Modern Supercritical Fluid Chromatography

Chemical Analysis

A SERIES OF MONOGRAPHS ON ANALYTICAL CHEMISTRY
AND ITS APPLICATIONS

Series Editor
MARK F. VITHA

Editorial Board
Stephen C. Jacobson
Stephen G. Weber

VOLUME 186

A complete list of the titles in this series appears at the end of this volume.

Modern Supercritical Fluid Chromatography

Carbon Dioxide Containing Mobile Phases

Larry M. Miller, J. David Pinkston, and Larry T. Taylor

Registered Office
John Wiley & Sons, Inc., 111 River Street, Hoboken, NJ 07030, USA

Editorial Office
111 River Street, Hoboken, NJ 07030, USA

For details of our global editorial offices, customer services, and more information about Wiley products visit us at www.wiley.com.

Wiley also publishes its books in a variety of electronic formats and by print-on-demand. Some content that appears in standard print versions of this book may not be available in other formats.

Library of Congress Cataloging-in-Publication data applied for

ISBN: 9781118948392

Cover design: Wiley
Cover image: Courtesy of Larry M. Miller

Set in 10/12pt Warnock by SPi Global, Pondicherry, India

Printed in the United States of America

V10015060_102819

Contents

Preface

0.1 Scope of the Book

Supercritical fluid chromatography (SFC) is more than 50 years old. Chapter 1 entitled "Historical Development of SFC" recaps over a much greater time-frame of the discovery of supercritical fluids and their development as a medium for chromatographic separation of both volatile and nonvolatile analytes. A real interest in SFC using either packed or open tubular columns began in the early 1980s when the first commercial preparative SFC instrument became available [1]. This development led to growing interest in the separation of stereoisomers which started with the pioneering work of Frenchman Marcel Caude and his research group in 1985 [2]. Thus, a wide variety of chiral separations were reported and applied near the turn of the century employing both analytical and preparative packed column (pcSFC) technology. SFC with open tubular columns (otSFC) also peaked in the 1980s but fizzled during the following decade. Interest in pcSFC is currently higher than ever before. For example, the technique is capable of generating peak efficiencies approaching those observed in gas chromatography (GC). On the other hand, pcSFC separations can achieve much higher efficiencies per unit time than in high performance liquid chromatography (HPLC). pcSFC has embraced a critical mass of separation scientists and technicians in terms of the number of workers in the field worldwide. Hundreds of supercritical fluid chromatographs currently are in use. Furthermore, pcSFC is (i) detector and environmentally friendly,

(ii) interfaceable with sample preparation, (iii) relatively economical in cost, and (iv) is a superior purification tool. Chapters 3 and 4 provide discussion of these critical developments that earlier had been referred to as *dense gas chromatography* [3]. Related work in the field currently uses both supercritical and subcritical mobile phase conditions to perform separations as well as purifications.

During the past 20 years, pcSFC has created a *bonafide* niche for itself as the go-to workhorse in chiral separations. Chapter 6 discusses in detail this topic. It has afforded many advantages for rapid separation of enantiomers over HPLC due to its greater separation efficiency per unit time. The advantages of pcSFC over HPLC which are also discussed later in the book, however, are practical but not fundamental. The greatest difference between pcSFC and pcHPLC is just simply the need to hold the outlet pressure above ambient in separations in order to prevent expansion (i.e. boiling) of the mobile phase fluid.

Enantiomeric separations are more compatible with ambient SFC than with high temperature HPLC because chiral selectivity usually favors decreasing temperature wherein the risk of analyte racemization is minimized. On the other hand, the risk of analyte thermal decomposition as in the GC of cannabis – related components is lessened. Furthermore, the straightforward search (primarily by trial and error) for a highly selective chiral stationary phase is a key step in the development of chiral pcSFC separations that address *industrial* applications. In this regard, a number of screening strategies that incorporate a wealth of stationary phases are discussed in the book that take advantage of short columns, small particles, high flow rates, and fast gradients.

Upon scale-up of analytical chromatography to preparative supercritical fluid separations as discussed in Chapter 8, the resulting decrease in solvent usage and waste generation relative to preparative scale HPLC is strikingly dramatic. SFC product can be routinely recovered at higher concentration relative to HPLC which greatly reduces the amount of mobile phase that must be evaporated during product isolation. Higher SFC flow rates contribute to higher productivity. The faster SFC process makes the separation cycle time significantly shorter such that it becomes practical as well as feasible to make purification runs by "stacking" small injections in short time windows without compromising throughput. Table 0.1 lists additional advantages of supercritical fluid chromatography.

Table 0.1 Advantages of supercritical fluid chromatography.

- High diffusivity/low viscosity yield greater resolution per unit time.
- Longer packed columns afford greater number of theoretical plates
- Low temperature reduces risk of analyte isomerization
- Scale-up of separation and isolation of fractions are facilitated

pcSFC (as most analytical techniques) has had a tortuous development history, but it appears that analytical and preparative scale chiral SFC are currently on the firmest foundation ever experienced with vendors that are strongly committed to advancing the technology. Extensive, new developments in achiral SFC and a much broader spectrum of applications outside the pharmaceutical area are already happening. Unlike reversed phase HPLC, the identification of the correct column chemistry is critical for the successful application of achiral pcSFC. Very different selectivity can be achieved depending on the column chemistry. Basic, neutral, and acidic compounds are well eluted on most columns that indicates the suitability of pcSFC for a broad range of chemical functionalities. The number of "SFC" columns for achiral purifications has also grown rapidly in the past three years. Activity in (i) agricultural and clinical research, (ii) environmental remediation, (iii) food and polymer science, (iv) petrochemicals, and (v) biological chemistry immediately come to mind. Additional Chapters 9–12 have been introduced into the book since writing began that reflect numerous additional applications of pcSFC such as pharmaceuticals, petroleum, food, personal care products, and cannabis. Additional advantages of SFC are listed in Table 0.2.

0.2 Background for the Book

While there have been numerous books published concerning SFC as both monographs and edited volumes, there appear to be only two texts that have had teaching as a major emphasis. One, published in 1990, was edited by Milton L. Lee (Brigham Young University) and Karin E. Markides (Uppsala University, Sweden) and written by a committee of peers is entitled "Analytical Supercritical Fluid Chromatography and Extraction" [4]. For chromatographic discussion, this book focused almost entirely on wall coated open tubular capillary column SFC (otSFC), which is not widely performed today having been replaced almost 100% by packed column SFC (pcSFC).

In the early days, otSFC and pcSFC coevolved and vigorously competed with each other as described in Chapter 1. otSFC lost ground and eventually faded

Table 0.2 Additional advantages using pcSFC.

- No pre-derivatization to achieve solubility and/or volatility
- Shorter cycle time with gradual gradient elution
- Faster separation facilitated by higher fluid diffusivity
- Reduced column diameter/particle size via lower fluid viscosity
- Less extreme chromatography conditions
- Routine normal phase chromatography

away, mainly as a result of poor chromatographic reproducibility issues in terms of flow rate, gradient delivery, pressure programming, and sample injection. The early systems were costly and not user friendly, which resulted in the technique being marginalized as too expensive and inefficient. While otSFC was capable of outstanding feats such as the separation of nonvolatile polymeric mixtures and isomeric polyaromatic hydrocarbons, most workers in the field would agree nowadays that the approaches used in otSFC are among the worst parameters to test with pcSFC.

Another book entitled "Packed Column SFC," published by the Royal Society of Chemistry and authored by Terry A. Berger [5] was published in 1995. Given that over 20 years have elapsed since the publication of Berger's book, the book presented here today provides ample references that reflect the current state-of-the-art as understood today. We have written our book that incorporates a more pedagogical style with the explicit intention of providing a sound education in pcSFC. Relatively new users of SFC in the early days were largely forced to rely on concepts developed for either HPLC (in the case of packed columns) or GC (in the case of open tubular columns), which were often inappropriate or misleading when applied to both otSFC and pcSFC. Our book addresses these deficiencies.

In this regard, a detailed discussion of current SFC instrumentation as it relates to greater robustness, better reproducibility, and enhanced analytical sensitivity is a focus of the book (Chapter 3). Originally, SFC was thought to be solely for low molecular weight, nonpolar compounds. Today, we know that SFC spans a much larger polarity and molecular mass range. Even though modern pcSFC books may be more adequately described as either "Carbon Dioxide-Based HPLC" (as Terry Burger once suggested) or "Separations Facilitated by Carbon Dioxide" (as suggested by Fiona Geiser) than "Packed Column Supercritical Fluid Chromatography," a change in nomenclature this drastic was not encouraged by attendees at several recent pcSFC conferences in both Europe and the United States. Suffice it to say, a change in nomenclature at this time is not suggested here. Nevertheless, this drastic shift in mindset and practice as suggested by Berger and Geiser during the last decade concerning both stationary phase and mobile phase has been a large reason for the current resurgence of pcSFC technology for problem solving at the industrial and academic levels worldwide. As proof, analytical scale achiral SFC is discussed in Chapter 6 along with ion pair SFC, reversed phase SFC, and HILIC-SFC.

While SFC has experienced much painful growth and disappointment during its evolution over 50 plus years, the "flame" has never been extinguished in the minds of a core group of separation scientists. A major reason for this mindset has been the near-annual, well-attended scientific meetings that have taken place in Europe and the United States over the past 25 years. Initially, the meetings were known as "the International Symposium on Supercritical Fluid

Chromatography and Extraction" wherein the focus was almost exclusively on capillary column SFC. Milton Lee at BYU and Karen Markides from University of Uppsala, Sweden served as hosts for the first meeting (1988) in Park City, UT. Subsequent meetings and approximate dates that have mostly been within the United States are listed in Table 0.3. Not shown in the table, but the youngest of us (DP) presented a poster at probably the earliest conference in this series called "SFC-87, Pittsburgh." Attendance was approximately 150.

These meetings were terminated soon after 2004 due to a lack of vendor commitment and support and user interest. In 2007, a series of new conferences with a different name ("International Conference on Packed Column Supercritical Fluid Chromatography") that gave attention to exclusively packed column Supercritical Fluid Chromatography was initiated first by Suprex Corporation, Pittsburgh, PA, then Berger SFC, and later by both Waters Corp. and Agilent. These meetings which now attract primarily industrial scientists, engineers, and academic colleagues from Europe and the United States are currently sponsored by the Green Chemistry Group. During the past 10 years the meetings have occurred annually and have alternated mostly between Europe and the United States (Table 0.4). To gain a greater world-wide audience the Green Chemistry Group has sponsored pcSFC meetings in China and Japan (i.e. 2016–2017, respectively). Additional meetings are scheduled in 2019 for both China and Japan.

pcSFC during the past 10 years has become a viable chiral chromatographic technique in the areas of pharmaceutical drug discovery and drug development. Chiral separations using carbon dioxide which incorporate a host of normal phase, silica-based stationary phases with principally ultraviolet and mass spectrometric online detection are now common. Nearly every pharmaceutical company in the United States, Asia, and Europe has multiple pcSFC instruments operating in a variety of laboratories. Interest in India, China, Korea, and the Pacific Rim, for example, is growing.

Table 0.3 Open tubular column SFC meetings.

SFC-1 (1988) – Park City, UT

SFC-2 (1989) – Snowbird, UT

SFC-3 (1991) – Park City, UT

SFC-4 (1992) – Cincinnati, OH

SFC-5 (1994) – Baltimore, MD

SFC-6 (1995) – Uppsala, Sweden

SFC-7 (1996) – Indianapolis, IN

SFC-8 (1998) – St. Louis, MO

SFC-9 (1999) – Munich, Germany

SFC-10 (2001) – Myrtle Beach, SC

SFC-11 (2004) – Pittsburgh, PA

Table 0.4 Packed column SFC meetings.

pcSFC 2007 – Pittsburgh, PA, USA
pcSFC 2008 – Zurich, Switzerland
pcSFC 2009 – Philadelphia, PA, USA
pcSFC 2010 – Stockholm, Sweden
pcSFC 2011 – New York City, USA
pcSFC 2012 – Brussels, Belgium
pcSFC 2013 – Boston, MA, USA
pcSFC 2014 – Basel, Switzerland
pcSFC 2015 – Philadelphia, PA, USA
pcSFC 2016 – Vienna, Austria
pcSFC 2017 – Rockville, MD, USA
pcSFC 2018 – Strasbourg, France

Currently activity centers around (i) development and application of mass-directed pcSFC, (ii) enhancement of robustness and sensitivity to meet various regulatory requirements, (iii) production of new polar stationary phases for separation of metabolomics and related biochemicals, and (iv) theoretical modeling of column physical properties dictated by employment of compressible polar modified mobile phase and stationary phase – bonded sub-2-µm particles.

There is rapidly growing interest in achiral pcSFC where the separation of highly polar compounds has been demonstrated. Applications to polymeric materials, natural products, water soluble analytes, surfactants, organic salts, fatty acids, lipids, organometallics, etc. are experiencing great success. Depending upon the nature of the stationary and mobile phases employed, a variety of separation mechanisms can be expected such as reversed phase pcSFC, ion pairing pcSFC, and aqueous promoted HILIC-pcSFC. Each mode of chromatography can be expected to augment the more popular normal phase pcSFC that has been used for decades and employs nonpolar mobile phases.

0.3 Audience for the Book

This book will be of interest to industrial, government, and academic users of pcSFC and is expected to be useful as a chemistry textbook in graduate-level separations courses. Laboratories looking to adopt SFC as part of their regular analytical tools will find this book useful as they learn fundamental principles behind technology and how pcSFC complements both HPLC and GC.

One's view of SFC today is entirely different from that of 25–30 years ago wherein (i) flow rates and gradient delivery were not reproducible, (ii) analytical UV sensitivity was not acceptable, and (iii) stationary phases were designed for reversed phase chromatography as opposed to normal phase chromatography. Today, SFC is considered to be primarily normal phase chromatography (i.e. a separation technique similar to HPLC) using mostly the same hardware and software developed for HPLC. The mobile phase is a binary or ternary mixture with CO_2 as the main component. The separation is usually performed with gradient elution where the composition of the mobile phase becomes more polar with time. Polar stationary phases such as bare silica, cyanopropylsilica, 3-aminopropylsilica, and 2-ethylpyridylsilica are routinely employed. pcSFC has numerous practical advantages relative to reversed phase HPLC such as higher speed, greater throughput, more rapid equilibration, and shorter cycle times. SFC yields lower operating cost and lower column pressure drop, and is orthogonal to reversed phase HPLC. Finally, compounds of interest can be isolated with a relatively small amount of solvent because CO_2 vaporizes away. This feature has become particularly important for preparative applications in which elution volumes can be large.

During this time period, a SFC system was introduced by Waters Corp. (Milford, MA, USA). The system featured the efficient cooling of the CO_2 pump heads by Peltier and the design of a dual stage back pressure regulator that was heated to avoid frost formation. In this case, separations with the Waters instrument were mostly identified as ultrahigh performance supercritical fluid chromatography (UHPSFC). A similar system like Waters was introduced in 2012 by Agilent which was a hybrid that allowed both UHPLC and SFC separations. Shimadzu has more recently introduced hardware that performs similar operations. This combined vendor news reenergized many workers in the SFC community and caused potential users of the technology to re-investigate the research potential of pcSFC. The instrumentation from these three vendors nowadays appears to represent the current methodology to perform analytical pcSFC which should enhance its acceptability by the separation scientists into the immediate future. UHPSFC via either vendor affords a high throughput approach for profiling analytes such as free fatty acids, acylglycerols, biodiesel, peptides, basic drugs, etc. via light scattering, UV, and Q-TOF-MS detection without the waste and uncertainty of sample preparation procedures. This more modern terminology is prevalent throughout this book. The older pcSFC instruments, while still useable in numerous laboratories are no longer being manufactured.

0.4 SFC Today

Being green is a good thing, but most people nowadays seemingly go for pcSFC because of its speed and fast method development rather than its environmental advantages. Experts in the field now readily agree that ultrahigh performance supercritical fluid chromatography (UHPSFC) has established itself as the preferred way of doing chiral and achiral analysis on both analytical and preparative scales. They also say that SFC will become the norm for small-scale purifications. Increased interest in (i) petrochemical and food industries, (ii) environmental air quality, (iii) biodiesel quality control, and (iv) protein separations can be expected in the not too distant future [6].

Much of the increased experimental capability alluded to above has been made possible by the introduction of pumping systems that deliver enhanced reproducible and accurate flow of CO_2 and modifier. In this case, separations are generally identified as UHPSFC.

Anyone with an interest in analytical and/or preparative scale pcSFC coupled to both spectroscopic and flame-based detectors will find this book beneficial. Subcritical fluid chromatography and enhanced fluidity chromatography as developed by Susan Olesik at the Ohio State University are also applicable here. Bonafide experience of the separation scientist in analytical or preparative scale SFC is not necessary for reading this book. Some knowledge of chromatographic principles is, however, desirable. With the introduction of more reliable instrumentation and eye-catching applications, a new generation of separation scientists and engineers are beginning to express much interest in the technology. Because the book is written with teaching in mind, the text could very well be the reference document on the desk of each person who is applying pcSFC.

Enjoy reading!
Larry M. Miller, J. David Pinkston, Larry T. Taylor
February, 2019

References

1 Berger, T.A., Fogleman, K., Staats, T. et al. (2000). The development of a semi-preparatory scale SFC for high-throughput purification of combi-chem libraries. *Journal of Biochemical and Biophysical Methods* 43: 87–111.

2 Mourier, P., Eliot, E., Caude, M. et al. (1985). Super- and sub- critical fluid chromatography on a chiral stationary phase for the resolution of phosphine oxide enantiomers. *Analytical Chemistry* 57: 2819–2823.

3 Giddings, J.C., Myers, M.N., McLaren, L.M., and Keller, R.A. (1968). High pressure gas chromatography of nonvolatile species. *Science* 162: 67–73.

4 Lee, M.L. and Markides, K.E. (1990). *Analytical Supercritical Fluid Chromatography and Extraction*. Provo, Utah: Chromatography Conferences, Inc.

5 Berger, T.A. (1995). *Packed Column SFC*. Cambridge: Royal Society of Chemistry.

6 Francotte, E. (2016). Practical advances in SFC for the purification of pharmaceutical molecules. *LCGC Europe* 29 (4): 194–204.

1

Historical Development of SFC

1.1 Physical Properties of Supercritical Fluids

In supercritical fluid chromatography (SFC), the mobile phase is ideally in the supercritical state. The meaning of the word supercritical (literally, above critical) is explained in Figure 1.1. The figure shows a phase diagram for a single (pure) component. Depending on the temperature (T) and the pressure (P), three different states of matter may be distinguished. These are gas (G), liquid (L), and solid (S) states. At the triple point (tp) all three of these phases may coexist. Above the critical point (cp) a difference between gaseous and liquid states can no longer be observed. This region is illustrated in Figure 1.1 by the dashed lines, which defines the supercritical fluid region and the material is referred to as a supercritical fluid (Schoenmakers, P.J. and Uunk, L.G.M., "Mobile and stationary phases for supercritical fluid chromatography," Private Communication.).

The supercritical fluid region is not a fourth state of matter. Crossing one of these dashed lines does not result in a phase change, whereas crossing a solid line does. Both condensation and evaporation are phase changes, during which the physical properties (e.g. density, viscosity, and diffusivity) change abruptly. On the other hand, a gas can also be transformed into a liquid in a manner indicated by the curved arrow in Figure 1.1. During this process, a phase

Modern Supercritical Fluid Chromatography: Carbon Dioxide Containing Mobile Phases,
First Edition. Larry M. Miller, J. David Pinkston, and Larry T. Taylor.
© 2020 John Wiley & Sons, Inc. Published 2020 by John Wiley & Sons, Inc.

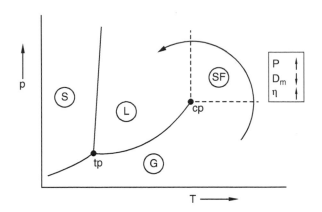

Figure 1.1 Phase diagram for a single pure component, illustrating areas in which solid (S), liquid (L), gaseous (G), and supercritical (SF) conditions occur. tp is the triple point and cp is the critical point. A gas can be transferred into a liquid by following the arrow. In doing so, the density, the viscosity, and the diffusion coefficient change continuously from gas-like to liquid-like values, but no phase change is observed. *Source:* Schoenmakers [1, p. 102].

change is *not* observed; yet, gas is transformed into a liquid. More generally stated, the physical properties of a pure compound show continuous rather than abrupt variations when passing through one of the dash lines.

The region of the phase diagram at temperatures and pressures higher than the critical temperature and critical pressure values was formally (and arbitrarily) designated as the supercritical fluid region by both the American Society for Testing and Materials (ASTM) and by the International Union of Pure and Applied Chemistry (IUPAC) (see Figure 1.2). This designation introduced what appears to be a fourth state of matter, the supercritical fluid. A second designation can be found in Figure 1.3 wherein two subcritical regions are identified along with the supercritical fluid region. Chester has cautioned that this format is an immense source of confusion among novices and even some experts [4]. In this diagram, the supercritical fluid region is formally defined as shown, however the apparent boundaries are not phase transitions, only *arbitrary definitions*.

The literature is full of statements regarding the transition between a liquid and a supercritical fluid phase or between a vapor and a supercritical fluid phase. *This is incorrect* according to Chester. A discontinuous phase change is predicted when the boiling line is crossed, but no discontinuous transitions or phase changes take place for isothermal pressure changes above the critical temperature or for isobaric temperature changes above the critical pressure. *There are no transitions into or out of a supercritical fluid state even though the supercritical fluid region is defined formally* according to Chester.

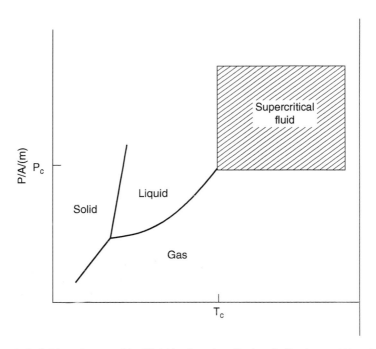

Figure 1.2 Definition of supercritical fluid by American Society for Testing and Materials (ASTM) and International Union of Pure and Applied Chemistry (IUPAC). *Source:* Smith [2]; ASTM [3]; Chester [4, vol. 2, p. 11, figure 2].

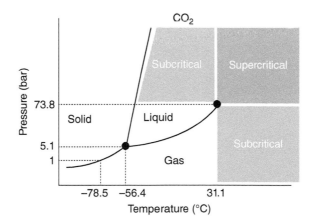

Figure 1.3 Misleading phase diagram for a single component supercritical fluid. *Source:* Laboureur et al. [5]. https://www.mdpi.com/1422-0067/16/6/13868. Licenced under CC BY 4.0.

In other words, it is possible to convert a liquid to a vapor, or a vapor to a liquid, without undergoing a discontinuous phase transition by choosing a pressure/temperature path that is wholly within the continuum. The required path simply goes around the critical point and avoids going through the boiling line. The distinction between liquid and vapor simply ceases for temperatures and pressures beyond the critical point [6]. As stated previously, Figure 1.1 is the accurate depiction of phase behavior. There is no fundamental difference between supercritical fluids and gases or liquids. Rather, a supercritical fluid may best be thought of as a very dense gas!

A more useful description of supercritical fluids for chromatographers is shown in Figure 1.4. In chromatography, multi-component supercritical mobile phases are frequently employed instead of a pure supercritical fluid. It is useful to continue thinking of the fluid phase behavior, but this requires one to expand the phase diagram to include the composition variation possible in a binary mobile phase [7]. Six general types of binary-mixture systems have been defined [8]. Some of the systems have large miscibility gaps rendering them useless for chromatography over much of their composition ranges. Type I mixtures, however, are the simplest and most widely used mixtures in liquid chromatography (LC). These are mixtures in which the two components are miscible in all proportions as liquids [7]. To consider the phase behavior of a

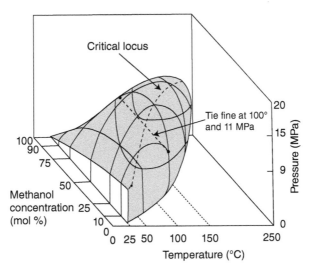

Figure 1.4 Two-phase *l-v* region of a binary mixture is a volume in a three-dimensional phase diagram. The Type I mixture CO_2-methanol is illustrated here. The two-phase region is the shaded interior of the figure. It has been cut off at 25 °C to show the isotherm, but actually extends to lower temperatures. *Source:* Adapted with permission from reference [4]. Reproduced with permission of American Chemical Society.

binary mixture it is necessary to add to the phase diagram a third axis representing the fluid composition. The two components here are arbitrarily called "a and b" except that "a" will be used to designate the more volatile component. The "a and b" choices are restricted to materials that together form a Type I binary mixture [8]. The reader is encouraged to consult reference 8 for a more complete interpretation of these plots.

In other words, there is no narrowly defined supercritical phase [4]. The behavior of a supercritical fluid may be very similar to that of a gas which would be the case just above the horizontal dashed line in Figure 1.1. Such a gas-like supercritical fluid possesses a relatively low density, a low viscosity, and high-diffusion coefficients. Just to the right of the vertical dashed line in Figure 1.1, a supercritical fluid may behave much more like a liquid. Such a liquid-like phase would show relatively high density, high viscosity, and low-diffusion coefficients. The most popular properties of supercritical fluids are listed in Table 1.1.

Whereas the physical properties of a liquid and solid are fixed, the physical properties of a supercritical fluid vary between the limits of a normal gas and those of a normal liquid by control of pressure and temperature as shown in Figure 1.5. Typically, supercritical fluids are used at densities ranging from 10 to 80% of their liquid density and at practical pressures for applications ranging from 50 to 300 atm. Under these conditions, the diffusion coefficients of supercritical fluids are substantially greater than those of liquids. Similarly, the viscosities of supercritical fluids are typically 10–100 times lower than liquids. These more favorable physical properties (as listed in Table 1.1) afford the advantages of supercritical fluids in chromatography and extraction applications.

"Supercriticality" is another term for a fluid that has reached a temperature higher than its critical temperature and a pressure higher than its critical pressure. Although rare, supercriticality exists in nature. For example, the atmosphere of the planet Venus is made of 96.5% carbon dioxide. Figure 1.5 pictorially compares the atmospheres of Venus (left) and Earth (right). At ground levels on Venus, the temperature is 735 K and its pressure is 93 bar. Therefore, in dealing with carbon dioxide, these conditions cause CO_2 to be supercritical on planet Venus.

Table 1.1 General properties of supercritical fluids.

- High diffusivity (gas-like)
- Low viscosity (gas-like)
- Zero surface tension
- Tunable solvent strength
- Nontoxic if CO_2

Figure 1.5 Comparison of Venus (CO_2) and Earth (Air) atmospheres. *Source:* Courtesy of NASA.

1.2 Discovery of Supercritical Fluids (1822–1892)

The phenomenon of the "critical state" was first described in 1822 by French engineer and physicist Charles Baron Cagnaird de la Tour when he noted the lack of discontinuity (i.e. disappearance of a meniscus) when passing between gaseous and liquid states in his famous cannon barrel experiment [9]. It was the work of the Irish chemist Dr. Thomas Andrews, Vice President of Queen's College in Belfast, Ireland in 1869 with CO_2, however, which is considered to be the first systematic study of a gas–liquid critical point [10]. It was also where matter was first referred to as a "supercritical fluid." It should be noted, nevertheless, that the general idea of the "critical state" was earlier and independently rumored by Mendeleeff in 1861 while working in Heidelberg with physicist Gustav Kirchoff where he discovered the principle of gases critical temperature. Mendeleeff's work went unnoticed, such that the discovery of critical temperatures is usually attributed to Thomas Andrews. Table 1.2 may be considered to contain a partial listing of the early studies wherein supercritical fluid behavior was demonstrated [11].

In 1879 and 1880, Hannay and Hogarth published the first account of the enhanced solvating properties of supercritical fluids [12] with an experimental apparatus earlier described by Andrews. Hannay's and Hogarth's original belief was that the ability to dissolve solid substances was a unique property of liquids. In their experiments, solutions of colored solids in liquids were heated through their critical points. When the liquids became gaseous, the solids were expected to precipitate and the fluids were predicted to become colorless. In practice, no such precipitation was observed, and the field of supercritical fluid extraction was born.

Table 1.2 Early studies of supercritical fluid behavior.

(1822) First report of supercritical fluid behavior
(1869) First measurement of critical parameters
(1879) First solvation of metal salts by gaseous fluids
(1879) First high pressure expt. via Hg column in mine shaft
(1892) Mercury column as high as the Eiffel Tower

Other investigators made similar observations at an earlier date [13]. In studying the solubility of inorganic salts such as cobalt(II) chloride, ferric chloride, potassium bromide, and potassium iodide in supercritical carbon dioxide, Hannay and Hogarth found a perfect continuity of liquid and gaseous states. Hannay summarized the findings later by stating that: "*The liquid condition of fluids has very little to do with their solvent power, but only indicates molecular closeness. Should this closeness be attained by external pressure instead of internal attraction, the result is that the same or even greater solvent power is obtained. The gas must have a certain density before it will act as a solvent, and when its volume is increased more than twice its liquid volume, its solvent action is almost destroyed*" [14].

Hannay's and Hogarth's experiments were largely based on transition metal salts and supercritical ethanol at temperatures too high to be convenient for a modern lecture demonstration. A lecture demonstration for supercritical fluids involving supercritical ethane with T_c 32.3 °C and blue dye, guaiazulene, has, however, been described for projection of an image of a high pressure silica capillary cell so as to be viewed by a large audience [15]. An excellent review of early studies regarding solubility measurements in the critical region was later provided by Booth and Bidwell [16].

Another informative, pictorial comparison of solvating properties appeared on the cover of Chemical and Engineering News (June 10, 1968 issue) that described the supercritical–liquid–gas inter-relationship (see Figure 1.6). Each of the enclosed glass vessels contained three spheres of unequal density and carbon dioxide. The temperature of the fluid on the far left is above the critical temperature. Thus, with no meniscus the condition was deemed supercritical fluid and at uniform density. Each of the spheres has a different density, thus, the high-density sphere in this bulb sunk to the bottom; while, the lowest density sphere rose to the top. The intermediate sphere density matched the supercritical fluid density and appeared to be suspended in the bulb.

Moving from left to right in the figure, there is a temperature decrease. As evidenced by the cloudiness in the second bulb, the CO_2 is at the critical temperature, and critical opalescence is predicted. At the third bulb from the left, temperature and pressure have decreased further, subcritical conditions exist, and a gas and liquid phase now appear. Two spheres floated on the liquid while the highest density sphere sank to the bottom. The temperature of the fourth bulb was thought to be lower than that in bulb #3. All three spheres floated on

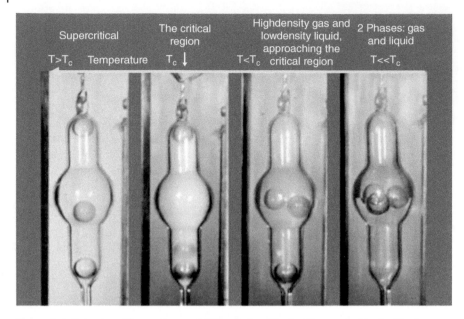

Figure 1.6 Behavior of four spheres of different densities in CO_2: supercritical, critical, subcritical, and liquid (0.92 g/mL). *Source:* Chemical & Engineering News, June 10, 1968, p. 105, Photo by Ray Rakow.

the liquid phase, which indicated that the liquid phase density had increased even more allowing even the greatest dense sphere to float on its surface.

During the late 1890s, numerous studies of high-pressure fluids and solubilization phenomena were recorded. For example, Amagat in 1879 performed high-pressure experiments using mercury columns that extended to the bottom of mine shafts [17]. Later, Cailletet (1891) used a mercury column from the top of the Eiffel Tower for high-pressure experiments [18]. By changing the density of the fluid through temperature and pressure variation, the solvation strength of a supercritical fluid was altered. An increase of the pressure caused the density of the supercritical fluid to increase thereby causing it to become more liquid-like. When the temperature was increased, the density of the supercritical fluid decreased, and the phase became more gas-like. Depending upon the density, the viscosities of supercritical fluids were thought to be similar to gases or intermediate between gases and liquids.

1.3 Supercritical Fluid Chromatography (1962–1980)

Considerable time passed before the previously described basic knowledge regarding supercritical fluids was utilized for SFC. It was first proposed in 1958 by James Lovelock while at Yale University [19]. He conceived the idea of using supercritical water, ammonia, sulfur dioxide, and carbon dioxide for chromatographic mobile phases in open tubular columns to increase the solvating power of the mobile phase in order to elute nonvolatile ionic substances, but he surprisingly did not attempt to practice the art of SFC. He, nevertheless, related his ideas concerning SFC to his coworkers Sandy Lipsky and Ray Landowne, who urged him to commit his thoughts and findings to paper. Lovelock, however, realized that the high liquid-like density, high gas-like diffusivity and low viscosity of fluids above the critical point would extend both GC and LC. Based upon these early discussions, he was reported to have had his research document notarized and witnessed soon after, but no patents were ever noted. He also suggested the name "critical state chromatography" for the separation. It was noted that acting as a solvent is a critical characteristic that differentiates the compressible mobile phases used in SFC from gas chromatography (GC).

Supercritical fluid chromatography was invented by gas chromatographers who initially explored gas at high pressures. Their hope was to elute compounds that could not be analyzed with GC because the analytes were prone to decompose at the temperatures needed to elute them. Ernst Klesper in 1962 employed a 30-inch long column packed with 33% Carbowax 20M on Chromosorb W (diatomaceous earth, GC packing, 180–250 μm), and is considered to be the first person to use higher gas pressures to elute, for example, porphyrin mixtures at lower temperatures than required for their elution using traditional GC conditions. Supercritical temperatures were maintained to enable the gas pressure to be continuously increased without it passing through vapor–liquid biphasic conditions [20]. In 1967, Sie and Rijnders were the first to use the term "supercritical fluid chromatography" for this new chromatographic technique. They also suggested the use of mobile phase pressure programming.

Later, Jentoft and Gouw were the first to employ mobile phase modifiers (i.e. methanol in n-pentane) to control retention in SFC. Figure 1.7 shows the pressure–density relationship for CO_2 in terms of reduced parameters (e.g. pressure, temperature, or density divided by the appropriate critical parameter) including the two-phase vapor–liquid region used in these earlier studies. This relationship is generally valid for most single-component systems. The isotherms at several reduced temperatures show the variation in density that can be expected with changes in pressure. Thus, the density of a supercritical fluid will be typically 100–1000 times greater than that of a gas at ambient temperatures. Consequently, molecular interactions increase due to shorter intermolecular distances.

The temperature ranges in supercritical fluid processes depend on the fluid used and reflect the respective critical values. The majority of supercritical work is done with CO_2, which has a T_c of 31.30°C (304.6 K). Critical properties of select compounds of differing polarity are shown in Table 1.3. The physical properties of CO_2 and other supercritical fluids are also summarized in the table.

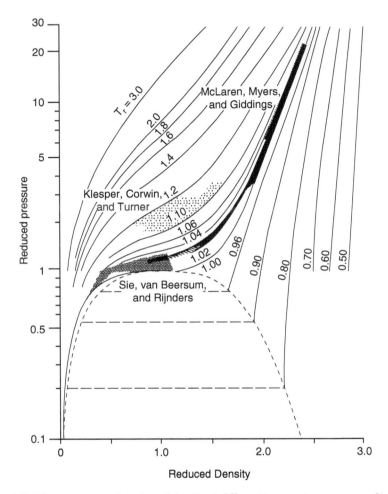

Figure 1.7 CO_2 pressure as a function of density at different temperatures expressed in terms of reduced parameters. The area below the dotted line represents the two-phase gas/liquid equilibrium region. The indicated areas refer to conditions at which the researchers performed their experiments during the initial studies of SFC. *Source:* Reproduced with permission of Milton L. Lee.

Klesper was born in Cologne, Germany and later finished his doctorate in inorganic chemistry in 1954 at the University of Hamburg. From there he moved to the United States, took an industrial position with the William T. Burnett Company in Baltimore, MD as a chief chemist. Thereafter he became a Research Associate at the Johns Hopkins University and an Assistant Professor at the University of Maryland. Subsequently he became associated with the Institut fur Macromolekulare Chemie at the University of Freiburg where he came into contact with polymer chemistry. He then joined A.H. Corwin's laboratory at Johns Hopkins University in Baltimore where the first experiments using

Table 1.3 Physical parameters of selected supercritical fluids.

Fluid	Dipole moment (debyes)[a]	T_c (°C)[a]	P_c (atm)[a]	ρ_c (g ml^{-1})[a]	ρ_{400} (g ml^{-1})[b]	ρ_1 (g ml^{-1})[a,c]
CO_2	0.00	31.3	72.9	0.47	0.96	0.71 (63.4 atm)
N_2O	0.17	36.5	72.5	0.45	0.94	0.91 (0 °C)
						0.64 (59 atm)
NH_3	1.47	132.5	112.5	0.24	0.40	0.68 (−33.7 °C)
						0.60 (10.5 atm)
n-C_5	0.00	196.6	33.3	0.23	0.51	0.75 (1 atm)
n-C_4	0.00	152.0	37.5	0.23	0.50	0.58 (20 °C)
						0.57 (2.6 atm)
$SF6$	0.00	45.5	37.1	0.74	1.61	1.91 (−50 °C)
Xe	0.00	16.6	58.4	1.10	2.30	3.08 (111.75 °C)
CCl_2F_2	0.17	111.8	40.7	0.56	1.12	1.53 (−45.6 °C)
						1.30 (6.7 atm)
CHF_3	1.62	25.9	46.9	0.52	1.15	1.51 (−100 °C)

[a] Data taken from references [21, 22].
[b] The density at 400 atm and T_r = 1.03 was calculated from compressibility data [23].
[c] Measurements were made under saturated conditions if no pressure is specified or were performed at 25 °C if no temperature is specified.

supercritical fluids as the mobile phase were performed in collaboration with Corwin and Turner as far back as 1962, which was interestingly well before the introduction of high performance liquid chromatograph (HPLC).

In Klesper's first reduction to practice a mixture of nickel porphyrins (etioporphyrin II, nickel etioporphyrin II, and nickel mesoporphyrin IX dimethylester) was separated with dichlorodifluromethane and monochlorodifluoromethane at pressures above 1000 and 1400 psia, respectively, as the mobile phase [20]. These particular mobile phases were chosen because of their low flammability and physiological inertness. The instrumentation was very simple but well suited to demonstrate the potential of employing supercritical fluids as the mobile phase in chromatography. At the time numerous supercritical fluid separations were presented and reported in the literature which were thought to be superior to existing gas and liquid chromatography reports. SFC was eagerly thought to be a future replacement for GC and LC of polar and high molecular weight components. See Table 1.4.

A diagram of the apparatus is shown in Figure 1.8. No mechanical pump or elaborate detector was used. The column was contained in a glass, high pressure gauge tube which allowed observation of the colored bands of porphyrins which moved down the column. The method was at that time called "high pressure gas chromatography" instead of SFC. The first publication by Klesper's group was a

Table 1.4 Early attraction of supercritical fluid chromatography (SFC).

- Capable of generating GC peak efficiencies
- Separations operated at higher flow rates than LC

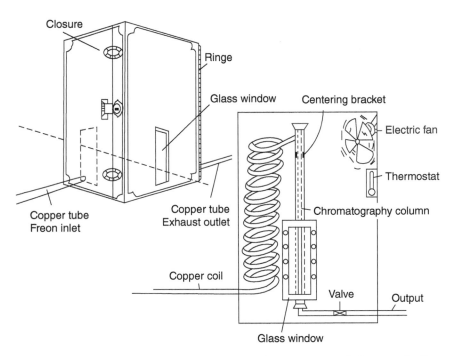

Figure 1.8 Diagram of apparatus used by Klesper to demonstrate proof of concept for first SFC. *Source:* Reproduced with permission of Milton L. Lee.

"Communications to the Editor" in the Journal of Organic Chemistry entitled "High Pressure Gas Chromatography Above Critical Temperatures [20]."

Klesper continued his research on polymers and SFC at the University of Freiburg in Germany [24]. The influence of temperature, pressure, and flow rate on the behavior of dimethylether and diethylether plus the separation of oligomers with UV-absorbing side groups via SFC using eluent gradients [25], and the elution behavior of styrene oligomer fractions in SFC [26] were just three of the early studies conducted by Klesper. By this time, Ernst Klesper was justifiably known in separation science as the "Father of Supercritical Fluid Chromatography" based upon his outstanding publication record in the field [27].

In 1968, Giddings et al. published an article concerning the influence of high pressures on retention in GC with carbon dioxide and ammonia as the mobile phase [28]. He stated that the use of high pressure in chromatography "would cause the convergence of gas chromatography with classical liquid chromatography." A decade later, Giddings also studied the general aspects of pressure-induced equilibrium shifts in size exclusion chromatography [29]. However, this new form of chromatography was generally noted as "dense gas" or "high pressure gas chromatography" rather than SFC.

Before Klesper and Giddings, other chromatographers were studying the use of high pressure gases and supercritical fluid mobile phases. The work of Riki Kobayshi at Rice University can be cited here. Kobayashi was also investigating high pressure equilibria by chromatography, but he was not using the increased solvent power of the media to elute nonvolatile materials. It was Sie, VanBeersum, and Rijnders in 1966 at the Shell Laboratories in Amsterdam, who first published the separation of C_7 to C_{13} n-paraffins, and soon after "supercritical fluid chromatography" was coined by the same workers in 1966–1967 [30]. Pressure programming and the use of mobile phase modifiers (i.e. methanol in n-pentane) were later (1969–1970) also demonstrated by Gouw and Jentoft to control retention [31, 32]. The prolific Amsterdam group published numerous papers during this time such as (a) the effect of mobile phase velocity and particle size on column plate height, (b) the use of porous polymeric stationary phases, and (c) numerous early useful reviews on SFC [33, 34], Table 1.5.

After the initial period of great interest in packed column supercritical fluid chromatography (pcSFC) in the 1960s, the progress of SFC slowed. In part, the slow development was due to several factors, such as (a) early experimental problems, (b) the lack of commercially available instrumentation, and (c) the fact that SFC development was over-shadowed by simultaneous developments in HPLC. A more striking reason surfaced in the late 1960s when several workers in the area suggested that the efficiency of crudely packed columns seemed to degrade with higher column pressure drops. Gouw and Jentoft postulated that decreasing density along the column axis acted like a decreasing temperature gradient in GC, causing a loss in efficiency [34]. It was, therefore, claimed from selected studies by various separation scientists that SFC packed columns could never produce more than ~20 000 theoretical plates, and at the time particles smaller than 5 μm were virtually ruled out in this research. Later it

Table 1.5 Milestones in the development of SFC.

(1996) First publication: C_7–C_{13} n-paraffins
(1967) Coined supercritical fluid chromatography
(1970) Convergence of high pressure GC and LC
(1970) Pressure programming and modifiers
(1974) Negative temperature programming
(1981) Open tubular column SFC
(1983) Density programming
(1984) Practical SFC with flame ionization detection

was learned that many of these problems stemmed directly from (i) inadequate use of home-made equipment related to back pressure control and (ii) the use of fixed restrictors instead of more effective back pressure regulation and variable restriction. Van Wasen et al. identified yet another reason around this time for the slow development of SFC [35]. They proposed that progress in SFC development was severely obstructed by the lack of physico-chemical knowledge regarding supercritical fluids. Unfortunately, this became a perception that existed for another 20 years.

The term "pressure drop" in chromatography refers to the decrease in pressure as the eluent is pumped through the column. The primary source of the system back pressure is the column, and it is proportional to the viscosity of the mobile phase. The magnitude of the pressure drop for the different types of chromatography increases in the order GC < SFC < HPLC. However, the effect of the pressure drop on the separation efficiency of the column is greatest in SFC because changes in pressure do not affect to any great extent the solvating abilities of liquids and gases. On the contrary, in SFC, the solvent properties of the supercritical fluid are greatly affected by the drop in pressure along the column. As the pressure of the supercritical fluid decreases, the density and consequently the solvent strength of the supercritical fluid decreases. As a result of the lower solvent strength, the peaks broaden instead of becoming narrower, and the total efficiency of the system decreases [36].

SFC in the 1970s experienced another dormant state. Research interest in SFC was limited; although some very important work was published by several groups during this period. For example, Schneider and Bartmann continued their research on the physicochemical aspects of supercritical fluids with studies on the density dependence of (a) retention factors, (b) binary diffusion coefficients, and (c) partial molar volumes [37]. The same co-workers also introduced SFC with negative temperature programming. More practical SFC investigations were introduced by Rogers and Graham regarding the separation of oligomers by a combination of pressure, temperature, and mobile phase programming [38]. Options afforded by supercritical fluids, nevertheless were beginning to be readily recognized (Table 1.6).

Table 1.6 Options afforded by supercritical fluids.

- Adjustable solvating power
- Greater and more variable diffusivity than liquid
- Lower and more variable viscosity than liquid
- Negligible surface tension
- Low temperature efficient separations
- Nontoxic mobile phase
- Analyte isolation achieved by pressure reduction

Table 1.7 Advantages: SFC vs. GC.

• Lower temperature
• Much higher molecular weight range
• Large programming range
• Selectivity tuning via pressure control
• Selectivity tuning via temperature control
• Universal mass detection via flame ionization

Hybrid chromatographic techniques made their first appearance in the late seventies as SFC was interfaced to mass spectrometry by Randall and Wahrhaftig [39]. In 1971, Novotny et al. published an important paper concerning the effects of temperature, pressure, eluent composition, flow rate, and type of stationary phase on retention factors [40]. It should be noted here that during the time period between 1962 and 1980, nearly all of the publications that related to SFC (which totaled about 90) involved the use of packed stainless steel columns. With the advent of capillary columns in 1980, SFC experiments were about to drastically change. Advantages of SFC versus GC (Table 1.7) were becoming very popular.

1.4 SFC with Open Tubular Columns (1980–1992)

A strong revival of interest in SFC occurred in the early eighties for two reasons. One important aspect was the introduction in 1981 of a commercial kit that converted a Hewlett-Packard model 1084 HPLC into an SFC system [41]. The resulting instrument, which was greatly due to the combined efforts of Berger, Gere, Lauer, and McManigill, was again for packed column SFC [42, 43]. A striking feature of this new instrumentation was that flow rate, mobile phase composition, column temperature, and column outlet pressure were all independently controlled. Unfortunately, there was no pressure programming and detection was exclusively UV.

The modified instrument, however, permitted operation with liquid CO_2 at the pump stage while maintaining supercritical conditions at the separation stage. Stationary phases were mostly standard, spherical silica bonded phase columns borrowed from HPLC. When the SFC-*incompatible* HP model 1090 HPLC system was introduced to the public, the HP 1084 model for SFC was withdrawn from the market, thus ending production of the **first** commercial SFC system that shocked the previously growing SFC market [43]. To partly fill the gap, it should be noted that in 1985, JASCO introduced a combined supercritical fluid extraction/supercritical fluid chromatography system which was similar to the modified but outdated HP 1084 system. The JASCO featured, however, the first electronic back pressure regulator specifically designed for SFC. User interest in this instrumentation, however, never lived up to expectations [44].

The introduction of open tubular wall coated columns in SFC by Lee and Novotny using pressure programming of pure CO_2 was reported in 1981 [45]. This development came at about the same time that Berger's group at Hewlett Packard reported the invention of fused-silica capillary columns which were capable of withstanding the high pressures necessary for open tubular column SFC (otSFC) [41]. otSFC experienced an explosive growth wherein the number of publications increased dramatically [46] because it was deemed by workers in the field to be more applicable (after considerable heated debate) for otSFC than for pcSFC. As a result, the otSFC technology was commercialized in 1986 through Lee Scientific (Salt Lake City, UT), which was later acquired by Dionex Corporation. The development of pcSFC was thus stymied and subsequently put on hold again as capillary columns dominated the development of both theory and instrumentation in SFC for the next five to seven years [47, 48] even though the earlier popular opinion regarding open tubular columns would be shown to be without merit.

The first workshop totally devoted to users of SFC was conducted in January 11–14, 1988 at Park City, UT. The meeting was sponsored by the State of Utah and Brigham Young University. The program attracted 175 participants from all over the world. The workshop schedule was demanding (8:00 a.m.– 10:00 p.m.) with 39 presentations of 20 minute duration. Each session was terminated by an hour long discussion of a particular aspect of SFC. At the beginning of the conference, Professor Ernst Klesper was presented a plaque for his pioneering work in SFC. His plenary lecture considered various gradient methods (temperature, pressure/density, and eluent composition) in SFC and their overall effect on resolution. The workshop closed on a high note of optimism regarding the potential of SFC, but everyone agreed that much research and developmental work needed to be performed.

In the early days, otSFC instruments resembled GC instruments, but the former used a syringe pump as a pressure source to change the CO_2 density and solvating power of the mobile phase. A homemade tapered fused silica fixed restrictor to limit flow through the small diameter chromatography column was employed. Detection was primarily via high temperature flame ionization. The simplicity of such an approach was for a time overwhelming such that most accompanying research activity replaced pcSFC usage. At one point, there were six to seven companies manufacturing and selling capillary SFC's and only one (JASCO) selling packed SFC columns [43].

Some of the immediate benefits of otSFC included (i) lower operational temperatures than GC, (ii) much higher molecular weight range than GC, (iii) large programming range primarily via pressure adjustment, (iv) selectivity tuning via temperature change, (v) universal mass spectrometric detection coupled with flame ionization detection, and (vi) amenable to very polar and some ionic analytes via pre-column derivatization [49, 50]. Figure 1.9 illustrates the separation of a polymeric material with three open tubular columns each with a different bonded phase. Pressure programming was employed in each case, yet separations were time consuming. Open columns possessed a high efficiency if operated under optimum conditions. In the Figure, separation in each case appears to be by oligomer unit. The time required to generate a greater plate number under this situation largely depended on both the inner

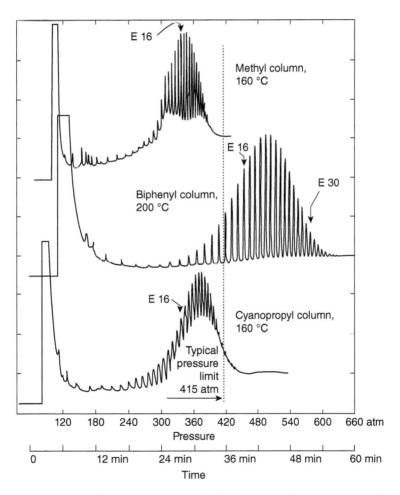

Figure 1.9 Separation of a polymeric material with three open tubular columns each with a different wall coated stationary phase. Pressure programming was employed in each case; yet separations were time-consuming. Separation in each case appears to be by oligomer unit. *Source:* Reproduced with permission of Wiley. Chester [52].

diameter of the column and the diffusivity of the analyte in the mobile phase which apparently was not the same as evidenced by the varying retention time of the E-16 oligomer. The slower the diffusion in the mobile phase, the smaller the inner diameter of the open tubular column had to be in order to obtain a given plate number. As diffusion in supercritical fluids is much slower than in gases, the inner diameter of open columns in otSFC was always much smaller than in GC. Consequently, a SFC column diameter of 50 μm was typically used.

The use of narrow open columns, however, imposed severe requirements on the instrumentation [47]. Sample injection, post column pressure/flow restrictors, and analyte detection devices were notable points of study. Flow split, timed split, and combined split injection techniques allowed the introduction of nanoliter sample sizes onto open columns with inner diameters below 50 μm. Total injection without splitting, on the other hand, was achieved by the use of a retention gap combined with a solvent venting exit. A number of fixed restrictor designs were also developed to control flow and pressure such as the polished integral tapered restrictor and the frit restrictor. The sensitivity of the detection devices required special consideration due to the reduced sample capacity of narrow-bore open columns. Furthermore, stationary phases in contact with supercritical fluids (i.e. otSFC) had to be highly cross-linked in order to impede column bleed. Thus, the employment of GC open tubular columns for otSFC where the mobile phase was supercritical CO_2 was not chromatographically successful. This feature dictated the use of columns that exhibited the best levels of deactivation attainable.

otSFC technology, in spite of the instrumental challenges, flourished for some time in both academic and industrial laboratories, but it had several disadvantages for wide-scale acceptance by the analytical community [46, 50]. Specifically, the technique, which relied mostly on nonpolar stationary phases, was traditionally limited to relatively nonpolar analytes because (i) pure CO_2 as the mobile phase exhibited poor solvating power similar to hexane, (ii) robust sample injection was not feasible, (iii) a wide polarity range of stationary phases was not available, (iv) high quality separations required ~60 minutes, (v) gradient CO_2 pressure and flow rate were coupled such that flow rate changed as mobile phase pressure changed, (vi) passive, fragile fused silica restrictors at the column outlet often plugged causing the separation to be aborted, and (vii) multi-gram scale-up of a successful separation was not feasible. Unfortunately, the technique was oversold as being appropriate for more polar analytes.

Still, in the 1980s, pcSFC was not very popular in spite of the fact that packed columns were thought by some separation scientists to be generally much easier to operate than open tubular narrow columns. Furthermore, columns packed with sub-10 μm particles were demonstrated to be more time efficient than contemporary, open columns. A fundamental problem, however, of packed columns in SFC continued to be the inherently high column pressure

Table 1.8 SFC was in a bad shape in 1980s.

- Fragile fixed restrictors/no flow control
- Pre-mixed cylinders of modified CO_2
- Helium-padded CO_2
- High temperatures/Long retention times
- Open tubular columns in/packed columns out
- No long-term vendor commitment
- Lacked HPLC figures of merit/Extension of GC
- Amenable only to nonpolar analytes

drop [51]. As pressure is an extremely important parameter in SFC, a marked change in the properties of the mobile phase along the column was anticipated to occur in the presence of a significant pressure drop. In contrast, the pressure drop across the column was negligible for otSFC.

Later in the decade, consensus was reached that both column types have their own unique advantages and disadvantages. History has, however, revealed that more highly significant technological developments have occurred in pcSFC than in otSFC, which accounts for the higher popularity and applicability for problem solving of the former than the latter. In the early days, capillary SFC and packed column SFC co-evolved and competed with each other. Capillary SFC lost ground and eventually faded away, mainly as a result of poor reproducibility issues. The early systems were costly and not user friendly, resulting in the technique being marginalized as too expensive and inefficient. The various features of SFC in the 1980s are listed in Table 1.8.

1.5 Rediscovery of pcSFC (1992–2005)

pcSFC underwent another renaissance in interest near the beginning of the 1990s when (i) the long-term limitations of otSFC became obvious and (ii) important progress in pressure gradient techniques involving mixed mobile phases was achieved for pcSFC. Today, it is generally conceded that dramatic differences exist between the instrumentation required for use of packed columns and open tubular columns in SFC. In fact, the use of open tubular instrumentation with packed columns often results in complete failure. otSFC resembles GC at high pressures, except that pressure (or density) programming in open tubular SFC has replaced temperature programming in GC. Because otSFC is still controlled using a passive fixed restrictor, flow varies with pressure, temperature, and mobile phase composition. Furthermore, the flow rate is typically measured in µL/min, and solutes are usually quantified with a flame ionization detector after the mobile phase has expanded to atmospheric pressure. In retrospect, otSFC has been demonstrated with a host of flame based detectors such as electron capture and nitrogen chemiluminescence. Few papers nowadays are published that deal with otSFC, and vendor support for otSFC technology has practically disappeared.

pcSFC with enhanced sophisticated instrumentation that allowed independent flow control under pressure gradient conditions became available in 1992. pcSFC was generally much easier to operate because it resembled HPLC wherein binary and ternary mobile phases were common. Then, it was realized that the composition of the mobile phase was almost always more important than mobile phase density programming in controlling retention, akin to HPLC in this regard. Flow rates were typically several mL/min. Pressure was controlled by an electronically controlled back pressure regulator (BPR) mounted at the end of the column. As previously noted, a fundamental problem of pcSFC during this period continued to be the inherently high column pressure drop [51]. As pressure was an extremely important parameter in pcSFC, a marked change in the properties of the compressible mobile phase along the column was predicted to occur in the presence of a significant pressure drop. Standard ultra-violet detectors with a high-pressure flow cell pre-BPR or a mass spectrometer post-BPR were used for detection and quantification.

Interest in pcSFC during this period was higher than ever because pcSFC was capable of generating peak efficiencies similar to those observed in GC and separations could be run at much higher flow rates than in HPLC. Furthermore, pcSFC was scalable, detector friendly, and economical. It afforded both rapid method development and a more environmentally acceptable chromatographic process than previously encountered with SFC [53]. pcSFC usually used (i) silica-bonded stationary phases, (ii) binary or ternary fluids, (iii) composition programming, and (iv) an ultraviolet detector.

Once modifiers were added to the CO_2 mobile phase, composition became even more important than CO_2 pressure or density in determining retention, unlike the situation that prevailed with otSFC. Packed columns are usually operated near the critical temperature of the fluid with flow control pumps and electronically controlled back pressure regulators mounted downstream of the column to obtain accurate flow rates and mobile phase composition. The combination of upstream flow control and downstream pressure control allowed both volumetric mixing of the main fluid and gradient modifier elution. pcSFC became viewed as an extension or subset of HPLC. Unfortunately, when one attempted to use inappropriate open tubular instrumentation with packed columns, the results were usually complete failure; whereas the use of packed column instrumentation with open tubular column instrumentation soon was realized to be entirely feasible.

With the availability of more reliable pcSFC instrumentation, a ready solution to several major analytical problems became economically and environmentally feasible. For example, pcSFC affords many advantages in comparison to

HPLC not the least of which is the rapid *separation of enantiomers* because chiral selectivity usually increases with decreasing temperature. Low temperature also reduces the risk of analyte racemization and thermal decomposition. Furthermore, CO_2 replaces petrochemically derived hydrocarbons that are used in the mobile phases of HPLC separation, resulting in a reduction in solvent utilization by as much as 90%. To successfully perform chiral pcSFC, a polar modifier must be added to the CO_2 mobile phase and in many cases operation across a gradient may exist during the separation.

pcSFC is ideal for preliminary rapid chiral screening. A fast chiral screen is used for rapid discernment of the most appropriate stationary phase and modifier. Finding a highly selective chiral stationary phase is a very key step in method development of a chiral separation. In addition, modifiers and additives usually play a significant role in enantiomeric recognition. A number of screening strategies have been reported that take advantage of short columns, high-flow rates, and fast gradients.

Preparative separations received considerable attention during the rediscovery of pcSFC [53–56]. The decrease in solvent use and waste generation when using CO_2 mobile phases compared to conventional liquid phases makes preparative SFC especially attractive for providing purified materials on a kilogram scale. Furthermore, the product is recovered in a more concentrated form relative to HPLC, thereby greatly reducing the amount of solvent that must be evaporated. The higher pcSFC flow rates also contribute to higher productivity relative to HPLC methods. The faster pcSFC process makes the separation cycle time significantly shorter such that it is viable to make purification runs by "stacking" small injections in short time windows without compromising the throughput. In this way, the utilization rate of expensive column material is much higher than with conventional preparative columns.

The unique properties of supercritical fluids have had a broad impact across the wider field of separation science through the development of so-called "unified chromatography" whose underlying principle was that there are no theoretical boundaries between mobile phases. Roger Smith pointed out in 1999 [48] that probably the most important idea that supercritical fluids have brought to separation science is a recognition that there is unity in the separation method and that a continuum exist from gases to liquids. Tom Chester noted during this period that when viewed from the perspective of the mobile phase, the perceived complexities of old and new modes of chromatography are not so complex after all [57]. Calvin Giddings also called attention to this notion over half a century ago and made the observation that as the column diameter decreases so do the differences between chromatographic techniques [28, 29]. As has hopefully been evident thus far in this chapter, knowledge of the history of SFC truly brought separation methods together near the close of the twenty-first century.

1.6   Modern Packed Column SFC

In the 1990s and in the early 2000s, pcSFC struggled to establish itself in other applications beyond preparative scale chiral separations. At the time, the technique's repeatability and robustness were still below those achieved in HPLC, and thus these issues hindered the implementation of SFC in both analytical and QC laboratories. SFC is experimentally different than LC, mainly because of the compressibility of the mobile phase. In this regard, Fornstedt has stated that SFC can be thought of as a "rubber variant" of HPLC where everything considered constant in HPLC varies in SFC [58]. The good news is that reluctance of analysts to use pcSFC has faded since the introduction by several major vendors of a new generation of SFC instruments dedicated to analytical purposes by several important manufacturers has happened. These new systems (first bullet point in Table 1.9) have benefitted from a novel automated back pressure regulator design and from ultra-high performance liquid chromatography (UHPLC) technology, which incorporates higher pressure limits and reduced void volumes. Improved performance, reliability, and full compatibility with most modern stationary phases (sub-3 µm core shell and fully porous sub-2 µm particles) have greatly broadened the application spectrum of this technique, making modern pcSFC with CO_2-based mobile phases competitive and complementary to UHPLC [58].

In modern pcSFC, the control of density is crucial. If not properly managed, the solvent strength differs between analyses leading to shifting retention times. The automated back pressure regulator (ABPR) is responsible for the pressure control, so proper design is of crucial importance. Changes in the temperature of the incoming CO_2 can also result in shifting retention times. In addition to these issues, the injection volume flexibility was limited in older generations of pcSFC instrumentation reflecting imperfections in the design of the partial loop injector for SFC. From a detection perspective, the UV noise level caused by differences in the refractive indices of mobile phase constituents needed to be minimized. Thus, the revival of interest in pcSFC in recent years has been predominantly the availability of state of the art instrumentation [58]. The ACQUITY system was commercially introduced in 2012 by Waters Corp. (Milford, MA, USA). To stress its differences from previous generations of pcSFC equipment, the term "convergence chromatography,"

Table 1.9 Reproducible flow and gradient delivery were keys to the re-emergence of SFC.

- Constant flow pump/variable restrictor. Linear velocity decreased with density increase. Efficiency remained constant
- Pressure controlled pump/fixed restrictor, mass flow increased with pressure increase. Efficiency decreased

originating from a statement of Giddings, was introduced by Waters [59]. Important features included the efficient cooling of the CO_2 pump heads by Peltier cooling and the design of a new dual stage ABPR that was heated to avoid frost formation. The maximum flow rate and pressure of the instrument was 4 mL/min and 413 bar, respectively. The column outlet was connected to a photo-diode array detector, which included a high pressure UV cell with a volume of 8 μL and a path length of 10 mm. The instrument could be controlled by typical software packages such as Empower or MassLynx.

The 1260 Infinity SFC/HPLC system was introduced by Agilent Technologies (Waldbronn, Germany), also in 2012 like the Waters system. The Agilent system is a hybrid that allows both UHPLC and pcSFC separations [60, 61]. Using switching valves and two pumps (one for SFC – Pump A and one for UHPLC – Pump B), the system can be modified based upon the needs of the user. Earlier collaboration (2010) between Aurora and Agilent Technologies resulted in the introduction of a dedicated analytical SFC module (1260 Infinity Analytical SFC system). The module is responsible for the compression of the incoming gaseous CO_2 with temperature control via a chilling liquid. The pre-compression requirement is removed from the other pump, which now functions only as a metering pump to control the flow of liquid CO_2. The BPR in this apparatus is a heated single stage device. Compared to traditional pcSFC systems, which employ a single pump to both compress the liquid CO_2 and meter the required flow, this approach has been shown to significantly reduce baseline noise. The maximum flow rates and pressures are 5 mL/min and 600 bar, respectively. The photodiode array is equipped with a micro flow cell with a volume of 1.7 μL and a 6 mm path length that is resistant to a pressure of 400 bar. The 1260 Infinity Analytical SFC system is controlled via ChemStation or OpenLab [62, 63].

One's view of SFC today is entirely different from that of 25 years ago when flow rates and gradient delivery were not reproducible, analytical UV sensitivity was not acceptable, and stationary phases were designed for reversed phase chromatography as opposed to normal phase. Today, SFC is considered to be primarily normal phase using mostly the same hardware and software developed for HPLC. The mobile phase is a binary or ternary mixture with CO_2 as the main component. The separation is usually performed as a gradient elution where the composition of the mobile phase is changed versus time. Polar stationary phases such as bare silica and 3-aminopropylsilica are routinely employed. pcSFC has numerous practical advantages relative to reversed phase HPLC such as higher speed throughput, more samples per day, more rapid equilibration, and shorter cycle time. SFC yields lower operating cost and lower column pressure drop, and it is orthogonal to reversed phase HPLC. Solvent consumption is low; therefore, waste generation is minimal. Finally, compounds of interest can be isolated in a relatively small amount of solvent because CO_2 vaporizes away.

The reluctance of analysts to use pcSFC has faded since the recent introduction of a new generation of instruments dedicated to analytical purposes by several important manufacturers [64]. These new systems benefit from a novel BPR design and are largely based on UHPLC technology. Improved performance, reliability, and full compatibility with the most modern stationary phases have greatly broadened the application spectrum of this technique. This text will draw heavily upon this advanced technology as it deals with a variety of analytical and preparative "real world" applications. Pedagogy as it relates to methods development involving a host of polar, nonpolar, and bipolar analytes will be emphasized. Applications to polymeric materials, natural products, polar water soluble analytes, surfactants, organic salts, fatty acids, lipids, organometallics, etc. are experiencing great success. Depending upon the nature of the stationary phase, reversed phase pcSFC, ion pairing pcSFC and aqueous promoted HILIC pcSFC can be expected to augment the more popular normal phase pcSFC that has historically been known for decades.

References

1 Schoenmakers, P.J. (1988). Open column or packed columns for supercritical fluid chromatography. In: *Supercritical Fluid Chromatography* (ed. R.M. Smith). London, UK: RSC Chromatography Monographs, The Royal Society of Chemistry Chapter 4.

2 Smith, R.M. (1993). *Pure and Applied Chemistry* 65: 2379–2403.

3 ASTM (1995). *Standard Guide for Supercritical Fluid Chromatography Terms and Relationships*, Designation E 1449-92, Annual Book of ASTM Standards., vol. 14.02, 905–910. Philadelphia, PA: American Society for Testing and Materials.

4 Chester, T.L. (2000). Unified chromatography from the mobile phase perspective. In: *Unified Chromatography*, ACS Symposium Series, 748 (eds. J.F. Parcher and T.L. Chester), 6–29. Cincinnati, Ohio: Procter and Gamble Company.

5 Laboureur, L., Ollero, M., and Touboul, D. (2015). Lipidomics by supercritical fluid chromatography. *International Journal of Molecular Sciences* 16 (6): 13868–13884.

6 Wells, P.S., Zhou, S., and Parcher, J.F. (2003). Unified chromatography with CO_2-based binary phases. *Analytical Chemistry* 73: 18A–24A.

7 Chester, T.L. (1999). The road to unified chromatography: the importance of phase behavior knowledge in supercritical fluid chromatography and related techniques, and a look at unification. *Microchemical Journal* 61: 12–24.

8 Konynenburg, P.H. and, Scott, R.L. "Critical lines and phase equilibria in binary van der Waals mixtures", *Philosophical Transactions of the Royal Society of London, Series A: Mathematical, Physical and Engineering Sciences*, 298 (1980) 265–273.

9 Charles Baron Cagnaird de la Tour (1822). Liquefaction of gases. *Annales de Chimie Physique* 21: 127.

10 Andrews, T. (1869). On the continuity of the gaseous and liquid states of matter. *Philosophical Transactions of the Royal Society* 159: 575–590.

11 Kelman, P. and Stone, H. (1970). *Mendeleeff: Prophet of Chemical Elements.* Engelwood Cliffs, NJ: Prentice Hall.

12 Hannay, J.B. and Hogarth, J. (1879). On the solubility of solids in gases. *Proceedings of the Royal Society of London* 29: 324–326.

13 Villard, P. (1897). Dissolution des liquides et des solides dans les gaz. *Zeitschrift für Physikalische Chemie* 23: 246.

14 Hannay, J.B. and Hogarth, J. (1880). On the solubility of solids in gases. *Proceedings of the Royal Society of London* 30: 178–484.

15 Banister, J. and Poliakoff, M. (1993). On the solubility of solids in gases; a lecture demonstration for supercritical fluids. *Journal of Supercritical Fluids* 6: 233–235.

16 H. S. Booth and, R.M. Bidwell "Solubility measurement in the critical region", *Chemical Reviews*, 44 (1949) 477–513.

17 Amagat, E.G. (1879). Researches on the compressibility of gases at elevated pressures. *Comptes Rendus des Seances de L'Academie des Sciences* 88: 336.

18 Cailetet, L. (1891). Description of a manometer in open air of 300 meters, established at the Eiffel Tower. *Comptes Rendus des Seances de L'Academie des Sciences* 112: 764.

19 White, C.M. (ed.) (1988). *Modern Supercritical Fluid Chromatography*, v. Heidelberg, Germany: Huthig.

20 Klesper, E., Corwin, A.H., and Turner, D.A. (1962). High pressure gas chromatography above critical temperatures. *The Journal of Organic Chemistry* 27: 700–701.

21 Braker, W. and Mussman, A.L. (eds.) (1980). *Matheson Gas Data Book*, 6e. Secaucus, NJ: Matheson.

22 Weast, R.C. (1984). *CRC Handbook of Chemistry and Physics*, 65e. Boca Raton, FL: Chemical Rubber Company.

23 Lewis, G.N. and Randall, M. (1961). *Thermodynamics*, 2e. (Rev. K.S. Pitzer and L. Brewer). New York: McGraw-Hill.

24 Leyendecker, D., Schmitz, F.P., and Klesper, E. (1984). Chromatography with sub- and supercritical eluents. *Journal of Chromatography* 315: 19–30.

25 Schmitz, F.P., Hilgers, H., Lorenschat, B., and Klesper, E. (1985). Separation of oligomers with UV-absorbing side groups by supercritical fluid chromatography using eluent gradients. *Journal of Chromatography* 346: 69–79.

26 Leyendecker, D., Schmitz, F.P., and Klesper, E. (1986). Ternary gradients by programming eluent composition and temperature at varying pressure. *Journal of High Resolution Chromatography and Chromatography Communications*: 525–527.

27 Wenclawiak, B.W. (1992). Ernst Klesper, the father of Supercritical Fluid Chromatography. *Fresenius' Journal of Analytical Chemistry* 344: 425.

28 Giddings, J.C., Myers, M.N., McLaren, L.M., and Keller, R.A. (1968). High pressure gas chromatography of nonvolatile species. *Science* 162: 67–73.

29 Giddings, J.C., Bowman, L.M., and Myers, M.N. (1977). Exclusion chromatography in dense gases: an approach to viscosity optimization. *Analytical Chemistry* 49: 243–249.

30 Sie, S.T., van Beersum, W., and Rijnders, G.W.A. (1966). High pressure gas chromatography and chromatography with supercritical fluids. 1. The effect of pressure on partition coefficients in gas-liquid chromatography with CO_2 as a carrier gas. *Separation Science* 1: 459–490.

31 Jentoft, R.E. and Gouw, T.H. (1970). Pressure program supercritical fluid chromatography of wide molecular weight range mixtures. *Journal of Chromatographic Science* 8: 138.

32 Jentoft, R.E. and Gouw, T.H. (1969). Supercritical fluid chromatography of monodisperse polystyrene. *Journal of Polymer Science, Polymer Letters* 7: 811–813.

33 Gouw, T.H. and Jentoft, R.E. (1972). Supercritical fluid chromatography. *Journal of Chromatography* 68: 303–323.

34 Jentoft, R.E. and Gouw, T.H. (1975). Analysis of polynuclear aromatic hydrocarbons in automobile exhaust by supercritical fluid chromatography. *Analytical Chemistry* 48: 2195–2200.

35 van Wassen, U. and Schneider, G.M. (1975). Pressure and density dependence of capacity ratios in sfc with CO_2 as the mobile phase. *Chromatographia* 8: 274–286.

36 Palmierl, M.D. (1988). An introduction to supercritical fluid chromatography. *Journal of Chemical Education* 65: A254–A259.

37 Bartmann, D. and Schneider, G.M. (1973). Experimental results and physico-chemical aspects of sfc with CO_2 as the mobile phase. *Journal of Chromatography* 83: 135–145.

38 Graham, J.A. and Rogers, L.B. (1980). Effect of column length, particle size, flow rate, and pressure programming rate on resolution in pressure programmed supercritical fluid chromatography. *Journal of Chromatographic Science* 18: 75–84.

39 Randall, L.G. and Wahrhaftig, A.L. (1978). Dense gas chromatography/mass spectrometer interface. *Analytical Chemistry* 50: 1703–1705.

40 Novotny, M., Bertsch, W., and Zlatkis, A. (1971). Temperature and pressure effects in SFC. *Journal of Chromatography* 61: 17–28.

41 Berger, T.A. (2015). *Supercritical Fluid Chromatography*, vii. Agilent Technologies, Inc.

42 Gere, D.R., Board, R., and McManigill, D. (1982). Supercritical fluid chromatography with small particle diameter columns. *Analytical Chemistry* 54: 730–736.

43 Berger, T.A. (2014). The past, present, and future of analytical supercritical fluid chromatography. *Chromatography Today* August–September: 26–29.

44 Saito, M. (2013). History of supercritical fluid chromatography: instrumental development. *Journal of Bioscience and Bioengineering* 115: 590–599.

45 Novotny, M., Springston, S.R., Peaden, P.A. et al. (1981). Capillary supercritical fluid chromatography. *Analytical Chemistry* 53: 407A–414A.

46 Chester, T.L. (1986). The role of supercritical fluid chromatography in analytical chemistry. *Journal of Chromatographic Science* 24: 225–229.

47 Peaden, P.A., Feldstedt, J.C., Lee, M. et al. (1982). Instrumental aspects of supercritical fluid chromatography. *Analytical Chemistry* 54: 1090–1093.

48 Smith, R.M. (1999). Supercritical fluids in separation science – (1) dreams, (2) reality and (3) future. *Journal of Chromatography A* 856: 83–115.

49 Later, D.W., Richter, B.E., Knowles, D.E., and Andersen, M.R. (1986). Analysis of various classes of drugs by capillary supercritical fluid chromatography. *Journal of Chromatographic Science* 24: 249–253.

50 Knowles, D.E., Richter, B.R., Wygant, M.B. et al. (1988). Supercritical fluid chromatography: a new technique for AOAC. *Journal of the Association of Official Analytical Chemists* 71: 451–457.

51 Rajendran, O., Krauchi, M., Mazzotti, M., and Morbidelli, M. (2005). Effect of pressure drop on solute retention and column efficiency in supercritical fluid chromatography. *Journal of Chromatography A* 10092: 149–160.

52 Chester, T.L. and Innis, D.P. (1993). Investigation of retention and selectivity in high-temperature, high-pressure, open-tubular supercritical fluid chromatography with CO_2 mobile phase. *Journal of Microcolumn Separations* 5 (5): 441–449.

53 Farrell, W.P., Aurigemma, C.M., and Masters-Moore, D.F. (2009). Advances in high throughput supercritical fluid chromatography. *Journal of Liquid Chromatography & Related Technologies* 32: 1689–1710.

54 Mangelings, D. and vander Heyden, Y. (2008). Chiral separations in sub- and supercritical fluid chromatography. *Journal of Separation Science* 31: 1252–1273.

55 Liu, Y., Berthod, A., Mitchell, C.R. et al. (2002). Super/subcritical fluid chromatography chiral separations with macrocyclic glycopeptide stationary phases. *Journal of Chromatography A* 978: 185–204.

56 Miller, L. and Potter, M. (2008). Preparative chromatographic resolution of racemates using HPLC and SFC in a pharmaceutical discovery environment. *Journal of Chromatography B, Analytical Technologies in the Biomedical and Life Sciences* 875: 230–236.

57 Chester, T.L. (1997). Chromatography from the mobile phase perspective. *Analytical Chemistry* 69: 165A–169A.

58 Fornstedt, T. (2015). Modern supercritical fluid chromatography – possibilities and pitfalls. *LC/GC North America* 33 (3): 166–174.

59 Tarafder, A., Hill, J.F., and Baynham, M. (2014). Convergence Chromatography Versus SFC – What's in a Name? *Chromatography Today* August–September: 34–36.

60 Alexander, A.J. (2012). SFC instrumentation modification to allow greater flexibility in method development by generating mixtures of solvents and modifiers on-line for mobile phase optimization. *Chromatographia* 75: 1185–1190.

61 de la Puente, M.L., Soto-Yarritu, P.L., and Burnett, J. (21011). Supercritical fluid chromatography in research laboratories: design, development, and implementation of an efficient generic screening for exploiting this technique in the achiral environment. *Journal of Chromatography A* 1218: 8551–8560.

62 Stevenson, R. (2013). State of the art separations with convergence chromatography. *American Laboratory* March: 36–38.

63 Swartz, M. (2012). HPLC Systems and Components Introduced at Pittcon 2012. *Chromatography Online*, April 30.

64 Novakova, L., Grand-Guillaume Perrenoud, A., Francois, I. et al. (2014). Modern analytical supercritical fluid chromatography using columns packed with sub-2 μm particles: a tutorial. *Analytica Chimica Acta* 824: 18–35.

2

Carbon Dioxide as the Mobile Phase

OUTLINE

2.1   Introduction to Carbon Dioxide

There is no mobile phase for packed column supercritical fluid chromatography (SFC) more suitable than CO_2. It is both a gas at room temperature and at atmospheric pressure that is produced when (i) people and animals breathe or (ii) certain fossil fuels are burned. It is also absorbed by plants in photosynthesis. It is (i) relatively nontoxic, (ii) does not support combustion, and (iii) is commercially and widely available. *Beverage grade* CO_2 is 99.9% CO_2 and is now mandated by the U.S. Food and Drug Administration as the mobile phase to be used in SFC provided the vapor is used and *not* the liquid. Regulations currently stipulate that the remaining 0.09% beverage gas can be made-up with low molecular weight hydrocarbons.

For comparison, *supercritical fluid grade* and *research grade* CO_2 are 99.998 and 99.999%, respectively (www.CO2Meter.com). Most contamination of these two grades occurs when the fluid is routinely packaged into smaller containers. Because carbonated soft drinks are now available worldwide, a distribution infrastructure exists that provides low-cost access to pure CO_2 in both high pressure cylinders and in larger volume Dewars. Higher pounds of CO_2 can cost as low as \$0.50/lb, but when delivered in bulk systems for mostly preparative chromatographic applications, the cost can be as low as \$0.05/lb. Carbon dioxide is (i) recyclable, (ii) abundant as water, (iii) highly diffusive, (iv) dissolves a

Modern Supercritical Fluid Chromatography: Carbon Dioxide Containing Mobile Phases,
First Edition. Larry M. Miller, J. David Pinkston, and Larry T. Taylor.
© 2020 John Wiley & Sons, Inc. Published 2020 by John Wiley & Sons, Inc.

large number of substances, and (v) easily removed by vaporization at reduced pressure. Both beverage and food grade CO_2 are therefore convenient and safe choices for any type of SFC.

Low critical parameters (**31.1 °C** and **7.36 MPa**) ensure that supercritical CO_2 is a safe solvent in, for example, bio-molecular separations, pharmaceutical applications, and a host of thermally labile systems. Despite general concerns, the use of CO_2 is unlikely to contribute significantly to an increase of CO_2 in the atmosphere. For example, a chromatographer driving the average American car to work 4.5 miles and back (~7 km) for example releases as much carbon dioxide into the environment as does a packed column instrument pumping liquid mobile phase at 2 mL/min for 24 hours. SFC, on the other hand, does not produce CO_2, but rather it *reuses* available CO_2. For comparison, all the CO_2 resulting from the burning of HPLC effluent is new CO_2. Carbon dioxide is today recognized as a "greenhouse" gas whereby recycling of carbon dioxide byproducts has been clearly demonstrated [1]. Much of the SFC work between 1984 and 1995 employed (i) capillary columns, (ii) flame ionization detection, (iii) isocratic or density programming, and (iv) 100% carbon dioxide as the mobile phase. Pressure or density programming that was popular in the early days with open tubular columns has now disappeared and been replaced by packed column SFC.

During the first 15 years of SFC experimentation, the solvent strength of carbon dioxide was grossly over-estimated. This misinformation came about during the mid-1960s, when Giddings predicted that chromatographic retention should be related to the Hildebrand solubility parameter (δ) of the solvent [2]. Unfortunately, estimated parameters for dense CO_2 were thought to be similar to that for isopropyl alcohol. Predicted large changes in solvent strength with pressure programming by mostly gas chromatographers were clearly over optimistic. Figure 2.1 compares Hildebrand solubility and Nile Red solvatochromic data for numerous solvents including CO_2. Today, it is generally accepted that even very dense CO_2 is known to act more like pentane or hexane than any alcohol solvent such as methyl alcohol or isopropyl alcohol [3]. Closer examination suggests several nonobvious polar attributes of carbon dioxide.

CO_2 in the gas phase exists as a linear nonpolar molecule with significant quadrupole moment. Because it is a charge-separated molecule with two polar carbon–oxygen bonds, it exhibits a significant nonzero bond dipole moment. Whether these features accurately characterize the solvent behavior of CO_2 has been extensively debated [4].

Because CO_2 has the potential to act both as weak Lewis acid and weak Lewis base, one may find it instructive to examine the solvent attributes of CO_2 from this perspective. Furthermore, strong theoretical and experimental evidence have indicated that CO_2 can either participate in both conventional and non-conventional hydrogen-bonding where CO_2 participates as a Lewis base or

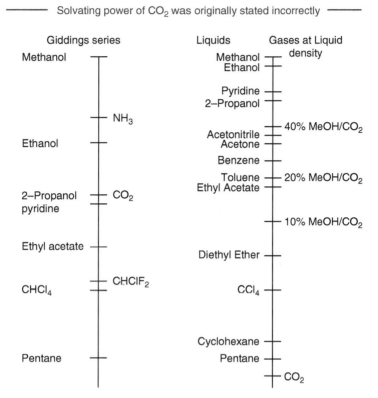

Figure 2.1 Solvating power of dense carbon dioxide erroneously predicted to be similar to isopropyl alcohol. *Source:* Berger [3]. Reproduced with permission of Royal Society of Chemistry.

nonconventional hydrogen-bond interactions where CO_2 participates as a weak Lewis acid. See Table 2.1 for a listing of CO_2 polar attributes.

One classic example that illustrates the polar nature of CO_2 is its higher solubility in H_2O compared to that of carbon monoxide (CO), which is both a polar molecule and contains a polar bond. Because CO is smaller in size than CO_2 and possesses a net dipole moment, one would assume that CO should be more soluble in water than CO_2. After all, a rule of thumb that many chemists first learn regarding solvation was that "like dissolves like." Thus, while polar solvents dissolve polar and ionic solutes, nonpolar solvents should dissolve nonpolar solutes. The higher solubility of CO_2 in H_2O has, therefore, been attributed to strong solute/solvent interactions despite the zero molecular dipole moment of carbon dioxide [5, 6]. Evidence for this observation came from molecular dynamic studies which identified the existence of a hydrogen bond between an oxygen atom in CO_2 and a hydrogen atom in H_2O. Such hydrogen

Table 2.1 Polar attributes of carbon dioxide.

- Weak Lewis acid carbon – weak Lewis base oxygen
- Low dielectric constant compared with hydrocarbons
- Nonzero dipole moment
- Charge separated molecule
- Solvent for thermally labile biomolecules
- Nonpolar molecule with polar bonds
---

bonds were not observed in a CO–H_2O system, indicating that site-specific, solute–solvent interactions are more important in the solvation of CO_2 by H_2O. Consequently, these studies also suggested considerable intermolecular interaction between the carbon (e.g. Lewis acid site) of CO_2 and the oxygen (e.g. Lewis base site) in H_2O.

2.2 Supercritical Carbon Dioxide

The first reported observation of the critical point of a substance was made in 1822 by Baron Cagniard de la Tour [7, 8]. The first reports of the solvating ability of a supercritical fluid were in 1879 by Hannay and Hogarth [9]. In 1967, Sie and Rijnders were the first to call the technique SFC [10]. The term was used when both fluid temperature and pressure exceeded their respective critical values. This working definition implied that the fluid needed to always be supercritical in order to display the characteristics of interest. This chromatographic assumption, unfortunately, was later found to be incorrect. "Supercritical" is a defined state. It possesses no physically unique characteristics, and thus, it possesses no physically unique characteristics, and thereby continues to be only three states of matter (i.e. solid, liquid, and gas).

Heating a liquid in a sealed tube has intrigued chemists and physicists for over 180 years, probably because the results are unexpected and almost counterintuitive. As Poliakoff and King noted, what one sees depends on how much liquid is in the tube [11]. If the tube is almost empty, the liquid evaporates quickly and one is left with a gas of moderate pressure. If it's almost full, the liquid expands rapidly to fill the whole tube and the pressure rises quickly. Put in just the right amount, and the meniscus, separating liquid from gas, grows faint and disappears abruptly. The contents of the tube in this case have passed through the critical point and have become supercritical.

If you heat the tube more slowly, the liquid begins to look opalescent as it approaches its critical point. As the opalescence increases, it grows more red and darker and goes completely black regardless of the substance. At the critical point, it becomes perfectly reflecting which prompted Poliakoff to comment: "you see your own eye staring back at you upon inspection." Heat a bit more and the fluid passes back through red to become completely transparent again. The change in temperature during this experiment is observed to be as small as one-hundredth of a degree.

Opalescence arises because the compressibility of the fluid at the critical point is infinitely high. When this happened, Poliakoff commented that the "microscopic thermal fluctuations that occur naturally in any fluid become strongly correlated, leading to large-scale, coherent density fluctuations which are very effective at scattering light." The high compressibility of the supercritical fluid and the correlated fluctuations also cause the speed of sound to drop to a minimum as the fluid passes through the critical point and the fluctuations attenuate the sound. When supercritical fluid of uniform density is rapidly cooled through the critical point, the contents of the tube separate into two phases. Gas bubbles and liquid droplets form with equal probability throughout the tube. According to Poliakoff [11], droplets begin to fall and the bubbles rise. "A veritable storm erupts which is quite different from the gradual disappearance of the meniscus when a liquid is heated."

Several interesting additional features of supercritical carbon dioxide have been reported [12]. The compressibility of the supercritical fluid is higher than the corresponding liquid. For example, the compressibilities of CO_2 and acetonitrile at 25 °C are, respectively, equal to 7.3×10^{-4} mL/atm and 4.5×10^{-5} mL/atm. On the other hand, turbulent flow is possible with supercritical fluids which gives rise to the notion that longer tubing will afford less extra band broadening. This hypothesis, however, has not been thoroughly documented [12].

The uniqueness of carbon dioxide is captured in Table 2.2 which lists the critical parameters for selected elements and molecules for comparison with CO_2 (e.g. critical temperature, critical pressure, and density at the critical point). The last column gives the solubility parameter under typical SFC conditions which measures the eluent strength and polarity. The Hildebrand parameter (δ_{SFC}) was calculated for each entry at a reduced pressure of 2 and at a reduced temperature of 1.02. For carbon dioxide, these conditions correspond to 145 atm and 37 °C.

For a solute to be soluble in a fluid, the solubility parameter of the solute and fluid should be similar. A study of the information in the table therefore strongly suggests that there are unique critical parameters related to CO_2 such as low critical pressure and low critical temperature when coupled with its solvating potential. This statement is strengthened by the following brief discussion which contrasts carbon dioxide and helium.

It may be argued that chromatography with helium as the carrier gas at pressures above 2.24 atm is SFC rather than GC. According to the definition given for SFC, this would indeed be the case (although GC conditions are likely to exist at the column outlet). This illustrates that the boundaries between GC and SFC are smooth. There is no phase change and there is nothing special

Table 2.2 Critical parameters for selected elements and molecules compared with CO_2.

Mobile phase	Critical temperature (°C)	Critical pressure (atm)	Critical density (gcm^{-3})	Calcd. solubility parameter (δ)
Group I				
Helium	−268	2.24	0.070	1.0
Hydrogen	−240	12.8	0.031	2.6
Neon	−229	27.2	0.484	4.2
Nitrogen	−147	33.5	0.313	4.7
Argon	−122	48.1	0.533	5.5
Krypton	−63.7	54.3	0.919	5.9
Group II				
Ethene	9.2	49.7	0.217	5.8
Xenon	16.6	57.6	1.113	6.1
CO_2	**31.05**	**72.9**	**0.466**	**7.5**
Ethane	32.2	48.2	0.203	5.8
N_2O	36.4	71.5	0.452	7.2
SF_6	45.5	37.1	0.738	5.5
NH_3	132.4	111.3	0.235	9.3
Diethy ether	193.5	35.9	0.265	5.4
n-Pentane	196.5	33.3	0.237	5.1
Group III				
n-Hexane	234.2	29.3	0.233	4.9
2-Propanol	235.1	47.0	0.273	7.4
Methanol	239.4	79.9	0.272	8.9
Ethyl acetate	250.1	37.8	0.308	5.7
Tetrahydrofuran	267.0	51.2	0.322	6.2
Acetonitrile	274.8	47.7	0.253	6.3
Water	374.1	217.6	0.322	13.5

about the supercritical area. Helium is nevertheless of no interest for SFC. The reason for this is seen in the table. At typical SFC conditions, helium's solubility parameter is not higher than 1.0. Moreover, the absolute values corresponding to the reduced parameters necessary to achieve SFC are highly impractical. For helium, a reduced temperature ($T_R = 1.02$) would correspond to a temperature of about −268 °C. At higher temperatures, the density and the solubility parameter of helium will decrease even further. Therefore, under GC conditions helium has virtually no solvating power and solutes would be eluted exclusively on the basis of their vapor pressure.

As stated previously, pure substances with critical temperatures below 0 °C have no solvating power. Therefore, the highest group of substances in Table 2.2 cannot be used as solvents for SFC. The second highest group of substances in the table contain a number of solvents which have been used for SFC. The critical temperatures of these solvents have been seen to be in the range 0 °C < T_C < 200 °C. All common SFC solvents thus appear to be in the second block of materials. The third group which rests at the bottom of the table contains some solvents used in HPLC for which T_C > 200 °C. The use of these pure substances for pcSFC appears to be impractical. Good SFC solvents, for practical purposes, should have critical temperature in the 0–200 °C range.

2.3 Solvating Power of Supercritical CO_2

The solvating power of supercritical CO_2 is highly dependent on its temperature and pressure. Surprisingly, solvating power *decreases* with rising temperature at low pressure; whereas at high pressures it *increases* in a straightforward fashion as measured by naphthalene solubility in carbon dioxide, for example,

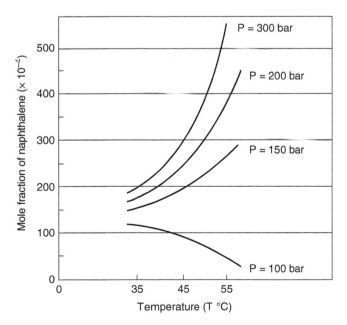

Figure 2.2 Solubility (mole fraction) of naphthalene in CO_2 as a function of temperature at various pressures [13].

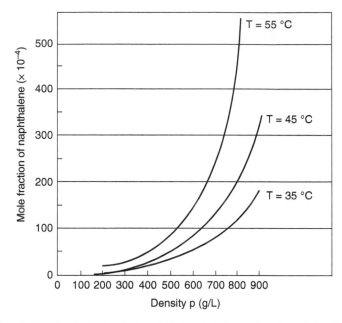

Figure 2.3 Solubility (mole fraction) of naphthalene in CO_2 as a function of density at various temperatures [13].

Figure 2.2. If we replace the parameter "pressure" by the parameter "density," the solubility–temperature relationship becomes much simpler as shown in Figure 2.3. This anomaly comes about because density decreases dramatically with an increase in temperature at low pressure; whereas, at higher pressure, changes in temperature have much less effect on density [14]. Thus, density, not pressure, to a first approximation, is proportional to the solvating power of the supercritical fluid (SF). The following trends are thus based upon many solubility measurements in the region from ambient conditions up to 1000 atm and 100 °C.

- Solvating power of a SF increases with density at a given temperature.
- Solvating power of a SF increases with temperature at a given density.

From the van der Waal's Law of Corresponding States one would expect that these temperature-pressure–density relationships would be applicable to all substances. For such a comparison, it is more convenient to define the pressure–temperature region in terms of reduced pressure ($P_{actual}/P_C = P_R$), reduced temperature ($T_{actual}/T_C = T_R$), and reduced density ($d_{actual}/d_C = d_R$) (where P and d may be in any units, but T must be in Kelvin for calculation purposes). The region just above $T_R = 1.0$ (304.05 K), $P_R = 1.0$ (72.9 atm) is the traditional operational supercritical region. At high values of T_R, the fluid density may be reduced to the point where solvent properties are no longer favorable if restricted by the pump to a relatively low pressure. Because of safety concerns arising from equipment limitation, reduced pressures above $P_R = 5$ (364 atm) or 6 (437 atm) may be difficult to achieve in the analytical laboratory. The region of condensed phase between $T_R = 1.0$ (31.05 °C) (304.05 K) and $T_R = 0.95$ (15.2 °C) (288.2 K) is termed the subcritical (or near critical) region.

Employment of a pressurized view cell affords a dramatic way to demonstrate solvating power. Figure 2.4 shows the effect of increasing pressure on solubility of tributyl phosphine nitric acid adduct. As the pressure is increased more analyte disappears, however, the stirring bar added to the experiment is unaffected as might be expected by the increase in pressure.

It is important to recognize how density (i.e. solvating power) changes with changes in pressure and temperature. As Figure 2.5 attests, a very small increase in pressure at a reduced temperature (T_R) between 1.0 and 1.2 results in a dramatic increase in density. Whereas the same change in pressure at $T_R > 1.5$ (183 °C) (456 K) hardly has an effect on the fluid density [15]. Density can be varied with the external pressure or with temperature, but it is linear only at high temperature. One should also note that the density of the SF practically never exceeds the density of the comparable liquid regardless of the pressure [13]. For reference, the density of liquid CO_2 is generally considered to be ~0.92 g/mL.

The solvent strength of a supercritical fluid (e.g. compressed gas) may be adjusted continuously from gas-like to liquid-like values and is well described qualitatively by the Hildebrand solubility parameter, which is the square root of the cohesive energy density. The parameter for gaseous carbon dioxide is

Dissolution of TBP-HNO$_3$ in SC-CO$_2$

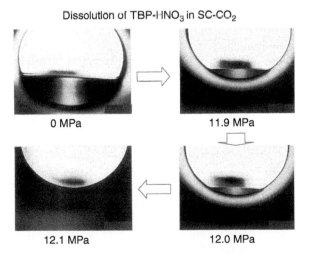

0 MPa   11.9 MPa

12.1 MPa   12.0 MPa

Figure 2.4 Dissolution of TBP–HNO$_3$ in supercritical CO_2 as a function of pressure [11].

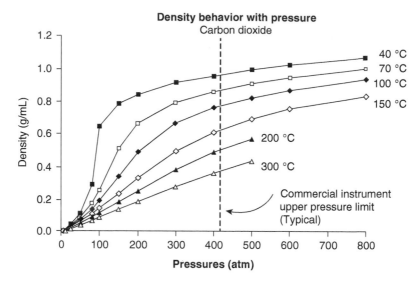

✦ Unlike gases or liquids, solvent strength (i.e. density) varies
   continuously in supercritical fluids with pressure

✦ Density can be varied with the external pressure or
   temperature. Linear only at high temperature

Figure 2.5 Solvent strength adjustment with density for pure fluids. Reprinted with permission from reference [15]. Copyright 2005 Marcel Deckker.

essentially zero, whereas the value for liquid carbon dioxide is comparable with that of a hydrocarbon [16]. At −30 °C, there is a large increase in δ upon condensation from vapor to liquid. Above the critical temperature, it is possible to tune the solubility parameter continuously over a wide range with either a small isothermal pressure change or a small isobaric temperature change. Unlike gases and liquids, solvent strength (i.e. density) varies continuously in a supercritical fluid with pressure. Density, on the other hand, can be varied with external pressure or temperature. It is linear however only at high temperature. The ability to tune the solvent strength of a supercritical fluid is its unique feature compared to conventional liquids, and it, as a result, can be practically used to extract and then recover many types of products [13].

Studies demonstrating the possibility of continuous phase change from liquid to gas led Hannay and Hogarth to inquire whether solvent action is characteristic only of the liquid phase, or whether gases might not also be capable of dissolving solids. Their work demonstrated that the solution of several salts in ethanol is not limited to the liquid state [9]. In particular, it was noted that an alcoholic solution of cobalt chloride above its critical point had essentially the same absorption spectrum as in the normal liquid state, thus indicating that the usual ionic condition persists even above the critical temperature of a solution. The concentration of solids in the supercritical phase was far higher than that could be accounted for by the normal volatility of the salts concerned.

Viewed from another perspective in this matter is (i) analyte solubility in the supercritical fluid as opposed to (ii) solvating power of the supercritical fluid. An established class of compounds that are highly miscible with carbon dioxide is fluorocarbons. Additional CO$_2$-philic materials are carbonyl-containing molecular systems such as poly (ether-carbonate) copolymers and sugar acetates. These facts [13] along with recent computational and spectroscopic studies surprisingly suggest a rather polar nature for carbon dioxide as a solvent. In other words, what kind of analytes dissolve in supercritical CO$_2$. Based upon the high affinity of CO$_2$ for molecules containing fluorine and carbonyl groups, it would appear that Lewis acidity is more strongly favored for CO$_2$ interaction than is Lewis basicity.

Solubility of analytes can be difficult to predict, however, because of purity issues. Even though impurities appear as minor constituents, their presence can have a major impact on solubility in supercritical carbon dioxide. Numerous constituents are involved that may differ significantly in molecular size, shape, structure, and polarity. Experimental solubility measurements on well-defined model systems are required to develop better predictive methods. The first understanding of the solvent power of carbon dioxide came from solubility data on 261 solid compounds by Francis in near critical carbon dioxide (900 psi, 25 °C) [17]. Table 2.3 lists some of the results from this early study. Although Francis studied solubility behavior in liquid carbon dioxide rather than supercritical carbon dioxide, the results may be applicable to high-density supercritical CO$_2$.

Table 2.3 Solubility of specific compounds in subcritical CO$_2$ [17].

	Weight Percent		Weight Percent
Esters		*Amines and Heterocyclics*	
Benzyl benzoate	10	Aniline	3
Butyl oxalate	M	*o*-Chloroaniline	5
Butyl phthalate	8	*m*-Chloroaniline	1
Butyl stearate	3	*N, N*-Diethylaniline	17
Ethyl acetate	M	*N, N*-Dimethylaniline	M
Ethyl acetoacetate	M	Diphenylamine	1
Ethyl benzoate	M	*N*-Ethylaniline	13

(Continued)

Table 2.3 (Continued)

	Weight Percent		Weight Percent
Ethyl chloroformate	M	N-Methylaniline	20
Ethyl maleate	M	α-Naphthylamine	1
Ethyl oxalate	M	2,5-Dimethyl-pyrrole	5
Ethyl phthalate	10	Pyridine	M
Methyl salicylate	M	o-Toluidine	7
Phenyl phthalate	1	m-Toluidine	15
Phenyl salicylate	9	p-Toluidine	7
Alcohols		*Phenols*	
t-Amyl alcohol	M	o-Chlorophenol	M
Benzyl alcohol	8	p-Chlorophenol	8
Cinnamyl alcohol	5	o-Cresol	2
Cyclohexanol	4	m-Cresol	4
1-Decyl alcohol	1	p-Cresol	2
Methyl alcohol	M	2,4-Dichlorophenol	14
Ethyl alcohol	M	p-Ethylphenol	1
2-Ethylhexanol	17	o-Nitrophenol	M
Furfuryl alcohol	4	Phenol (MP 41°C)	3
Heptyl alcohol	6	β-Methoxyethanol	M
Hexyl alcohol	11	Phenylethanol	3
Carboxylic Acids		*Nitriles and Amides*	
Acetic acid	M	Acetonitrile	M
Caproic acid	M	Acrylonitrile	M
Caprylic acid	M	Phenylacetonitrile	13
Formic acid	M	Succinonitrile	2
Isocaproic acid	M	Tolunitriles (mixed)	M
Lactic acid	0.5	Acetamide	1
Laurie acid	1	N, N-Diethyl acetamide	M
Oleic acid	2	N, N-Dimethylacetamide	M
		Formamide	0.5

M, miscible

A typical solubility study is described here with a series of coumarins [18]. Solubilities of the basic coumarin and four monosubstituted derivatives (7-hydroxy, 7-methoxy, 7-methyl, 6-methyl, and 4 hydroxy-coumarin) in supercritical CO$_2$ were measured from 50 to 35°C and pressures of 8.5–25 MPa as shown in Figures 2.6 and 2.7. In general, monosubstituted coumarin derivatives were less soluble than the basic coumarin in CO$_2$.The degree of solubility tends to increase in the order: 7-methyl, 7-methoxy, 7-hydroxy, and 4-hydroxycoumarin. However, in the case of 6-methylcoumarin, solubility was unusually high relative to the basic coumarin in CO$_2$ over the entire range of experimental conditions. Furthermore, the disubstituted coumarin derivatives were extremely less soluble than the basic coumarin and monosubstituted coumarins. For each

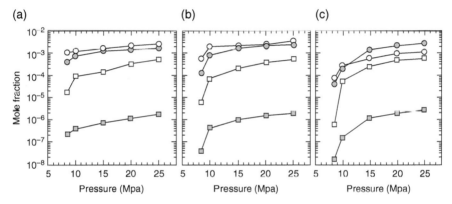

Figure 2.6 Compared solubilities of coumarin, 7-methylcoumarin, 4-hydroxycoumarin, and 7-hydroxycoumarin in supercritical CO$_2$ at 35, 40, and 50 °C, respectively [18].

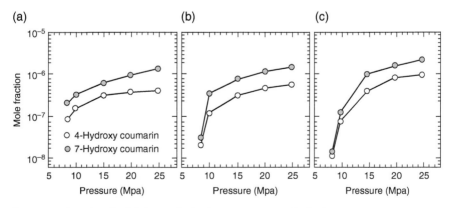

Figure 2.7 Compared solubilities of 4-hydroxycoumarin and 7-hydroxycoumarin in supercritical CO$_2$ at 35, 40, and 50 °C [18].

coumarin derivative, optimum equilibrium conditions which gave maximum solubility in CO_2 were reported.

The reader is referred to similar solubility studies of mesotetraphenylporphyrin coupled with supercritical pentane and toluene phases for comparison [19]. Supercritical toluene was the preferred solvent for this porphyrin because relatively large masses of analyte were solubilized at moderate pressures.

Counter to the previous studies, Johnston has noted that solubility differences among various solids are governed primarily by vapor pressure and only secondarily by solute–solvent interactions in the supercritical phase at a given density [20]. This concept is best illustrated via a normalized enhancement factor, which is the solubility normalized by the solubility in an ideal gas. Here, the vapor pressure effect is removed which provides a means to focus on solute–solvent interactions. Unfortunately, accurate enhancement factors are known for only a limited number of solids because vapor pressures are often inaccurate or unavailable in the range of 10^{-6} to 10^{-2} Pa.

To understand solvent strength more effectively, an alternate approach which focusses on interactions in the SF phase was considered. Spectroscopic measurement of solvatochromic shifts provided such an approach. Solvatochromic parameters are influenced by the local environment near the solute and thus describe solvent strength more effectively than bulk properties such as solubility parameter [21, 22]. Suffice it to say more traditional experimental measurements of solid solubilities are time-consuming and may have large uncertainties [23].

To allow comparison with conventional liquid solvents, the pressure dependent solvating powers of supercritical fluids have therefore been studied using spectroscopic methods such as the solvatochromic method, which uses selected probe molecules to determine the polarity or polarizability of the fluid [21, 22]. The solvent strength for several fluids expressed within the Kamlet–Taft π^* scale of solvent polarity/polarizability compared with several conventional liquids are shown in Figure 2.8. The π^* scale was deliberately designed to exclude hydrogen bonding effects and thus focused on the polarity/polarizability of the solvent exclusively. The measurements confirmed that polarity or polarizability increases as fluid density increases. More polar fluids such as ammonia show a large change in solvent strength as density increases, whereas less polar fluids such as SF_6 and C_2H_6 show smaller effects. Xenon shows poorer solvent strength than CO_2 even though xenon has a much higher critical density. Both CO_2 and N_2O show similar solvent strength over the entire density regime even though the latter is a more polar molecule.

In summary, there appears to be no material with critical parameters as mild as those of CO_2 and with solvating power as high as CO_2. The π^* values for liquid methylene chloride, benzene, ethanol, carbon tetrachloride, hexane, and perfluorohexane are also included in Figure 2.8 for comparison. At liquid-like

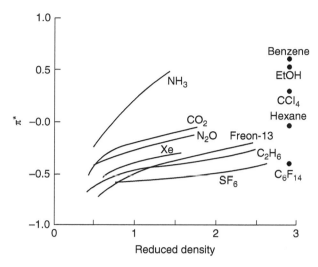

Figure 2.8 π^* solvent polarizability parameter for various supercritical fluids as a function of reduced density at a reduced temperature of 1.03. The single π^* values for benzene, ethanol, carbon tetrachloride, hexane, and perfluorohexane are given for comparison [22].

densities, CO_2 exhibits solvating properties greater than hexane. As previously discussed, one should also note that solvent properties of conventional liquids are fixed and appear as a dot in the figure, whereas those of supercritical fluids are variable. Searching for a CO_2 substitute such as SF_6, H_2O, ammonia, or fluoroform, seldom produced a positive result, and in many cases safety issues with several of these solvents became insurmountable [24].

An SF exhibits other physicochemical properties intermediate between those of a liquid and a gas. Mass transfer relative to a liquid is rapid in SF's. For pressures between 50 and 500 atm, *diffusivity* of supercritical CO_2 varies between 10^{-4} and 10^{-3} cm²/sec. Similarly, *viscosities* of SFs are 10–100 times lower than liquids. In other words, the viscosity and diffusivity of a SF begin to approach those of a liquid as pressure and density are increased. Diffusivity will increase with an increase in temperature. These changes in viscosity and diffusivity are most pronounced in the region around the critical point. It should be noted, however, from a chromatographic viewpoint that at even high external pressures (e.g. 300–400 atm), viscosity and diffusivity of SF's differ by one to two orders of magnitude from normal liquids (see Figures 2.9 and 2.10).

The faster rates of diffusion and lower viscosity relative to liquids translate into (i) greater optimum chromatographic mobile phase velocity, (ii) shorter analysis times, (iii) higher efficiency per unit time for chromatography, and (iv) the ability to use longer packed columns and/or smaller bonded packed particles without dramatically higher precolumn pressures. Additional properties of SF's that vary

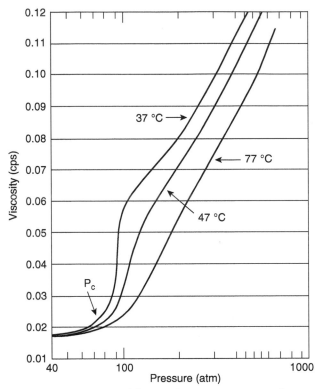

Viscosity behavior of CO_2 at various temperatures and pressures

Figure 2.9 Viscosity behavior of CO_2 at various temperatures and pressures. Greatest change in viscosity with pressure occurs near the critical point. Reprinted with permission from reference [13]. Copyright 1996 John Wiley & Sons.

widely over a broad range of temperatures and pressures around the critical point are (i) thermal conductivity, (ii) partial molar volume, and (iii) heat capacity.

Carbon dioxide, however, is not perfectly ideal in all aspects. For example, mixtures of CO_2 and H_2O are corrosive and CO_2 is not inert with respect to primary and secondary aliphatic amines such as dimethylamine. Many fluoro-forms (such as Freon-23 and Freon-134A) have physical/chemical properties that recommend them for use as the main fluid in SFC. In practice, however, none of these, show a significant difference from simple carbon dioxide. Xenon has been used as a mobile phase in SFC when performing infrared detection. But the high cost precludes this fluid from all but the most important investigations. Ammonia was suggested as the primary fluid in SFC as far back as 1968 with mixed results. Giddings reported [25, 26] the elution of some very polar biological solutes using ammonia although there was significant difficulty when others attempted to repeat the work.

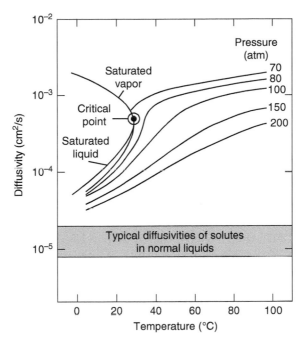

Figure 2.10 Diffusivity of CO_2 varies with temperature at various pressures. Greatest change in diffusivity with temperature is near the critical point. Reprinted with permission from reference [13]. Copyright 1996 John Wiley & Sons.

2.4 Solvating Power of Modified CO_2

The solvent strength of carbon dioxide varies markedly with the density of the fluid. The solvent strength at a density less than $0.25 \, g/cm^3$ is less than that of perfluorinated alkanes; at a density of $0.98 \, g/cm^3$ it exceeds that of hexane. Even at high densities near $1.0 \, g/mL$, CO_2 has limited ability to dissolve polar molecules [27]. The solvating power of SF CO_2 can, however, be enhanced by the addition of a small amount of miscible, polar compound to the primary fluid [28] such as methanol.

This second component is referred to as a modifier or cosolvent. Methanol is currently the most common modifier for supercritical CO_2 [29]. It is important to note that mobile phase composition is more important than mobile phase density regarding chromatographic retention changes. However, when the mobile phase exists as both vapor and liquid in the column, separations are destroyed. One is more likely to encounter a two-phase liquid/vapor region when the column pressure is either too low, temperature is too high, or modifier concentration is too great. The two-phase region, however, is easily avoided by simply maintaining the column pressure above the two-phase region at the temperature used. For example, ~120 bar for a mixture of $MeOH/CO_2$ at $60\,°C$ would be required.

Table 2.4 Polar modifiers which enhance solvating power of CO_2.

Modifier	T_C (°C)	P_C (atm)
Methanol	239.4	79.9
Ethanol	243.0	63.0
2-Propanol	235.1	47.0
Tetrahydrofuran	267.0	51.2
Acetonitrile	275.0	47.7

Adding Modifier Raises T_c and P_c.
The Effect on T_c is Larger than on P_c.

T_c and P_c of CO_2 are altered by the incorporation of modifier into the SF. The T_c of different CO_2/modifier mixtures lies between those of the two pure components (i.e. CO_2 and modifier). The P_c value, for comparison, usually shows a maximum at intermediate composition between that of either pure CO_2 or the pure liquid modifier. Table 2.4 lists the critical pressure and temperature for several pure, common modifiers. T_c is much higher for these modifiers than the T_c for CO_2. Critical pressure for the compounds is comparable to that of CO_2 (31 °C).

The critical constants of the mixture can be measured experimentally but may be more easily approximated as the arithmetic mean of the critical temperatures and pressures of the two components as follows where X = mole fraction of either modifier or CO_2 in the mixture and T_c' and P_c' are the critical parameters for the modifier/CO_2 *mixture*.

$$T'_c = X(CO_2)T_c(CO_2) + X(m)T_c(m)$$

$$P'_c = X(CO_2)P_c(CO_2) + X(m)P_c(m)$$

Most polar organic liquids (i.e. alcohols, ethers, tetrahydrofuran, dimethyl sulfoxide, chloroform, and methoxyethanol) have solubility parameters that are larger than that of CO_2, so that they may be used to increase extraction yields or to decrease pressure and solvent requirements in SFC or SFE. A summary of the large increase in analyte solubility that may be obtained with the addition of a simple cosolvent to CO_2 is given in Table 2.5. The key to predicting cosolvent effects is to calculate the important physical and chemical interactions between the solute and the cosolvent.

The complexing agent tributyl-phosphate increases the solubility of hydroquinone by a factor of 300. An anionic surfactant, aerosol–OT forms aggregates and reverse micelles in supercritical ethane, which can solubilize ionic substances such as tryptophan at levels of 0.4 wt%. The use of complexing

Table 2.5 Effects of co-solvents including simple liquids, complexing agents, and surfactants versus solubility in supercritical fluid at 35 °C [20].

Solute	Co-solvent	$\dfrac{y}{y\text{binary}}$
acridine	3.5% CH$_3$OH	2.3
2-aminobenzoic acid	3.5% CH$_3$OH	7.2
cholesterol	9% CH$_3$OH	100
hydroquinone	2% tributyl-phosphate	>300
hydroquinone	0.65% AOT, W$_o$ = 10*6% octanol	>200
tryptophan	0.53% AOT, W$_o$ = 10*5% octanol	>>100

agents and surfactants provides an opportunity to extend supercritical technology to hydrophilic substances such as biomolecules.

The use of modifier mixtures offers great flexibility since the modifier and its concentration are easily and widely variable, and the solvent power can be altered to fit the analysis. Modifiers can be introduced into chromatographic systems in primarily two ways. First, premixed tanks of modified fluid can be purchased with a variety of fluid and modifier concentrations. There is limited flexibility when premixed tanks of modified fluid are used because each time a change in the composition of the mobile phase is desired, another cylinder must be purchased. Furthermore, vapor–liquid equilibrium changes have been documented in liquid aluminum tanks from which most SFC mobile phases are provided. In other words, as the liquid volume in a premixed cylinder is depleted, the total gas volume above the liquid increases. Because the primary fluid (CO_2) has a much higher vapor pressure than the modifier (methanol), it preferentially moves into the vapor phase thus occupying the newly created volume. A concentration of 5.1% v/v methanol in the cylinder containing the remaining modified fluid was revealed after 26.3 lb liquid (CO_2) had been removed from the filled cylinder [30].

A second (and more convenient) way to add modifier is to use a two-pump system wherein one pump delivers the pure fluid and the other pump delivers the liquid modifier. After efficient mixing, the two-fluid systems are mixed in a volume/volume ratio to form the mobile phase. While the incorporation of modifier enhances the fluid's solvating power, the effect on mass transfer may also be altered, especially if the modifier and the solute interact. Solute binary diffusion coefficients in pure CO_2 are more than an order of magnitude greater than in normal organic solvent or water. The typical operating range in SFC is 5–50% modifier although the greatest speed advantage of SFC over HPLC is at the lower end of the scale. Near 100% modifier, Berger has stated that SFC is indistinguishable from HPLC [31].

As an illustration of the effect of modifiers on physical properties of CO_2-based fluids, diffusion coefficients have been measured for acridine, phenanthrene, and benzoic acid in pure CO_2 and methanol-modified CO_2 as a function of temperature (Table 2.6), basic acridine and benzoic acid showed considerably lower diffusion coefficients in the modified fluid [32]. This indicates likely association between the methanol and the analyte. In the presence of a modifier, phenanthrene being neither acidic nor basic, showed no change in diffusion coefficient relative to pure CO_2. For the best chromatography or extraction, the mobile phase should have high solvating power, high diffusivity, and low viscosity. Because all three conditions cannot be met together, chromatographic parameters such as pressure, temperature, and composition must be optimized.

With supercritical CO_2 at typical SFC operating conditions (70–450 atm) the range of Hildebrand solubility parameters achieved overlaps with the solubility parameters of several common solvents such as Freons, pentanes, methylene chloride, THF, toluene, and methyl and isopropyl alcohols. Even though the Hildebrand solubility parameter is at best only a first approximation to the actual phenomena involved, it provides the SFC user with the useful rule of

Table 2.6 Addition of modifier lowers diffusivity: diffusion coefficients for Acridine, Phenanthrene, and Benzoic acid ($10^9 D_{12}$, m^2/s). Reprinted with permission from reference [32]. Copyright 1990 Elsevier.

Temperature (°C)	D_{12}	
	Pure $CO_2{}^a$	CO_2/MeOH^b
	Benzoic Acid	
35	8.70 ± 0.9	5.75 ± 0.42
45	10.5 ±0.9	6.13 ± 0.51
55	11.6 ± 1.1	6.69 ± 0.60
	Acridine	
35	6.37 ± 0.6	5.65 ± 0.60
45	7.88 ± 0.9	6.96 ± 0.40
55		7.78 ± 0.90
	Phenanthrene	
35	7.18 ± 0.5	7.77 ± 0.40
45	8.87 ± 0.7	9.36 ± 1.10
55	9.43 ± 1.2	10.0 ± 1.20

^{a} 173 bar.
^{b} 3.5 mol %.
Phenanthrene does not H-bond with MeOH.

Table 2.7 Role of modifiers in pcSFC.

- Cover active sites on solid support
- Swell or solvate he stationary phase
- Reduce analyte retention
- Improve peak symmetry
- Increase density of CO$_2$ mobile phase

---

thumb that if a sample has appreciable solubility in one or more of the previously mentioned organic solvents, there is a reasonable chance that it will migrate in supercritical CO$_2$.

This obviously excludes many polar molecules that are of great interest to pharmaceutical and biotechnology laboratories that have zero or limited solubility in organic solvents. It should be noted here that modifiers have numerous roles in addition to solubility enhancement. Table 2.7 captures the most notable chromatographic features. For example, the adsorbed film on the stationary phase when using a mixture of MeOH and CO$_2$ is at least three monolayers thick. At maximum total adsorption (CO$_2$ + MeOH) was 9% of the adsorbed film on an octyldecyl (ODS) silica column and 28% of the adsorbed film on silica. Total adsorption was ODS (23 µmol/m^2); Silica (29 µmol/m^2).

2.5 Clustering of CO$_2$

Many years ago early observations of SF behavior were placed on a stronger foundation by the experimental work of Berger and Deye who characterized SFs using solvatochromic dyes that changed color when the polarity of their solvation spheres changed [33, 34]. Pure CO$_2$ was reported to produce an apparent solvent strength similar to hexane. In a subsequent study, Berger and Deye measured the solvating strength of *modified* supercritical fluids using Nile Red (E$_{NR}$) and a dye named E$_{t30}$, Figure 2.11. The pertinent results are shown in Figure 2.12. Both solvent strength and hydrogen bonding were predicted to increase from upper left to lower right in the figure. The middle line indicates the solvent strength of MeOH/CO$_2$ mixtures. It is important to note that small additions of MeOH caused large increases in solvent strength. The E$_{t30}$ scale on the x-axis is very sensitive to hydrogen bonding. The Nile Red scale on the y-axis is less dependent on hydrogen bonding. The nonlinearity between modifier concentration and solvent strength was thought to be due to "clustering" of MeOH molecules.

Within a cluster, modifier concentration and density are higher thus creating microenvironments of locally high polarity compared to the nominal bulk composition and density. Clustering is more prevalent when the compressibility is

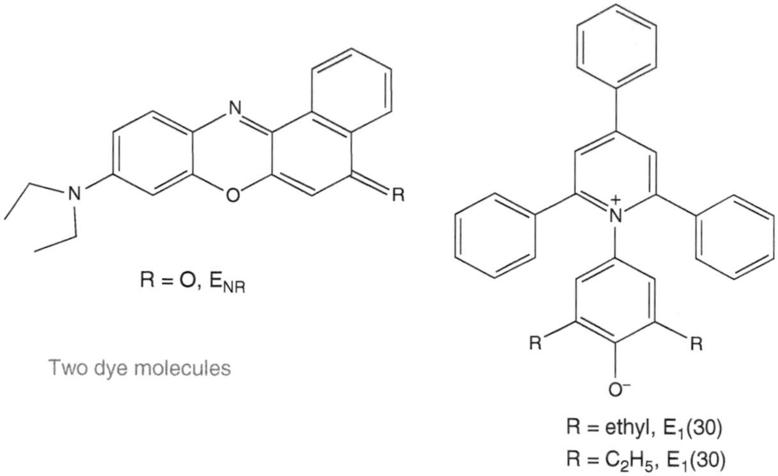

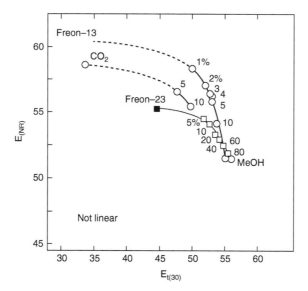

R = O, E_{NR}

Two dye molecules

R = ethyl, $E_1(30)$
R = C_2H_5, $E_1(30)$

Figure 2.11 Measurement of binary CO_2 – modifier solvating power [33, 34].

Figure 2.12 Comparison of transition energies from solvatochromic data for mixtures of methanol and three supercritical fluids [34].

large – for instance, around the critical point [35]. Polar solutes interact with the polar clusters of modifier. Thus, a polar modifier may increase markedly the solubility of a polar solute, but may not affect the solubility of a nonpolar solute. The larger the difference in solvent strength between the pure supercritical fluid and the modifier, the more dramatic the solvent strength shift upon addition of the modifier. For example, 5% MeOH/Freon13 (i.e. more extensive clustering) is predicted to be a stronger solvent than 5% MeOH/CO_2 (less extensive clustering) with both dyes. In the highly compressible near critical region, the solvent clusters arrange themselves about the solute due to the attractive intermolecular forces. For example, at 308.5 K and 79.8 bar, the partial molar volume of naphthalene in supercritical carbon dioxide at infinite dilution is $-7800\, \text{cm}^3$/mol, which corresponds to the condensation of about 80 solvent molecules around a solute molecule [36]. This large degree of condensation corresponds to macroscopic clusters that extend over many coordination shells.

Clustering in supercritical fluid mixtures was later studied with a different dye and acetone as the modifier [37]. Solubility data were measured with and without cosolvent, and then regressed using an augmented van der Waals-density-dependent local composition model to determine the relevant binary interaction energies. The solute, phenol blue, was used as a probe to estimate local concentrations by measuring the effect on the absorption wavelength. The bulk concentrations of the cosolvents n-octane, acetone, ethanol, and methanol ranged from 0 to 100 mole percent and the pressure ranged from 80 to 300 bar. At bulk concentrations of 3.5 and 5.25%, the local compositions were greater for methanol and ethanol, which form stronger hydrogen bonds with phenol blue than does acetone. In all cases, the local compositions exceed the bulk concentrations because the cosolvent interacts more strongly with the solute than it does with carbon dioxide. The local compositions decrease toward the bulk values as the pressure increases, because the isothermal compressibility decreases. The effect of the attractive forces on the structure of the fluid decreased as the isothermal compressibility decreased because the molecules became less mobile.

Clustering is more common and extensive when fluid and modifier are highly dissimilar in polarity [38]. For example, more extensive clustering is predicted for Freon/MeOH than for CO_2/MeOH. A molecular model for CO_2/modifier mobile phase has been suggested. (i) A unique feature of both pure and modified supercritical fluids is that solvent molecules cluster about solute when there is large compressibility (i.e. critical point). (ii) There is preferential solvation of the solute by the modifier rather than by the fluid. The local concentration of modifier near the solute thus exceeds the average bulk value. (iii) The mobile phase is heterogeneous. (iv) Solute partitions between the small cluster, the absorbed film and the purchased stationary phase.

Numerous ^{1}H NMR studies of hydrogen-bonding in methanol mixtures with supercritical CO_2 have been reported. A large body of experimental data was generated over a wide range of pressure and temperature using high pressure online spectroscopy [39, 40]. The study showed success in quantitatively predicting hydrogen bonding. Investigations into the dynamics of hydrogen bonding in CO_2–ethanol mixtures have revealed that hydrogen bonds between CO_2 and ethanol molecules are limited both in number and lifetime. Each NMR spectrum consisted of three peaks, attributable to the OH, CH_2, and CH_3 moieties. The chemical shifts of all three peaks changed with variable concentration, temperature, and pressure. The effect of these variables on the CH_2 and CH_3 shifts are the same and can be attributed to non-hydrogen bonding effects. The shift of the OH peak, relative to the other two, reflects purely hydrogen bonding effects.

In a somewhat related study, near-infrared absorption spectroscopy of supercritical CO_2–ethanol for molar fractions from 0.05 to 1 in ethanol in the temperature range 40–200 °C and at two different constant pressures P = 15 and 20 MPa were obtained [41]. A strong dependence of the degree of hydrogen bonding as a function of temperature and ethanol molar fraction was observed. The degree of hydrogen bonding in these mixtures was predicted to have a strong influence on the dissolution of polar solutes in ethanol–CO_2 mixtures. The study focused in the 4000–7800 cm^{-1} region where it is possible to detect the contribution of the OH stretching overtone.

References

1 Welch, C.J., Wu, N., Biba, M. et al. (2010). Green analytical chromatography. *Trends in Analytical Chemistry* 29: 667–680.

2 Giddings, J.C., Myers, M.N., McLaren, L., and Keller, R.A. (1968). High pressure gas chromatography of nonvolatile species. *Science* 162: 67–73.

3 Berger, T.A. (1995). Solubility was based upon calculated Hilldebrand solubility parameters. In: *Packed Column SFC* (ed. R.M. Smith), 56. Royal Society of Chemistry.

4 Raveendran, P., Ikushima, Y., and Wallen, S.L. (2005). Polar attributes of supercritical carbon dioxide. *Accounts of Chemical Research* 38: 478–485.

5 Sato, T., Matubayan, N., Nakahara, M., and Hirata, F. (2002). Which carbon oxide is more soluble? *Ab Initio* study on carbon monoxide and dioxide in aqueous solution. *Chemical Physics Letters* 323: 257–262.

6 Brogle, H. (1982). CO_2 as a solvent: its properties and applications. *Chemistry & Industry* 12: 385.

7 Cagniard de la Tour, C. Nouvelle note sur les effets qu`on obtient par l`application simultanee de la chaleur et de la compression a certains liquides. *Annales de Chimie Physique* 22 (1823): 410–415.

8 Cagniard de la Tour, C. (1822). Expose de quelques resultats obtenu par l`action combinee de la chaleur et de la compression sur certains liquides, tells que l`eau, l`alcool, l ether sulfuriquee et l`essence de petrole rectifee. *Annales de Chimie Physique* 21: 127–132.

9 Hannay, J.B. and Hogarth, K. (1879). *Proceedings. Royal Society of London* 29: 324.

10 Sie, S.T., van Beersum, W., and Rijnders, G.W.A. (1966). High pressure gas chromatography and chromatography with supercritical fluids. 1. The effect of pressure on partition coefficients in gas-liquid chromatography with CO_2 as a carrier gas. *Separation Science* 1: 459–90.

11 Poliakoff, M. and King, P. (2001). Phenomenal fluids. *Nature* 412: 125.
12 Brown, M.W. (1987). Neither liquid nor gas, odd substances find use. *New York Times* (19 May).
13 Taylor, L.T. (1996). *Supercritical Fluid Extraction.* New York, NY: Wiley Chapter 2.
14 Schoenmakers, P.J. and Uunk, L.G.M. (1987). Supercritical fluid chromatography – recent and future developments. *European Chromatography News* 1 (3): 14–18.
15 Chester, T.L. and Pinkston, J.D. (2005). Supercritical Fluid Chromatography in *Ewing's Analytical Instrumentation Handbook, 3rd Edition,* Cazes, J., Editor, Marcel Dekker, New York.
16 Sengers, J.V. and Sengers, A.L. (1968). The critical region. *C&EN* 10: 104–118.
17 Francis, A.W. (1954). Ternary systems of liquid carbon dioxide. *The Journal of Physical Chemistry* 58: 1099.
18 Choi, Y.H., Kim, J., Noh, M.J. et al. (1997). Effect of functional groups on the solubility of coumarin derivatives in supercritical carbon dioxide. In: *Supercritical Fluids: Extraction and Pollution Prevention,* ACS Symposium Series, 670 (eds. M.A. Abraham and A.K. Sunol), 110–118.
19 Bergstresser, T.R. and Paulaitis, M.E. (1987). Solubility of meso-tetraphenylporphrin in two supercritical fluid solvents. *Supercritical Fluids: Chemical and Engineering Principles and Applications, ACS Symposium Series, 329,* (eds T.G. Squires and M.E. Paulaitis), University of Deleware, Iowa, 138–148.
20 Johnston, K.P., Kim, S., and Combes, J. (1989). Spectroscopic determination of solvent strength and structure in supercritical fluid mixtures: a review. *Supercritical Fluid Science and Technology, ACS Symposium Series, 406,* Symposium sponsored by American Institute of Chemical Engineers Washington, DC, 52–70.
21 Bush, D. and Eckert, C.A. (1996). Estimation of solid solubilities in supercritical carbon dioxide from solute solvatochromic parameters. *Supercritical Fluids: Extraction and Pollution Prevention, ACS Symposium Series, 670* (eds. M.A. Abraham and A.K. Sunol), *Symposium Sponsored by* Division of Industrial and Engineering Chemistry, ISSN 0097-6156, 37–50.
22 Frye, S.L., Yonker, C.R., Kalkwarf, D.R., and Smith, R.D. (1987). Application of solvatochromic probes to supercritical and mixed fluid solvents. *Supercritical Fluids: Chemical and Engineering Principles and Applications, ACS Symposium Series, 329,* (eds T.G. Squires and M.E. Paulaitis), Sponsored by Division of Fuel Chemistry, Iowa State University, Iowa, 29.
23 Booth, H.S. and Bidwell, R.M. (1949). Solubility measurement in the critical region. *Chemical Reviews* 44: 477–513.
24 Howard, A.L., Yoo, W.J., Taylor, L.T. et al. (1993). Supercritical fluid extraction of environmental analytes using trifluoromethane. *Journal of Chromatographic Science* 31: 401–408.
25 Giddings, J.C., Myers, N.N., and King, J.W. (1969). Dense gas chromatography pressures to 2000 atmospheres. *Journal of Chromatographic Science* 7: 276–283.
26 Giddings, J.C., McLaren, M.N., and Myers, L. (1968). Dense gas chromatography of nonvolatile substances of high molecular weight. *Science* 159: 197–199.

27 Yonker, C.R. and Smith, R.D. (1988). *Supercritical Fluid Extraction and Chromatography, ACS Symposium Series, 366* (eds. B.A. Charpentier and M.R. Sevenants), American Chemical Society, Washington, DC, Sponsored by Division of Agricultural and Food Chemistry, 161.

28 Saito, M. (2013). History of supercritical fluid chromatography: instrumental development. *Journal of Bioscience and Bioengineering* 115: 590–599.

29 Yang, Y. (2010, 2010). Mobile phase modifiers for SFC: influence on retention. In: *Encyclopedia of Chromatography*, vol. 10 (ed. J. Cazes), 1519–1522. Greenville, NC: East Carolina University.

30 Schweighardt, F.K. and Mathias, P.M. (1993). Impact of phase equilibria on the behavior of cylinder-stored CO_2-modifier mixtures used as supercritical fluids. *Journal of Chromatographic Science* 31: 207–211.

31 Berger, T.A. (2015). Instrumentation for analytical scale supercritical fluid chromatography. *Journal of Chromatography A* 1421: 172–183.

32 Smith, S.A., Shenai, V., and Matthews, M.A. (1990). Diffusion in supercritical mixtures: CO_2 + cosolvent + solute. *Journal of Supercritical Fluids* 3: 175–179.

33 Deye, J.F., Berger, T.A., and Anderson, A.G. (1990). Nile red as a solvatochromic dye for measuring solvent strength in normal liquids and mixtures of normal liquids with supercritical and near critical liquids. *Analytical Chemistry* 62: 615–622.

34 Berger, T.A. and Deye, J.F. (1992). Use of solvatochromic dyes to correlate mobile phase solvent strength to chromatographic retention in SFC. *Supercritical Fluid Technology, ACS Symposium Series, 488* (eds F.V. Bright and M.E.P. McNally), Sponsored by Division of Analytical Chemistry, ISSN 0097-6156, 132-142.

35 Kaufman, J.F. (1996). Spectroscopy of solvent clustering. *Analytical Chemistry News & Features* 68: 248A–253A.

36 Kim, S. and Johnston, K.P. (1987). Clustering in supercritical fluid mixtures. *AIChE Journal* 33: 1603–1611.

37 Eckert, C.A., Ziger, D.H., Johnston, K.P., and Kim, S. (1986). Solute partial molar volumes in supercritical fluids. *The Journal of Physical Chemistry* 90: 2738.

38 Johnston, K.P., McFann, G.J., Peck, D.G., and Lemen, R.M. (1989). Design and characterization of the molecular environment in supercritical fluids. *Fluid Phase Equilibria* 52: 337–346.

39 Maiwald, M., Li, H., Schnabei, T. et al. (2007). On-line ^{1}H NMR spectroscopic investigation of hydrogen bonding in supercritical and near critical CO_2–methanol up to 35 MPa and 403 K. *Journal of Supercritical Fluids* 43: 267–275.

40 Reiser, S., McCann, N., Horsch, M., and Hasse, H. (2012). Hydrogen bonding of ethanol in supercritical mixtures with CO_2 by ^{1}H NMR spectroscopy and molecular simulation. *Journal of Supercritical Fluids* 68: 94–103.

41 Lafrad, F., Idrissi, A., and Tassaing, T. (2014). What is the extent of ethanol molecule aggregation in ethanol-supercritical carbon dioxide mixtures? An FTIR investigation in the full molar fraction range. *Journal of Supercritical Fluids* 94: 65–70.

3

Instrumentation for Analytical Scale Packed Column SFC

Modern Supercritical Fluid Chromatography: Carbon Dioxide Containing Mobile Phases,
First Edition. Larry M. Miller, J. David Pinkston, and Larry T. Taylor.
© 2020 John Wiley & Sons, Inc. Published 2020 by John Wiley & Sons, Inc.

3.1   Introduction

This chapter covers the instrumentation for contemporary, analytical-scale, packed column SFC. Figure 3.1 shows a simple schematic diagram of a modern SFC instrument. The pumps in this system are operated under flow control. The system pressure is regulated using a "downstream" (postcolumn) back-pressure regulator. The primary components of the system that are described in this chapter are shown on the diagram. See Chapter 1 for a short summary of instrumentation for open-tubular SFC. Detection in SFC is described in Chapter 4. SFC columns, stationary phases, mobile phases, and method development are treated in Chapter 6.

At first glance, instrumentation for packed column SFC is surprisingly similar to that for its cousin, high performance liquid chromatography (HPLC). Both require high pressure pumps for pumping and mixing a variety of mobile phase components. Both use high pressure "inline" injection and columns packed with small particles, housed (usually) in thermostated column compartments. A wide variety of similar detectors (or at least similar in appearance and function) are available for SFC and HPLC. But there are critical differences in virtually all these components, and SFC requires new instrumental components which are not generally used in HPLC. This chapter focusses on the different and unique aspects of SFC. Note that "SFC" is used throughout this chapter, despite the fact that the mobile phase in contemporary "supercritical" fluid chromatography is not always supercritical. However, whether the mobile phase is supercritical or liquid/subcritical, the same instrumental considerations are required to maintain the mobile phase in a single phase during the chromatographic separation. Namely, the column outlet must be held at or above a particular pressure dictated by the mobile phase composition. This is necessary for a good separation, and for sensitive condensed-phase detection.

3.2   Safety Considerations

All of the safety considerations that exist for HPLC systems also exist for SFC systems. Yet SFC systems exhibit additional safety concerns. Care must be taken with any system that incorporates compressed gasses. Compressed gas can expand rapidly, even explosively, if a component or fitting fails. Chester and Pinkston [1] recommend that the burst pressure rating of a component in SFC should be at least four times the desired maximum operating pressure. Temperature can also play a role. Maximum pressure ratings are often determined at ambient temperature. Higher operating temperatures may reduce the maximum operating pressure. Consult the vendor when in doubt. A major

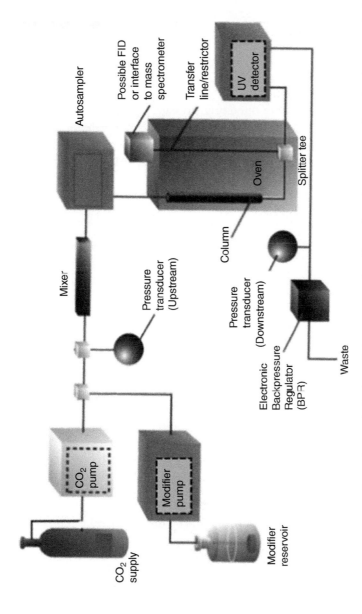

Figure 3.1 Schematic diagram of a modern packed column SFC instrument. The effluent from the column is split in this diagram to the "UV Detector" where condensed-phase detection occurs, and to a "Possible FID" (FID; flame ionization detector) where low-pressure/gas phase detection occurs. This lower pressure detection can also be an interface to mass spectrometric detection (see Chapter 4).

manufacturer of SFC and other popular chromatographic instrumentation states that "Components *[in our SFC instrumentation]* have a minimum of 3.5x safety factor for primary burst stresses. This complies with ASME BPVC Section VIII, Division I, (a popular standard worldwide due to its conservativeness)" (J.D. Pinkston, Discussion with major manufacturer of chromatographic equipment about pressure ratings of components in SFC instrumentation. 2016). (Text in italics is added by this author.) This vendor also noted that "The burst disks/pressure relief valves in the system are rated to about 25% greater than the maximum operating pressure, so technically the systems could never get to even 2x the maximum operating pressure" (J.D. Pinkston, Discussion with major manufacturer of chromatographic equipment about pressure ratings of components in SFC instrumentation. 2016).

While Joule–Thompson cooling due to fluid expansion can help reveal a small leak in a SFC system, a larger leak can result in a cryogenic-burn hazard to unprotected skin. Exercise proper care and use proper personal protection equipment (i.e. gloves, at a minimum).

The effluent from a SFC system often contains a mixture of organic solvents and decompressed CO_2. Vent the effluent to a solvent trap, followed by a chemical fume hood system.

3.3 Fluid Supply

3.3.1 Carbon Dioxide and Other Compressed Gases

Carbon dioxide is by far the most widely used primary mobile phase components for SFC, but other fluids have also been used in SFC, such as pentane [2, 3], nitrous oxide [4], Freons and sulfur hexafluoride [5], and others. As discussed in Chapter 2, CO_2 has many advantages, such as being nonflammable, inexpensive, and relatively inert, among others. The purity of the carbon dioxide required for SFC can vary greatly, depending on the particular application and mode of detection. For example, flame ionization detection requires ultra-high-purity, instrument-grade CO_2 that can be very expensive. Instruments that require a supply of liquid CO_2 (rather than using recompression of gaseous CO_2 drawn from the supply tank) generally require higher purity CO_2. In contrast, instruments which are designed to condense gaseous CO_2 and which use ultra-violet absorbance, mass spectrometric, or other more selective modes of detection can use very inexpensive, lower purity grades of CO_2, such as "beverage grade." Evaporative-light-scattering and charged-aerosol detection, while nonspecific, are also compatible with inexpensive grades of CO_2, because impurities in liquid CO_2 are usually volatile and not detected.

Table 3.1 Comparison of estimated cost of HPLC and SFC mobile phase components.

Hexane, HPLC-grade, 95%	$88/L[a]
Ethanol, Analytical-Grade, 96%	$118/L[a]
Methanol, HPLC-grade, >99.9%	$29/L[a]
Acetonitrile, HPLC-grade, >99.9%	$88/L[a]
Water, HPLC-grade (<0.000 3% evap. residue)	$21/L[a]
CO_2 – SFC/SFE (99.999 9%) grade/eductor tube	$16/L[b]
CO_2 – Research grade (99.999%), vapor from cylinder	$6.8/L[b]
CO_2 – Pure Clean grade (99.995%), vapor from cylinder	$3.0/L[b]
CO_2 – Instrument Grade (99.99%), vapor from cylinder	$2.0/L[b]
CO_2 – Instrument Grade (99.99%), vapor from Dewar	$2.0/L[b]

[a] From the U.S. website of a major international supplier of HPLC-grade solvents, July 2016.
[b] At 6.08 MPa and 22.4 °C, quote from a major international gas supplier, July 2016.

Laboratories equipped with one or a few analytical scale SFC systems often use individual cylinders of CO_2. While convenient and requiring a low initial capital investment, this is usually the most expensive option in terms of cost/ unit mass. Labs equipped with many SFC systems, and especially those equipped with preparative scale SFC, benefit from larger CO_2 supply systems. Large fluid-supply tanks (sometimes many tons) are usually located in a remote location, and automatically notify the supplier when a refill is needed. The most common approach with these large systems is to use food/beverage grade CO_2 and gas recompression. The gas or fluid is carried to the laboratory via high purity stainless steel tubing. Important practical considerations and advantages and disadvantages of various types of fluid-supply installations have been reviewed by Pradines [6]. It's important to note that one of the primary advantages of SFC over HPLC is the greatly reduced cost of the primary component of the mobile phase. This is especially true for preparative-scale separations. Table 3.1 compares the estimated cost of various HPLC and SFC fluids. Note that the cost of some, in particular acetonitrile, have varied considerably and have been far higher, at times, than estimated here.

3.3.2 Mobile Phase "Modifiers" and "Additives"

Organic modifiers are usually high purity "HPLC-grade" solvents, low in UV-absorbing impurities. As described in Chapter 6 on method development, the most common modifier is methanol, and others are similar, small organics of moderate polarity. Degassing of these solvents is not required for SFC because of the high pressure maintained within inline detectors (such as UV/VIS absorbance detectors). Mobile-phase additives (discussed in detail in Chapter 6) are generally "ACS Reagent" grade or better. Additives range from small organic acids, to small bases [7, 8] – often small amines, but even including ammonium hydroxide [9], to volatile salts [10, 11], and even water [12, 13]. Most additives are dissolved in the modifier at low levels (mMolar to low-percent levels). The modifier and additive are therefore pumped as a single mixture, as described in the next section. One important consequence of this convenient arrangement is that the additive concentration in the mobile phase rises in unison with the modifier concentration during a mobile-phase gradient.

3.4 Fluid Delivery – Pumps and Pumping Considerations

3.4.1 Pump Thermostating

Pumping systems in analytical-scale SFC often look similar to their HPLC cousins, and there are indeed many similarities. But the similarities are only skin deep. The differences are described in detail by Berger [14]. Carbon dioxide, even when compressed and/or cooled to the liquid state, is a very compressible fluid as compared to traditional liquids. Its compressibility decreases as liquid CO_2 is cooled, so the pump heads of most CO_2 pumping systems are cooled to 5–10 °C. Some cooling systems use Peltier coolers, while others use a cooled fluid circulated through or around the pump head. The Peltier designs are more convenient, but may contribute more to the cost of the instrument, and may not have the cooling capacity of the circulating fluid designs.

3.4.2 Fluid Pressurization and Metering

Instrumentation suppliers have gone to great pains to develop pump-control software and firmware which incorporate dynamic compressibility compensation for CO_2. Built-in compressibility tables and empirical real-time feedback helped provide a smooth and accurate flow of CO_2. Despite these measures, pump-related noise was shown to be one cause of baseline noise when using some detection approaches, such as UV/VIS absorption [15]. Some recent designs have separated the compression stage of pumping from the flow metering or delivery stage [16]. This, along with other measures, has reduced noise in SFC related to pumping, pressure regulation, mobile-phase temperature regulation, and UV/VIS detector design to the point where ASTM noise in SFC/UV equals that of HPLC/UV (0.006 mAU per the specification of a major manufacturer) [17, 18]. Figure 3.2 illustrates the effect of one of these improvements [17].

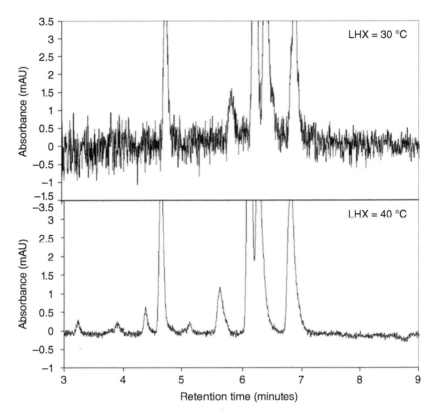

Figure 3.2 Example of the effect of instrumental parameters on UV detection noise in SFC. This is an example of the effect of temperature control of the column effluent entering the UV detector. Top chromatogram was acquired with a pre-UV-cell heat exchanger set at 30 °C, while the bottom chromatogram was acquired with the heat exchanger at 40 °C. Analytes are warfarin breakdown products. The samples injected were identical in the two chromatograms. Chromatographic conditions: 5–35% 2-propanol modifier added to CO_2, flow rate of 2 mL/min, column temperature of 40 °C, 100 bar outlet pressure. Column: 3 mm × 100 mm, 1.8 μm RX-SIL. 20 Hz. *Source:* Reprinted with permission from reference [17]. Copyright Elsevier, 2014.

3.4.3 Modifier Fluid Pumping

Mobile phase modifiers are usually organic liquids, or mixtures of organic liquids, as described above and in Chapter 6. Correspondingly, modifier pumps are generally traditional HPLC pumps. However, as it is often desirable to explore the effects of a number of modifiers during method development, especially in chiral SFC, SFC systems designed for chiral method development allow automated switching from one modifier fluid to another, some with up to six available fluids.

3.4.4 Pressure and Flow Ranges

Pressure limits of analytical scale SFC pumping systems are $\sim$30–60 MPa, and maximum flow rates vary from 3 to 5 mL/min of combined fluid (CO_2 and modifier). One modern system provides up to 3 mL/min at a maximum pressure of 41.4 MPa, or up to 4 mL/min with a maximum pressure of 29.3 MPa. Another lists a maximum flow rate of 5 mL/min with a maximum pressure of 60 MPa. While these operating parameters are sufficient for analytical-scale SFC, they do not allow crossover to semipreparative separations, as some earlier designs allowed. Some practitioners have argued for the need for higher maximum operating pressures to allow for longer columns, smaller particle sizes, higher flow rates, or all the three [19].

3.4.5 Fluid Mixing

Proper fluid mixing may be the most important "unseen" or unconsidered (from a users' perspective) aspect of fluid delivery in SFC. Instrument developers and researchers may choose dynamic (active) or static (passive) mixers. Each has advantages and limitations. Mixers that ensure optimum mixing may have larger dead volumes, and these dead volumes become a disadvantage when a high speed mobile-phase gradient (as in moving from 5 to 60% modifier concentration over 30 seconds or less) is required for high speed analytical separations [20]. This issue becomes quite important when a high-speed gradient is required with a narrow bore column (such as 2-mm internal diameter), and its associated lower flow rates. The user must be aware of the potential and consequences of gradient delay, and be prepared to move to a lower dead volume, though potentially less effective, mixer when these issues are encountered.

3.5 Sample Injection and Autosamplers

Much like pumps and other components of SFC systems, injectors and autosamplers resemble their HPLC counterparts. But there are some important differences in injection in SFC which researchers and instrument manufacturers have spent much effort in characterizing and addressing. Probably the most obvious difference is that the sample injection solvent in SFC is often a "chromatographically strong" solvent. It is important that the sample be injected in a solvent in which it is soluble. This frequently means that a short-chain alcohol, such as methanol, is used as the sample injection solvent, and this injection solvent is therefore the same, or very similar, to the mobile phase modifier. In practice, this results in tighter limitations on injection volume than is commonly seen in HPLC. Large injection volumes using a "chromatographically strong" injection solvent prevent focusing of the analytes on the head of the column. The analytes can be "washed" into the column, resulting in broad chromatographic peaks, and reducing the potential chromatographic resolving power of the chromatographic system. Enmark et al. provided an in-depth study which detailed this effect, as well as two other possible injection-solvent effects on peak distortion in SFC – the presence of retained cosolvents in the injection solvent mixture, and "viscous fingering" of the injection solvent [21]. The bottom line is that injection volume with a 4.6-mm-i.d. column used under common SFC conditions is often limited to 10 μL or less. Similarly, 2-mm-i.d. columns may be limited to 2 μL or less. The constraints on injection volume are rarely a significant limitation for analytical-scale separations, but they can become a severe limitation in preparative-scale separations, as discussed in Chapter 7.

Some researchers have addressed the "strong injection solvent" issue in SFC by using an injection-solvent mixture that approximates the solvent strength of the mobile phase [22]. For example, De Pauw et al. [23] used an injection-solvent mixture of hexane/ethanol/isopropanol 85/10/5 (% vol.) to match the CO_2/methanol mobile phase in terms of elution strength and polarity. By using this mixture and lowering the injection volume from 2.5 to 0.4 μL, and the pre- to extra-column volume from 7.4 to 4.4 μL, the apparent plate count was increased from 7600 to 21 300 for a weakly retained compound (k' = 2.3) on a 2.1 × 150 mm column, as shown in Figure 3.3.

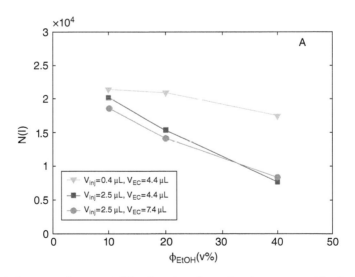

Figure 3.3 Apparent plate count (N) as function of sample solvent composition (volumetric EtOH-concentration, mixed with hexane and IPA), with the EtOH/IPA ratio always equal to 2, for different injection volumes (Vinj) and extra-precolumn volumes (V_{EC}). Probe analyte is testosterone, a weakly retained compound (k' = 2.3). Measured on a 2.1 mm × 150 mm Zorbax HILIC RRHD column (packed with 1.8-μm particles) at a flow rate of 1 mL/min. *Source:* Reprinted with permission from reference [23]. Copyright 2015, Elsevier.

A potentially more problematic issue, at least with regard to injection volume reproducibility and accuracy are concerned, is directly related to the gaseous nature of CO_2 at atmospheric pressure and ambient temperature. During full- or partial-loop injection using traditional HPLC technology, the injection valve loop is filled with the sample solution with the valve in the "load" position. The valve is then switched to the "inject" position, where the injection-valve loop is placed in series with the mobile-phase supply line between the mobile-phase mixer and the chromatographic column. The mobile phase moves through the loop and carries the sample solution from the loop to the column, and the chromatographic separation then proceeds without issue.

However, at this point, the injection valve is commonly switched back to 'load" position to prepare for the next sample injection. The loop is filled with mobile phase, which, in SFC, consists mostly of CO_2. This aliquot of mobile phase rapidly decompresses, with an audible "puff," when the injection-valve loop comes into communication with atmospheric pressure. The decompression of mobile phase from the loop can even expel mobile-phase modifier into the laboratory if the modifier concentration in the mobile phase is high. Placing a narrow bore restrictor (e.g. 50- or 75-μm-i.d. fused silica tubing) between the loop outlet and an appropriate waste vessel can prevent this potential problem. Regardless of how the decompression is accomplished, the injection-valve loop is filled with atmospheric pressure air or CO_2 after the decompression. If proper care is not taken when filling the loop with the next sample, poor injection volume accuracy and reproducibility can result [24]. This is believed to be due to very small gas bubbles which are not easily expelled from the injection-port loop during filling, and which displace sample solvent. By flushing the injection-port loop with copious amounts of liquid before sample loading, Chester and Coym demonstrated excellent injection volume reproducibility and accuracy [24]. They suggest that the small bubbles are gradually flushed from the system.

3.6 Tubing and Connections

3.6.1 Tubing

3.6.1.1 Stainless Steel Tubing

The most widely used connecting tubing in SFC is narrow-bore stainless steel. The most common internal diameter tubings when using 4.6- to 2-mm-i.d. columns are 178 μm (0.007 inch), 170 μm (0.0067 inch), 127 μm (0.005 inch), and 120 μm (0.0047 inch). The importance of using appropriate internal-diameter tubing was demonstrated by De Pauw et al. [23], as illustrated in Figure 3.4. Using a 2.1 mm × 150 mm column and a detector with a reduced internal dead volume of only 0.6 μL, a move from 120-μm internal diameter tubing to 250-μm

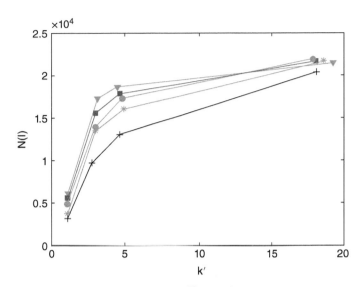

Figure 3.4 Influence of tubing ID (65 (■), 120 (▼), 170 (●), and 250 µm (✳)) on extra-column band broadening using a 0.6 L flow cell at 1 mL/min. The figure shows apparent plate count (N) vs. retention (k′). For reference, the 1.7 µL flow cell with 250 µm tubing is given as well (+). The experiments were performed on a 2.1 × 100 mm Zorbax HILIC column, back pressure at 130 bar, injection volume of 0.8 µL consisting of 10/5/85 EtOH/IPA/hexane. The used compounds (in order of elution) are aspirine (k~1), testosterone (k~3), β-estradiol (k~5) and chlorthalidone (k~18). *Source:* Reprinted with permission from reference [23]. Copyright 2015, Elsevier.

i.d. tubing resulted in a drop from 17 300 to 14 000 apparent theoretical plates for testosterone, a weakly retained compound (k ~ 3).

The components of modern analytical-scale SFC systems are located in close proximity, and, under normal operating conditions (e.g. 2-mL/min mobile phase flow rate, measured as a condensed fluid, for a 4.6-mm-i.d. column), the pressure drop due to the connecting tubing is small relative to that within the chromatographic column. However, as flow rate is increased beyond this level for higher speed separations, the pressure drop within the connecting tubing may become significant. Berger has published an extensive and careful study of extra-column band broadening in SFC [25]. By replacing 170-µm-i.d. tubing with shorter lengths of 120-µm-i.d., as well as using a low-dead-volume UV/VIS detection cell, he was able to achieve peak fidelity of >0.95 with peaks of k = 2, using a 3-mm-i.d. × 100-mm long column packed with 1.8-µm particles. A reduced plate height as low as 1.87 was achieved.

3.6.1.2 Polymeric Tubing

Some HPLC systems use polymeric (PEEK or polyetheretherketone) tubing, which is flexible and easy to cut with a clean, square end. It is especially useful when contact between the analytes and metal is to be avoided. PEEK tubing can be used to 100 °C, and 178 μm (0.007 inch) and 127 μm (0.005 inch) are rated to 48.3 MPa (8000 psi) [26]. But PEEK is not compatible with dichloromethane, dimethylsulfoxide, and tetrahydrofuran. These authors have successfully used PEEK tubing in analytical-scale SFC systems with CO_2 mobile phase with common modifiers and additives. Note that connections must be made with stainless steel unions, nuts, and ferrules, not with PEEK unions, nuts, and ferrules.

A note is included here about the use of fused-silica tubing in packed column SFC systems. Researchers may at times like to use fused-silica tubing with internal-diameter ranging from 125 to 50 μm, especially when interfacing with detectors or in other specialized uses. Baker and Pinkston found that undeactivated fused silica tubing is compatible with pure CO_2 mobile phase, but will develop small leaks, and will eventually fail when modified CO_2 mobile phase (containing, for example, methanol modifier) is used. However, deactivated fused silica tubing will resist this degradation, and can be used successfully for many months with modified CO_2 mixtures [27].

3.6.2 Connections

Coupling stainless steel or PEEK tubing to another piece of tubing or to other components of the chromatographic system is best accomplished with near-zero-dead-volume stainless steel nuts, ferrules, and unions. Such fitting components supplied for HPLC are rated to withstand the temperatures and pressures encountered in analytical-scale SFC. Vendors have introduced a variety of "finger-tight" column connection fittings that provide secure and convenient near-zero-dead-volume connections between tubing and columns.

3.7 Column and Mobile Phase Temperature Control

The temperature of the chromatographic separation has long been known to have more impact on selectivity and efficiency in SFC than in HPLC [28, 29]. This effect has been exploited to varying degrees by SFC instrument manufacturers. Not surprisingly, therefore, a wide variety of column-oven styles have been used in SFC over the years. These range from GC-style forced-air convection ovens to traditional HPLC-style "column heaters." Each approach has advantages and disadvantages. The HPLC-style column compartments are relatively inexpensive and compact. Some provide enough space for a coil of mobile-phase inlet tubing, or other form of heat exchanger, to heat the incoming mobile phase to the column temperature before the mobile phase enters the column. This reduces cooling of the column by the incoming mobile phase, and a resulting axial temperature gradient. The HPLC-style column heaters have a limited range of available temperatures, generally ranging from slightly above ambient temperature to a maximum of 60 °C or 80 °C. Changing the column temperature can also be relatively slow (as compared to the GC-style oven), especially when lowering the temperature. Nevertheless, this approach to temperature control has been widely used, and has been successful for arguably the majority of modern SFC separations. These include separations of most pharmaceutical mixtures, where elevated temperatures are not desirable, and separations where a temperature gradient is not needed to adjust selectivity.

The GC-style convective ovens are generally more expensive, but are also larger. Their size allows larger columns, automated column-switching devices, and even multiple, coupled columns for high-resolution separations [19]. These ovens also allow higher, and sometimes lower, temperatures than their HPLC-style counterparts. Higher temperatures can be useful for high-resolution separations of thermally stable analytes. For example, Pinkston et al. have performed high resolution SFC separations of small polymers (alkyl ethoxy propoxy branched chain block copolymers) at temperatures up to 200 °C [19]. (A note of caution here – many columns are not compatible with such high temperatures. See Chapter 6.) Many GC-style ovens also allow subambient operation. This can be especially useful for separations where the selectivity between a critical pair is inversely related to temperature. Such is the case, for example, in some chiral separations. These authors have performed enantiomeric separations at column temperatures as low as −50 °C [30]. Such column temperatures would not be practical in traditional normal phase HPLC, because the viscosity of traditional normal phase mobile phase rises steeply as temperature drops below ambient.

One factor driving the use of larger column ovens in recent years is the use of automated column screening to find the most promising phase for a separation. This has become a fairly common practice in chiral SFC (see Chapter 5) where predicting the most promising phase is challenging without some empirical data. Some modern SFC systems now accommodate up to twelve 4.6 mm × 250-mm columns, and can screen all these, along with various modifiers, at a range of temperatures and downstream pressures, in a fully automated fashion [31]. These modern, larger ovens are also designed to cool and heat more rapidly than their predecessors.

3.8 Chromatographic Column Materials of Construction

Stainless steel is the material of choice for SFC columns because it can withstand the pressures and temperatures used in SFC. Column manufacturers go to great lengths to achieve a smooth, polished inner column bore for efficient column packing, and such surfaces can be achieved with stainless steel. While stainless steel can withstand the temperatures and pressures of SFC, the other components of the column (ferrules, nuts, and frits) must do so as well. Columns manufactured for HPLC may contain polymeric components which may not be suited for SFC, so users must be cautious when choosing a column. Column packing materials and phases are treated in Chapter 6.

3.9 Backpressure Regulation

The backpressure regulation (BPR) system may be the single most critical component of a SFC system. So much depends on the BPR that designs are sometimes patented or published, but are more often closely guarded trade secrets. The great majority of conventional SFC systems use pumping systems operated under flow control, and independent control of system pressure using a "downstream" (postcolumn) mechanical BPR under computer control, as shown in Figure 3.1. Proper backpressure regulation is a critical aspect of virtually any SFC separation. It is generally accepted that the mobile phase should be in a single phase for efficient and effective chromatography. So, at a minimum, the mobile phase must be maintained at a pressure sufficient to prevent boiling of the mobile phase throughout the chromatographic system. Moreover, mobile-phase density and solvating power vary with mobile phase pressure, especially at lower mobile-phase modifier concentrations. The BPR system must therefore be accurate and reproducible to ensure reproducible chromatography. When operating with little or no mobile-phase modifier, the influence of the mobile phase pressure on solvating power can be used to the chromatographer's advantage to obtain a mobile-phase strength gradient, in much the same way as obtained by varying the mobile phase composition. So the backpressure regulation system should also be able to provide a smooth and reproducible backpressure gradient.

Yet there are even more requirements of this critical component of the SFC system. The refractive index of the SFC mobile phase typically varies with pressure. So small variations in the mobile phase pressure can increase baseline noise, and reduce overall signal-to-noise ratios, when using high pressure (preBPR) spectroscopic detection, such as the common ultraviolet/visible (UV/VIS) absorbance detector. Instrument manufacturers have therefore expended great effort to reduce these minor variations in mobile phase backpressure. These improvements have resulted in reduced baseline noise, and corresponding improvements in signal-to-noise ratios, matching those of contemporary HPLC/UV systems [18].

It is often desirable to introduce the full mobile phase flow into low-pressure (postBPR) detectors, such as atmospheric pressure ionization mass spectrometry (API-MS), evaporative light scattering (ELSD), and corona charged aerosol detection (CAD) [32]. Total flow introduction generally improves sensitivity and lowers limits of detection. Yet this approach requires yet another feature of the BPR system: low internal dead volume. Chromatographic band broadening increases, and chromatographic fidelity decreases, as the internal dead volume of the BPR increases. Anecdotal reports have described significant variation in the internal dead volume of the BPRs of modern SFC systems, ranging from only a few microliters to many hundreds of microliters. While these have not been verified, some instrument suppliers do indeed recommend interfacing to low-pressure detectors using a preBPR split, avoiding any influence of the BPR dead volume on chromatographic band broadening, rather than using total flow introduction (through the BPR). This is likely due to the use of a BPR with a comparatively large dead volume in these vendors' SFC systems.

Finally, the BPR must be designed so that solutes do not precipitate within the BPR, which can lead to erratic behavior and carryover for post-BPR detection or fraction collection in preparative-scale SFC. The BPR must be "self-cleaning," rugged, and affordable. It must also be designed so that it can be manufactured reproducibly and precisely. We'll describe how some of the published or patented BPRs stack up against these myriad requirements in the following paragraphs.

3.9.1 Passive Flow Restriction

In describing specific styles of BPR systems, let's begin with the simplest approach, the passive flow restrictor. The most common passive flow restrictor is narrow bore fused silica or stainless steel tubing, though laser-drilled pinholes have also been explored [33]. A variety of modifications of the narrow-bore fused silica restrictor were implemented to achieve better performance [34, 35], as shown in Figure 3.5. Passive flow restrictors were widely used in open-tubular SFC, as described in Chapter 1 [36]. In that implementation, the mobile phase was pumped under pressure control, and the flow was controlled by the dimensions and design of the flow restrictor. Gradient elution was generally performed using pressure or density programming, and the mobile phase flow rate therefore increased over the course of the separation due to the fixed nature of the flow restrictor. Attempts to vary the resistance of an otherwise passive flow restrictor have been described by Berger and Toney [37], Pyo [38], and Li and Thurbide [39]. Most commonly, the authors used resistive heating of the restrictor to vary its resistance and the resulting flow rate on column.

The passive flow restrictor is still alive and well in one contemporary interface for low-pressure detectors such as mass spectrometric, evaporative light scattering, and the corona charged aerosol detector. This is the preBPR split interface, where a portion of the mobile phase is split from the main flow after the column and before the BPR. This portion is directed to the detector, as shown in Figure 3.1. (Figure 3.1 shows a "possible flame ionization detector," unusual in modern SFC because it is compatible only with pure CO_2 or $CO_2/$

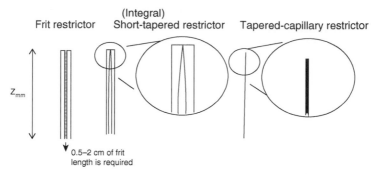

Figure 3.5 Illustration of the last 2 mm of the three most common flow restrictors used in open-tubular SFC: frit (left), integral ("short tapered") (middle), and tapered (right). The illustrations were drawn after observation under a microscope and are to scale. The length of the frit-filled portion of the frit restrictor is typically 1–3 cm. The length of the tapered region of the tapered restrictor is typically 2–4 cm. These restrictors prevented analyte nucleation and detector spiking, which was often observed with narrow bore linear restrictors. Larger bore (50–75 μm) linear restrictors are often used in modern packed column SFC as an interface to atmospheric pressure ionization mass spectrometry or evaporative light scattering detection. *Source:* Figure reprinted with permission from reference [36]. Copyright 1993, Wiley.

water mobile phases. But the preBPR split interface is similar for the more common detectors listed above.) The transfer line from the split point to the detector is generally a narrow bore tube that limits the flow to the detector and ensures that the majority of the flow is directed to the BPR for effective control of the system pressure. The mobile-phase pressure within the narrow-bore transfer line drops as the mobile phase approaches the exit of the transfer line. If the dimensions of the transfer line are not chosen carefully (i.e. internal diameter and length), the pressure within the transfer line can drop to a point where the mobile phase essentially boils before the exit. Analytes can fall out of solution and produce poor peak shapes. When modifiers are used, the mobile phase can separate into a segmented flow of a CO_2-rich phase and a modifier-rich phase. This results in a noticeable "sputtering" and flow instability at the inlet to the low pressure detector, often resulting in poor detector performance and high baseline noise.

3.9.2 Active Backpressure Regulation

One of the first active BPRs described in SFC was that of Klesper and colleagues [40, 41]. The system was based upon a Tescom valve incorporating a tapered pin in a hole. This regulator was capable of up to 69 MPa (10 000 psi). The pin was driven using bicycle gears, a chain, and a stepper motor and was capable of pressure programming. Klesper and colleagues were able to complete a great deal of ground-breaking research in SFC (most often with small alkanes as the primary mobile-phase component) using this design. The first commercial SFC, the Hewlett Packard 1084, used the same Tescom valve as Klesper, but the HP instrument did not perform pressure programming [42].

A later "vibrating pin in a hole" design was patented by JASCO [43]. The frequency of vibration was used to control pressure. Clogging could be easily cleared by backing the pin away from the hole, but the vibrating pin led to rapid, small pressure variations, and significant baseline noise in preBPR spectroscopic (UV/VIS) detectors. A later design reduced this baseline noise [44].

A series of BPRs were based upon metal diaphragms pushed against a seat with a strong solenoid. The diaphragms were generally manufactured from stainless steel, and were sometimes coated with gold for sealing and corrosion control. The seat was made of Hastaloy, later coated in diamond. Still later designs from Berger Instruments contained a platinum seat and a Vespel diaphragm [42]. These designs are said to have had relatively low dead volume (a few µL) such that peak broadening was not an issue for postBPR detection or fraction collection.

Berger et al. have patented similar BPR designs [45]. The diaphragm and seat materials include PEEK and Vespel. The newest version has a dead volume of <7 µL and allows direct flow introduction detection with very little chromatographic peak band broadening. The BPR drive stepper motor has a very fine relationship to pressure, and results in a pressure "noise" as low as 5 KPa (0.05 bar) at 20 MPa (200 bar). This provides a dramatic reduction in UV baseline noise (<0.1 mAU at 40 Hz, or as low as 0.006 mAU with a filter of 2.5 Hz [42]) and brings SFC/UV sensitivity in line with HPLC/UV sensitivity for trace analysis [18].

Other manufacturers attempted to address some of the BPR challenges by using an active BPR, designed to control pressure over a specific range, followed by a passive mechanical BPR or even a narrow-bore tube (typically on the order of 10–20 cm in length and 175 µm in internal diameter) over which much of the pressure drop occurs. This approach reduces the pressure differential across the active BPR, and reduces the mechanical requirements for the active BPR, as well as wear. This was the case with Gilson SFC instruments. A mechanical nozzle, downstream of the primary BPR, extended the lifetime of this BPR. The Gilson instrument was capable of pressure programming, but the programming range was limited due to the second, downstream nozzle. (As an aside, the Gilson instruments also had the advantage of being easily converted to HPLC mode.) Other vendors, such as Waters, have improved on this combination of active and passive BPRs for durability and precision of control.

BPR design continues to be an active area of research for instrument manufacturers, trying to come still closer to the "ideal" BPR.

3.10 Waste Disposal

Waste disposal has long been, and continues to be, an important part of managing analytical HPLC instrumentation. Mixed organic and aqueous/organic waste must be collected, stored, and disposed. One common concern in laboratories performing HPLC is the containment of organic solvent vapors emanating from the waste collection vessels. Ultimately, these wastes are most commonly transported off-site and incinerated. Many of the same considerations apply to SFC, yet the waste disposal requirements are usually far less onerous in SFC than in HPLC. After decompression of a modified CO_2 mobile phase, the waste stream consists of a modifier-rich gas (mostly CO_2) and a CO_2-rich polar organic liquid phase. This "segmented flow" mixed stream is typically directed to a collection vessel, where the organic liquid is collected. The gas phase is directed further to a chemical fume hood for disposal. The amount of organic solvent that much be disposed is generally far less in SFC than in HPLC, a welcomed benefit to the chromatographer. A second benefit of this arrangement is that solvent vapors are directed to the fume hood, along with the waste gas stream.

3.11 Conclusion

In summary, despite the similarities in HPLC and SFC instrumentation, we've shown that there are important differences – differences which are critical to successful SFC. Researchers were quick to take advantage of SFC, as described in Chapter 2, showing faster and higher resolution separations than in HPLC, and expanding the range of applications. Despite these successes, many challenges remained. Over the years, scientists and instrument manufacturers took aim at these challenges, and have improved many aspects of SFC instrumentation, such as injection reproducibility, pumping accuracy and reproducibility, extra-column volume, column and mobile phase temperature control, detector signal-to-noise, and BPR reliability and longevity. We hope we've made these advances evident here.

References

1 Chester, T.L. and Pinkston, J.D. (2004). Supercritical fluid chromatography instrumentation. In: *Analytical Instrumentation Handbook*, 3e (ed. J. Cazes), 759–760. CRC Press.

2 Smith, R.D., Fjeldsted, J.C., and Lee, M.L. (1982). Direct fluid injection interface for capillary supercritical fluid chromatography-mass spectrometry. *Journal of Chromatography A* 247 (2): 231–243.

3 Leyendecker, D., Leyendecker, D., Lorenschat, B. et al. (1987). Influence of density on the chromatographic behaviour of lower alkanes as mobile phases in supercritical fluid chromatography. *Journal of Chromatography A* 398 (0): 89–103.

4 Cammann, K. and Kleiböhmer, W. (1990). Supercritical fluid chromatography of polychlorinated biphenyls on packed columns. *Journal of Chromatography A* 522 (0): 267–275.

5 Kopner, A., Hamm, A., Ellert, J. et al. (1987). Determination of binary diffusion coefficients in supercritical chlorotrifluoromethane and sulphurhexafluoride with supercritical fluid chromatography (SFC). *Chemical Engineering Science* 42 (9): 2213–2218.

6 Pradines, A. (2012). Developments of SFC from analytical to preparative scale chromatography at Sanofi in Toulouse. In: *SFC 2012 – 6th International Conference on Packed Column SFC*, (ed. L. Miller and L. Taylor). (3–5 September 2012). Brussels, Belgium: The Green Chemistry Group.

7 Ashraf-Khorassani, M., Fessehaie, M.F., Taylor, L.T. et al. (1988). *Journal of High Resolution Chromatography* 11: 352.

8 Lesellier, E. and West, C. (2015). The many faces of packed column supercritical fluid chromatography – a critical review. *Journal of Chromatography A* 1382 (0): 2–46.

9 Hamman, C., Schmidt, D.E. Jr., Wong, M., and Hayes, M. (2011). The use of ammonium hydroxide as an additive in supercritical fluid chromatography for achiral and chiral separations and purifications of small, basic medicinal molecules. *Journal of Chromatography A* 1218 (43): 7886–7894.

10 Steuer, W., Baumann, J., and Erni, F. (1990). Separation of ionic drug substances by supercritical fluid chromatography. *Journal of Chromatography A* 500 (0): 469–479.

11 Pinkston, J.D., Stanton, D.T., and Wen, D. (2004). Elution and preliminary structure-retention modeling of polar and ionic substances in supercritical fluid chromatography using volatile ammonium salts as mobile phase additives. *Journal of Separation Science* 27 (1-2): 115–123.

12 Liu, J., Regalado, E.L., Mergelsberg, I., and Welch, C.J. (2013). Extending the range of supercritical fluid chromatography by use of water-rich modifiers. *Organic & Biomolecular Chemistry* 11 (30): 4925–4929.

13 Li, J. and Thurbide, K.B. (2008). A comparison of methanol and isopropanol in alcohol/water/CO2 mobile phases for packed column supercritical fluid chromatography. *Canadian Journal of Analytical Sciences and Spectroscopy* 53 (2).

14 Berger, T.A. (2015). Instrumentation for analytical scale supercritical fluid chromatography. *Journal of Chromatography A* 1421: 171–183.

15 Berger, D. Supercritical fluid chromatography (SFC): a review of technical developments. *LC GC Europe* 20 (3): 164–166.

16 Terry, A., Berger, K.D.F., Wikfors, E.E., et al. (2010). Compressible fluid pumping system for dynamically compensating compressible fluids over large pressure ranges United States.

17 Berger, T.A. (2014). Minimizing ultraviolet noise due to mis-matches between detector flow cell and post column mobile phase temperatures in supercritical fluid chromatography: effect of flow cell design. *Journal of Chromatography A* 1364 (0): 249–260.

18 Berger, T.A. and Berger, B.K. (2011). Minimizing UV noise in supercritical fluid chromatography. I. Improving back pressure regulator pressure noise. *Journal of Chromatography A* 1218 (16): 2320–2326.

19 Pinkston, J.D., Marapane, S.B., Jordan, G.T., and Clair, B.D. (2002). Characterization of low molecular weight alkoxylated polymers using long column SFC/MS and an image analysis based quantitation approach. *Journal of the American Society for Mass Spectrometry* 13 (10): 1195–1208.

20 Regalado, E.L. and Welch, C.J. (2015). Pushing the speed limit in enantioselective supercritical fluid chromatography. *Journal of Separation Science.*

21 Enmark, M., Asberg, D., Shalliker, A. et al. (2015). A closer study of peak distortions in supercritical fluid chromatography as generated by the injection. *Journal of Chromatography A*: 131–139.

22 Fairchild, J.N., Hill, J.F., and Iraneta, P.C. Influence of Sample Solvent Composition for SFC Separations. *LC GC North America* 31 (4): 326–333.

23 De Pauw, R., Shoykhet Choikhet, K., Desmet, G., and Broeckhoven, K. (2015). Understanding and diminishing the extra-column band broadening effects in supercritical fluid chromatography. *Journal of Chromatography A* 1403: 132–137.

24 Coym, J.W. and Chester, T.L. (2003). Improving Injection Precision in Packed-Column Supercritical Fluid Chromatography. *Journal of Separation Science* 26: 609–613.

25 Berger, T.A. (2016). Instrument modifications that produced reduced plate heights <2 with sub-2 μm particles and 95% of theoretical efficiency at k = 2 in supercritical fluid chromatography. *Journal of Chromatography A* 1444: 129–144.

26 PEEK Tubing Specifications 2016 Available from: http://www.westernanalytical.com/pdf/cat_512.pdf.

27 Baker, T.R. and Pinkston, J.D. (1998). Development and Application of Packed-Column Supercritical Fluid Chromatography/Pneumatically Assisted Electrospray Mass Spectrometry. *Journal of the American Society for Mass Spectrometry* 9 (5): 498–509.

28 Fields, S.M. and Lee, M.L. (1985). Effects of density and temperature on efficiency in capillary supercritical fluid chromatography. *Journal of Chromatography A* 349 (2): 305–316.

29 De Pauw, R., Choikhet, K., Desmet, G., and Broeckhoven, K. (2014). Temperature effects in supercritical fluid chromatography: a trade-off between viscous heating and decompression cooling. *Journal of Chromatography A* 1365: 212–218.

30 Pinkston JD. 2000.

31 Welch, C.J., Biba, M., Gouker, J.R. et al. (2007). Solving multicomponent chiral separation challenges using a new SFC tandem column screening tool. *Chirality* 19 (3): 184–189.

32 Pinkston, J.D. (2005). Advantages and drawbacks of popular supercritical fluid chromatography/mass spectrometry interfacing approaches – a user's perspective. *European Journal of Mass Spectrometry* 11 (2): 189–197.

33 Huang, E., Henion, J., and Covey, T.R. (1990). Packed-column supercritical fluid chromatography – mass spectrometry and supercritical fluid chromatography – tandem mass spectrometry with ionization at atmospheric pressure. *Journal of Chromatography A* 511 (0): 257–270.

34 Chester, T.L., Innis, D.P., and Owens, G.D. (1985). Separation of sucrose polyesters by capillary supercritical-fluid chromatography/flame ionization detection with robot-pulled capillary restrictors. *Analytical Chemistry* 57 (12): 2243–2247.

35 Cortes, H., Pfeiffer, C.D., Richter, B.E., and Stevens, T.S. 1988. Chromatography columns with cast porous plugs and methods of fabricating same. US Patents 4,793,920.

36 Pinkston, J.D. and Hentschel, R.T. (1993). Evaluation of flow restrictors for open-tubular supercritical fluid chromatography at pressures up to 560 atm. *Journal of High Resolution Chromatography* 16 (5): 269–274.

37 Berger, T.A. and Toney, C. (1989). Linear velocity control in capillary supercritical fluid chromatography by restrictor temperature programming. *Journal of Chromatography A* 465 (3): 157–167.

38 Pyo, D.J. (2007). Temperature controlled restrictor for packed capillary column supercritical fluid chromatography. *Journal of Liquid Chromatography & Related Technologies* 30 (20): 3085–3092.

39 Li, J.J. and Thurbide, K.B. (2009). Novel pressure control in supercritical fluid chromatography using a resistively heated restrictor. *Canadian Journal of Chemistry-Revue Canadienne de Chimie* 87 (3): 490–495.

40 Klesper, E. and Hartmann, W. (1978). Apparatus and separations in supercritical fluid chromatography. *European Polymer Journal* 14 (2): 77–88.

41 Klesper, E. and Schmitz, F.P. (1988). Gradient methods in supercritical fluid chromatography. *The Journal of Supercritical Fluids* 1 (1): 45–69.

42 Berger, T.A. (2015). Instrumentation for analytical scale supercritical fluid chromatography. *Journal of Chromatography A* 1421: 171–183.

43 Saito, M., Yamauchi, Y., Kashiwazaki, H., et al. (1991). Pressure control apparatus and apparatus for effecting extraction chromatographic separation, and fractionation by employing the same. US Patents 4,984,602.

44 Kanomata, T., Okubo, K., and Horioka, S. (2011). Pressure control apparatus for supercritical fluid. US Patent 8,915,261.

45 Berger, T. A., Colgate, S., and Casale, M. (2013). Low noise back pressure regulator for supercritical fluid chromatography, US Patent 8,419,936.

4

Detection in Packed Column SFC

OUTLINE

Modern Supercritical Fluid Chromatography: Carbon Dioxide Containing Mobile Phases,
First Edition. Larry M. Miller, J. David Pinkston, and Larry T. Taylor.
© 2020 John Wiley & Sons, Inc. Published 2020 by John Wiley & Sons, Inc.

4.1   Introduction

The best chromatographic separation provides no benefit without some form of detection of the separated components. The detector is the interface between the chromatographic separation and the mind of the researcher, providing the data that the scientist ultimately converts to information. This chapter is devoted to this important aspect of SFC. Nearly all modes of detection in SFC are familiar to practitioners of the more well-established methods of HPLC and GC. Yet, there are often significant differences in the detectors themselves and/or in the modes of interfacing these detectors to the SFC column. Without a clear understanding of these differences and details, detection in SFC is often impossible.

There are two approaches to the detection in SFC: predecompression and postdecompression. In the former mode, detection is accomplished within a compressed, dense, usually liquid-like (or even liquid) fluid. The detector is usually placed between the SFC column and the backpressure regulator. The first section of this chapter describes the most popular predecompression detectors, along with a few less common, yet interesting, modes of predecompression detection.

In postdecompression detection, detection takes place after the backpressure regulator, or, at a minimum, after a passive pressure restrictor, as described in Chapter 3. The pressure within the detection zone is always lower than that at the outlet of the chromatographic column. In fact, the pressure within the detection zone is often at or near atmospheric pressure. Detection takes place in a gas, or in a mixture of a gas and liquid droplets. The modes of interfacing the SFC column to a postdecompression detector are sufficiently varied that a section of this chapter is devoted to the advantages and limitations of these interfacing approaches. The last section of this chapter is devoted to postdecompression detectors, ranging from nonspecific detectors such as the evaporative-light-scattering detector, to more informative detectors, such as mass spectrometry.

4.2   Predecompression Detection (Condensed-Fluid-Phase Detection)

4.2.1   UV/VIS Absorbance

Ultraviolet/visible light absorbance is a mainstay of routine SFC applications, especially in quality control operations. Carbon dioxide, the primary mobile phase component in the great majority of SFC separations, does not absorb appreciably in the UV/VIS region above ~190 nm [1], so it is well suited for UV/VIS detection. Modifiers (most typically short-chain alcohols) and other mobile-phase additives, may affect the lower UV/VIS spectral limit, of course. See Table 4.1 for a list of the UV cutoffs of common SFC modifiers [2]. Diode-array detectors are common, and cover a spectral range of 190 to ~750 nm. The design of SFC UV/VIS detectors is quite similar to the design perfected in HPLC, with a few important differences. The flow cell must be capable of withstanding the pressure of the mobile phase. But such "high pressure" flow cells have become routinely available with the proliferation of atmospheric pressure LC/MS instruments, where the mobile phase is still under relatively high pressure until it is nebulized into the electrospray or atmospheric pressure chemical ionization source.

Another more fundamental difference in the SFC UV/VIS detector is the importance of carefully controlling and preventing density variations of the CO_2-dominated mobile phase in the UV/VIS absorbance flow cell. Density variations result in fluctuations in refractive index, which result in variations in the light flux incident on the detector and create significant baseline noise. This is the fundamental basis for the fact that early SFC/UV instrumentation was often deemed not suitable for trace analysis and pharmaceutical quality

Table 4.1 UV/VIS cutoffs for common SFC mobile-phase modifiers.

Mobile-phase modifier	UV/VIS cutoff (nm)
Methanol	205
Ethanol	210
2-Propanol	205
1-Butanol	215
Acetonitrile	190
Tetrahydrofuran	212
Acetone	330
Ethyl ether	215
Methyl-*tert*-butyl ether (MTBE)	210
Ethyl acetate	256
Propylene carbonate	220
Dichloromethane	233
Hexane	195
Methylethylketone	329
N,N-dimethylformamide	268
Dimethylsulfoxide	268

Source: Data taken from reference [2].

control, where peaks at 0.05% of the intensity of the largest major peak required quantification. UV detector noise in early SFC instruments ranged up to 0.5 milli-absorbance units (mAU) [3]. For comparison, ASTM standard E1657-98 calls for a noise of <0.02 mAU for trace analysis with HPLC/UV. Researchers and SFC instrumentation developers have devoted significant time and energy to improving this situation. The refractive indices of most common HPLC solvents range from ~1.33 to ~1.39. While the refractive indices of these fluids vary with density, there is relatively little variation of their density, and thus in their refractive indices, as a function of temperature, pressure, or even composition within the normal limits of HPLC. In contrast to HPLC solvents, the density of CO_2-based solvents varies more significantly with temperature, pressure, and composition. As a consequence, the refractive index of CO_2 can vary significantly (from 1.06 to 1.24 [3]) under typical SFC operating conditions. Figure 4.1 illustrates this phenomenon [4]. Most of the improvements in SFC/UV baseline noise come from reductions in pressure fluctuations due to pumping and due to backpressure regulation, and to density variations due to temperature changes as the mobile phase enters the detector flow cell. Pumping and backpressure regulation in SFC are described in Chapter 3. For example, reference 3 describes improvements in SFC/UV noise due to the separation of the compression and the metering strokes in SFC pumping. New backpressure regulators (BPRs) with lower pressure oscillations were also shown to improve SFC/UV noise [5]. Modern BPRs can control back

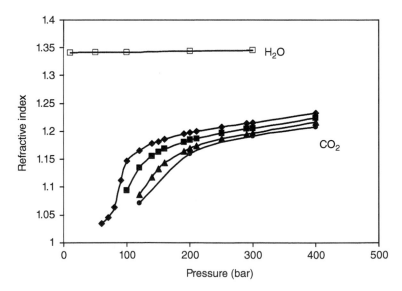

Figure 4.1 Refractive index vs. pressure curves: water at 40 °C (□), pure carbon dioxide at 40 °C (♦), 50 °C (■), 60 °C (▲), and 70 °C (●). *Source:* Reprinted from reference [4] with permission. Copyright 2003, Lord & Taylor.

pressure to <±0.05 bar at 200 bar, meaning that very little noise arises from backpressure regulation [5, 6].

Finally, proper mobile-phase cooling/thermostating before the mobile phase enters the flow cell, or insulating the mobile phase from the UV detector optical bench, has reduced density variations as the mobile phase enters the detector flow cell. When the mobile phase is relatively warm (e.g. 40 °C or higher), small variations in pressure from pumping or backpressure regulation can cause much larger variations in refractive index. Many UV/VIS detector designs therefore include a heat exchanger before the flow cell, to cool the mobile phase, or at least to minimize temperature changes of the mobile phase inside the flow cell. In other UV/VIS detectors, the optical bench is warmed, and may be warmer than the mobile phase exiting the column oven. In this case, the flow cell may be insulated from the optical bench in order to minimize temperature changes of the mobile phase inside the flow cell.

4.2.2 Fluorescence Detection

Fluorescence detection in HPLC is known to have superior selectivity and sensitivity compared to UV absorbance detection, especially for targeted analysis. The analyst specifies an excitation and an emission wavelength, and only analytes that fluoresce under those specific conditions are detected. Not only is this approach more selective, but also improved signal-to-noise ratio and response can improve sensitivity by a factor of 10–100. These same advantages apply to SFC. As in UV detection, the fluorescence detection cell in SFC must be capable of withstanding the pressure of the SFC mobile phase. In addition, the same concerns about the influence of small variations in pressure and temperature on mobile phase density, refractive index, and S/N apply in fluorescence detection. It is therefore not surprising that the advantages of fluorescence detection in SFC were first demonstrated in capillary open-tubular SFC, very soon after its invention [7]. Detection was performed on column, thereby avoiding the requirement of using a fluorescence detector with a high-pressure flow cell. Passive backpressure regulation and fluid supplied via a high pressure syringe pump avoid the small variations in mobile phase density and refractive index seen in earlier versions of packed column SFC [7].

Smith et al. brought the selectivity and sensitivity advantages of fluorescence detection to packed column SFC [8]. Sensitivity for propranolol was eightfold better with fluorescence detection than with UV absorbance. The authors used a passive restrictor to drop the mobile-phase pressure and protect the fluorescence detection cell. Other groups worked to devise sensitive fluorescence flow cells that could withstand the mobile phase pressure. Nomura et al. designed a square cross-section quartz flow cell, and used the SFC/

fluorescence system to selectively detect polyaromatic hydrocarbons in fuel oils [9]. Another example of SFC/fluorescence detection is the work of Kanomata et al. of Jasco, who patented a novel high pressure flow cell for SFC [10]. Note that modifiers in SFC can affect fluorescence detection. These effects may include absorbance of the excitation or emission wavelengths, as well as quenching. This is a similar situation to that in HPLC. Fluorescence detection is not widely used, but is available in commercially available SFC instruments.

4.2.3 Electrochemical Detection

Electrochemistry in supercritical fluid media is an active field of research [11, 12], so it is not surprising that electrochemical detection for SFC has been explored, though its use is relatively infrequent. Many analytes that are readily eluted in SFC are also electrochemically active, and electrochemical processes are known to occur in supercritical and subcritical fluids, so SFC/ECD should provide the same sensitive and selective detection to which the practitioners of HPLC/ECD are accustomed. But pure CO_2 is nonpolar and resistive, making conventional ECD challenging. Di Maso et al. used ultra-microelectrodes, with their low residual ohmic effects and reduced requirements for added supporting electrolyte, to first demonstrate SFC/ECD [13]. Mixtures of CO_2 and mobile-phase modifiers are quite amenable to ECD [14]. Dressman and Michael explored SFC/ECD with both pure and modified CO_2 mobile phases [15, 16]. They found that ECD of ferrocene with pure CO_2 provided similar performance to flame ionization detection in terms of limit of detection, linearity of response, and peak shape. Furthermore, the FID was completely inoperable at 3% methanol or acetonitrile modifier, while the ECD functioned well with modified CO_2 mobile phase [15]. They expanded their range of applications in later work and were able to use voltammetry to resolve closely eluting and coeluting mixtures of substituted phenols [16]. While the ECD worked well with modified fluids, they found that the electrode had to be refurbished after ~3 days of separations.

Señoráns demonstrated the impressive selectivity and sensitivity of SFC/ECD by determining tocopherols and vitamin A in vegetable oils, without the tedious clean-up (i.e. removal of triglycerides, isolation of unsaponifiables) required for conventional detection [17]. Improved longevity of the ECD cell in SFC/ECD was demonstrated by Toniolo et al. [18] through the use of a porous electrode supported on a moist perfluorinated ion-exchange polymer membrane. Ferrocene was used as the probe analyte, and its detection by SFC/ECD was superior to that provided by SFC/UV by all measures.

Despite these clear advantages for electrochemically active analytes, SFC/ECD remains relatively uncommon and is not typically available from commercial SFC instrumentation vendors.

4.2.4 Other Less Common Condensed Phase Detectors

4.2.4.1 Flow-Cell Fourier Transform Infra-Red Absorbance (FTIR) Detection

IR absorbance is a well-tested mainstay of organic structure elucidation. FTIR detection is more universal than UV/VIS absorbance, because virtually all organics absorb in the IR. Unfortunately, sensitivity is a real issue for FTIR detection. HPLC/FTIR, while not commonplace, has been used by many research groups over the years [19]. So it's no surprise that SFC/FTIR has also been explored. Much of this exploratory work was performed in the late 1980s and early 1990s, when both open-tubular and packed column SFC were popular. Supercritical CO_2 is much more transparent in the IR than are traditional HPLC organic solvents. Shafer and Griffiths first described in-line SFC/FTIR, and showed that the IR-transparency of CO_2 allowed a 1-cm path-length flow cell, approximately one order of magnitude greater than those used in HPLC/ FTIR [19]. They were able to obtain library-searchable IR spectra with approximately 3 µg of analyte. However, pressure programming resulted in a shift in the baseline signal. Olesik et al. also developed an early high pressure flow cell for online SFC/FTIR [20], and studied the limits of detection using simple probe analytes. Other groups were quite active in online SFC/FTIR, including a collaboration between Taylor [21–26] at Virginia Tech and Vidrine [27] at Nicolet. Morin et al. used optical path and detector improvements to provide searchable spectra of tens-of-ng of analyte on column [28]. Other flow-cell improvements allowed Wieboldt et al. to drop this important level to 10 ng [29].

A number of innovations were described by Raynor and colleagues. Among these were the addition of a make-up fluid to the effluent of a 50-µm-i.d. open tubular SFC column in order to minimize the band-broadening resulting from the online IR flow cell [30], as well as the exploration of xenon as a mobile phase [31]. While supercritical Xe had wonderful IR optical transparency, its high cost limited its use to high value applications and to 50-µm-i.d. open tubular SFC.

As with most SFC work, more recent SFC/FTIR research has involved larger packed column SFC, and has focused more on specific applications. Higher flow rates and larger on-column injection masses made FTIR interfacing more straightforward. For example, Auerbach et al. used SFE/SFC/FTIR to study surfactants in washing detergents [32]. FTIR allowed these scientists to determine various functional-group classes of detergent actives in this complex mixture.

The richness of IR absorbance spectra provides good selectivity, but, as mentioned earlier, IR absorbances are relatively weak, so online chromatography/IR detection suffers from poor sensitivity, which has limited its acceptance in general. Some of the disadvantages of online IR absorbance detection are alleviated by the off-line approach where the chromatographic effluent is deposited on a surface, the mobile-phase components are removed, and IR microscopy is used to examine the residue. The ease of mobile-phase elimination in SFC makes this approach especially attractive, but the complexity of the deposition instrumentation is a serious drawback. Jinno [33] as well as Taylor and Calvey [34] reviewed the advantages and disadvantages of both interfacing approaches. The deposition approach will be discussed in Section 4.4. Unfortunately, neither approach is available in modern commercial SFC instrumentation.

4.2.4.2 Online Nuclear Magnetic Resonance (NMR) Detection

Researchers working with HPLC/NMR were naturally drawn to SFC/NMR for the same advantage discussed in the above section about SFC/FTIR, namely spectral "transparency." Interfacing in SFC/NMR is similar to that in HPLC/NMR, using small microcoils wrapped around the flow cell to enhance sensitivity. ^{1}H-NMR is the most widely practiced and the most sensitive of the popular NMR methods, and CO_2 does not contain protons. So suppression techniques that are required for HPLC solvents are not required for a pure CO_2 mobile phase. Allen et al. were the first to describe online coupled SFC/NMR [35]. They demonstrated no need for ^{1}H-suppression, and full spectral acquisition in separating a model mixture.

The group of Bayer, Albert, and colleagues in Tubingen, Germany, has driven the majority of the research in online SFC/NMR [36–41]. They designed NMR probes that could withstand the pressure of the SFC system, and could operate at up to 100 °C [37]. Their initial publication demonstrated spectral line widths of 1.5 Hz in flowing SFC effluent under supercritical conditions. The authors stated that ^{1}H-NMR spectra acquired under flowing supercritical conditions were of the same quality as those acquired in liquids. A later publication [39] provides more detail about the nature of the spectra acquired under supercritical conditions, as well as a summary of a number of interesting applications. In particular, the proton spin-lattice relaxation times, T1, are approximately doubled due to the supercritical measuring conditions. Also, the ^{1}H NMR signals are shifted up field as pressure/density rises if a mobile-phase pressure gradient is used in the chromatographic separation. Applications of SFC/NMR include the analysis of plasticizers, polyacrylates, vitamins, and natural products [36, 39, 40].

A somewhat more fundamental application of the SFC/NMR system was described by Fischer and Albert in a collaboration with Gyllenhaal and Vessman of AstraZeneca R&D [41]. This group studied the reaction of amines with CO_2 to form carbamates under SFC conditions. This possibility has been the subject of much controversy over the years, with some groups demonstrating acceptable SFC chromatography of certain amines and others not [42]. Fischer et al. examined the structures of amines using NMR under SFC conditions, and, in cases where reaction occurred, of the corresponding carbamate [41]. They used a variety of primary and secondary aromatic amines as probes, and found that the propensity for reaction of the amine to form a carbamate depends on the stereochemistry of the substituent on the amine. Primary amines often reacted, while metoprolol, a 2-isopropylamino alcohol compound, for example, did not react. These reactions are reversible when the pressure of CO_2 drops, which explains why the intact amines are detected in atmospheric-pressure detectors.

4.2.4.3 Refractive Index (RI) Detection

RI is commonly used in HPLC, especially in preparative-scale HPLC, because it is a nearly universal detection method. The sensitivity of RI detection is relatively poor compared to UV and other HPLC detectors, with limits of detection frequently in the microgram or hundreds-of-nanograms range. But this is not an issue in many applications using 4.6-mm-i.d. and larger columns and large injection volumes. Low sensitivity is even less of an issue in preparative-scale separations in which injected quantities can range into the milligram or even gram scale. Yet RI detection in SFC faces a significant challenge. As discussed in Chapter 2, the refractive index of CO_2 can vary significantly with relatively small changes in temperature and pressure, especially near the critical point. These variations result in high baseline noise in SFC/RI, unless precautions are taken to minimize density/refractive index fluctuations. Hirata et al. actually showed that the lower RI of CO_2 results in higher detector response in SFC/RI than in HPLC/RI [43]. These researchers diverted approximately 1/1000th of the effluent of a preparative-scale SFC separation to a capillary cell RI detector and carefully controlled the pressure and temperature of this cell. The RI detector operated in parallel to peak collection on the main effluent stream. They showed that SFC/RI was appropriate for preparative-scale SFC, with a LOD of 30 µg of n-hexadecanol injected on column. This represented 30 ng in the capillary flow cell. The linear dynamic range of the RI detector was two orders of magnitude.

4.3 Postdecompression Detection (Gas/Droplet Phase Detection) – Interfacing Approaches

The remainder of the most commonly used detectors in SFC operate at atmospheric or lower pressures, in sharp contrast to the condensed-phase detectors described above. As the reader might imagine, moving a chromatographic peak from solution in a high-pressure, condensed phase to an atmospheric or lower-pressure detector, with good chromatographic fidelity, is fraught with potential peril. Nucleation of analytes into particles, deposition of analytes on surfaces, and formation of segmented, liquid-rich and gas-rich phases are among the issues that can cause havoc or that, in some cases, can be exploited. It is therefore not surprising that a variety of methods have been used to interface SFC to these lower-pressure detectors. Pinkston summarized the most popular interfacing approaches [44]. In short, they consist of the following:

1) A split of the effluent after the column, but before the backpressure regulation device (BPR). A fraction of the effluent is directed to the lower-pressure detector, while the remainder of the effluent is directed to the BPR in order to maintain pressure control of the mobile phase.
2) Movement of the entire effluent through the BPR device before introduction to the lower-pressure detector. This places demands on the BPR, as described below. A variety of pressure-regulation approaches may be used.

Here is a more detailed description of the advantages and limitations of these interfacing approaches.

4.3.1 Pre-BPR Flow Splitting

This is the most common interfacing approach for modern SFC/MS instruments, and is illustrated schematically in Figure 4.2. A fraction (in the range of 1–20%) of the mobile phase is directed to the lower-pressure detector through a near-zero-dead-volume tee and a transfer line and restrictor. The transfer line dimensions are chosen such that the remainder of the mobile phase continues to the BPR. Without sufficient flow to the BPR, the postcolumn system pressure cannot be regulated.

Pre-BPR flow splitting is straightforward and allows control of the postcolumn pressure using the instrument-control software of the SFC instrument. It also provides good chromatographic fidelity if the transfer line/restrictor dimensions do not allow a large pressure drop before the exit to the detector. If the dimensions of the transfer line are too long or too wide, precipitation and deposition of analytes within it can occur. Transfer line dimensions that are too

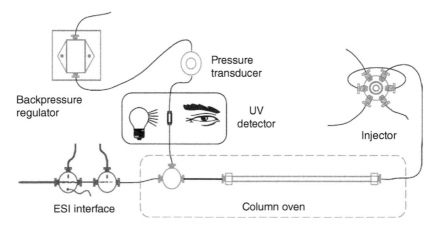

Figure 4.2 Schematic diagram of the preBPR split interface for SFC/MS. A variety of atmospheric pressure ionization sources can be used successfully. *Source:* Reprinted from reference [44] with permission. Copyright 2005, Sage Publishing.

long or too wide can also allow the mobile phase to boil and transition to segmented phases (one modifier-rich, one CO_2-rich) that move into the detector. This can result in unstable, erratic, pulsating detector response and jagged, noisy peaks. The required dimensions of the transfer line vary with total flow rate, the expected range of modifier concentration, and the desired downstream pressure. For guidance, a 1-m-long, 125-µm-i.d. transfer line provides acceptable results for a total CO_2/methanol flow rate of 2 mL/min operated with a downstream pressure of 200 bar. Some vendors advocate the addition of a low flow (e.g. 50–100 µL/min) of organic solvent to the split flow in order to minimize any deleterious effects.

Figure 4.3 is an example of the use of pre-BPR splitting for SFC/MS by Bolanos et al. [45]. The authors used the high spectral acquisition rate of time-of-flight

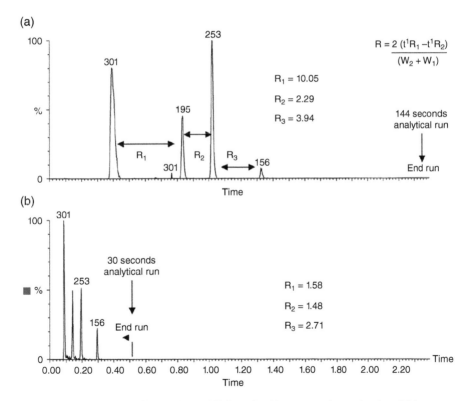

Figure 4.3 Comparison of (a) traditional SFC method (144 seconds run time) and (b) ultra-fast run (30 seconds run time) on a four-compound QC standard mixture (caffeine, pyridine, proprietary compound, and sulfanilamide). For the ultra-fast run, all compounds are baseline resolved in less than 18 seconds, but there is some loss of peak resolution due to a steep methanol modifier gradient. *Source:* Reprinted from reference [45] with permission. Copyright 2004, Elsevier.

mass spectrometry (TOFMS) to preserve the chromatographic profile of a high-speed SFC separation. Four peaks are baseline resolved, with a resolution of 1.5 or better, in an 18-s portion of this separation. The TOFMS provided a spectral acquisition rate of 10–20 Hz.

While pre-BPR splitting provides good chromatographic fidelity, this approach also has a few disadvantages. Only a fraction of the effluent is directed to the lower-pressure detector. For mass sensitive detectors (such as evaporative light scattering detection [ELSD] or atmospheric pressure chemical ionization [APCI]), this results in a lower response than if the full effluent is directed to the detector. Another issue is that the split ratio varies with mobile-phase pressure and with any change in the restriction of the transfer line from the split point to the detector. The former is predictable [46], while the latter is not.

4.3.2 Total Flow Introduction (Post-BPR Detection)

In many cases, the drawbacks of pre-BPR flow splitting are inconsequential. In others, such as when a wider dynamic range or better sensitivity with a mass-sensitive detector is needed, they are not. Chiral SFC is becoming increasingly important. When precise measurements of enantiomeric ratios are required, the variation in split ratio mentioned above might be a concern. Therefore, a variety of "total-flow-introduction" interfaces have been investigated. Here, we touch on the characteristics of three of these: the mechanical BPR interface, the pressure-regulating-fluid approach, and an interface employing no active backpressure regulation.

4.3.2.1 BPR Requirements for Total-Flow Introduction Detection

With total flow introduction detection, there is no split ratio about which to be concerned. Yet the requirements of the BPR system are stringent: the dead volume of the BPR, any pre-BPR pressure measuring device, and any connecting tubing must be low enough to minimize band broadening. A few microliters are acceptable for total SFC flow rates of 1–3 mL/min. Also, the BPR system must not contain any unswept dead volume, which can trap analytes and contribute to poor peak shapes. Third, the BPR system must not change the composition of the mobile phase in a way that impedes the proper operation of the detector. This will become more obvious during the discussion of the pressure-regulating-fluid interface. Finally, and perhaps most importantly, the detector itself must be able to accept the total flow of effluent from the SFC system without compromising its operation. Flows from 4.6-mm-i.d. SFC columns can easily be 3 mL/min or more under typical operating conditions. Of course, a significant fraction of this effluent is CO_2, which usually assists in the nebulization process for detectors that require droplets for proper operation, such as atmospheric pressure ionization mass spectrometry and evaporative light scattering detection.

4.3.2.2 Total Flow Introduction with Mechanical BPR

As discussed in Chapter 3, some contemporary SFC systems contain BPRs that meet the dead-volume requirements described above, but others have dead volumes in the tens or even hundreds of microliters, and are not designed to preserve the chromatographic separation as the effluent flows through the BPR system. Even when the BPR has an adequately low internal dead volume, any large-dead-volume components (i.e. some pressure transducers, large i.d. tubing) must be removed from the flow path between the column and the detector.

The total flow introduction approach with a mechanical BPR has the advantage of allowing full control of the system pressure by the instrument-control software. Of course, it also provides the improved limits of detection and the dynamic range advantages discussed above for the appropriate detector. There is an important similarity between this interface and the preBPR flow splitting approach: the dimensions of the transfer line between the mechanical BPR and the detector must not allow the transition to two phases (i.e. boiling) or precipitation of analytes due to excessive pressure drop. Yet, this is not as great a concern with the total flow introduction interface, simply because the flow through the transfer line is usually larger than with the pre-BPR flow splitting approach, and the majority of the pressure drop occurs near the exit of the transfer line.

4.3.2.3 Total Flow Introduction – Pressure-Regulating-Fluid (PRF) Interface

Backpressure in condensed phase chromatographic systems is most commonly controlled with a mechanical device, as described above and in Chapter 3. But this is not the only possibility. Pressure can also be controlled with a fluid set at a particular pressure. Figure 4.4 illustrates this approach, the "pressure-regulating-fluid" (PRF) interface. The mechanical BPR and pressure transducer are replaced with a single near-zero-dead-volume chromatographic tee. In the tee, the chromatographic effluent is mixed with a low flow of CO_2-miscible fluid (the pressure-regulating fluid) from a pump operated under pressure control. This mixture is then directed to the detector. The dimensions of the transfer line from the tee to the detector dictate the available ranges of chromatographic effluent flow, PRF flow, and post column pressure. The principles of this interface were described by Chester and Pinkston [47].

The PRF interface has a number of advantages. First, the pressure control point, the tee, can be placed very near the detector. The pressure is actively controlled nearly to the outlet of this interface. This avoids transfer through long lines with no active pressure control, and minimizes the possibility of phase transition (i.e. boiling) of the mobile phase and poor mass transfer of analytes. Second, it provides excellent chromatographic fidelity, with very low

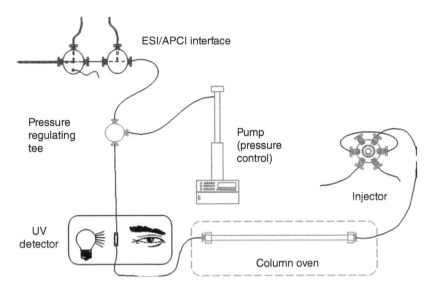

Figure 4.4 Schematic diagram of the pressure-regulating-fluid interface for SFC/MS. A variety of atmospheric pressure ionization sources can be used successfully. *Source:* Reprinted from reference [44] with permission. Copyright 2005, Sage Publishing.

dead volume and minimal extra-column broadening. Third, the entire sample is delivered to the ion source without any splitting. With a mass-sensitive detector, such as the ELSD, this can result in lower detection limits and greater dynamic range. Finally, the pressure-regulating fluid can be chosen to enhance detector response. For example, electrospray ionization (ESI) requires an effluent that is capable of forming droplets, and pH or solvent–mixture adjustment can often enhance ionization. In fact, SFC/ESI-MS with pure CO_2 as the mobile phase results in no signal. So when the mobile phase contains little or no organic modifier, a PRF fluid can be chosen that provides this necessary component. For example, an 80:20 mixture of methanol and water containing a low level (~1 mM) of ammonium acetate was found to provide excellent results with SFC/ESI-MS [48].

Unfortunately, despite providing excellent results, the PRF interface is not very user-friendly, and is not available in a commercial unit. Most commonly, the downstream pressure is no longer under control of the system software, but under the control of an independent pump. This requires occasional manual intervention and programming, or requires the user to establish a control link between the SFC system software and the PRF pump. Second, the dimensions of the transfer line from the PRF control-point tee to the detector dictate a range of available flow rate/pressure combinations. For example, a 3–5 cm length of 0.0025-inch (62.5-μm) i.d. PEEK tubing provides a backpressure of 200 bar for mobile phase flow rates ranging from 1.5 to 3 mL/min and PRF fluid flow rates ranging from ~50 to ~700 µL/min. As one might imagine, when performing a mobile-phase gradient, the PRF fluid flow and the mobile-phase modifier flow (not the total mobile-phase flow, which is generally held constant) vary in an inverse manner. If one requires flow rates beyond these ranges, one must install a transfer line of different dimensions. Third, the addition of the PRF fluid increases the solvent load on the detector. The total mobile-phase and PRF fluid flows may exceed the capacity of the detector. This requires a change in the transfer line dimensions (see above). Finally, the PRF approach requires an additional pump capable of providing a pulse-free flow of solvent under pressure control, such as a high pressure syringe pump. Reciprocating pumps induce a slight pulsation in the PRF flow, and this often results in detector noise [47].

When care is taken in setting up the total-flow-introduction interface with mechanical BPR, satisfactory results can be obtained. Because of its ease of use and reasonable performance, this interface has become the most commonly used total-flow-introduction interfacing approach. However, in our experience, this interface does not produce the ultimate chromatographic fidelity provided by the PRF interface, especially if the internal dead volume of the BPR is significant (i.e. more than a few microliters). Figure 4.5 shows a head-to-head comparison of these two interfaces in the separation of two isomers. While the isomers are clearly resolved with the PRF interface (4.5B), the smaller isomer is only a shoulder on the larger peak when the mechanical BPR interface is used (4.5A).

4.3.2.4 Total Flow Introduction Without Active Backpressure Regulation

The simplest method is sometimes the best, and total flow introduction with a passive BPR (i.e. a length of narrow bore tubing) is certainly simple. In this approach, the column is essentially directly coupled to the detector. Importantly, because there is no active downstream pressure control, this interface only works well under a specified combination of conditions. First, the flow conditions and transfer line dimensions must be such that phase transition does not occur in the column or transfer line. In other words, the transfer line/restrictor must be sufficiently restrictive to prevent boiling of the mobile phase until the effluent enters the detector. Second, even when the pressure and temperature remain in the one-phase region for the mobile phase mixture, the selectivity of the chromatographic separation must not be greatly affected by variation in the downstream pressure. In fact, these conditions are often met in contemporary SFC. Flow rates, and resulting column pressures, are relatively high (1–7 mL/min for a 2-mm-i.d. column, resulting in downstream pressures well above 150 bar with a 50-cm-long, 125-µm-i.d. transfer line). Column temperature is usually held between 40 and 60 °C. Mobile phases usually contains a significant level of organic modifier (10–50%) and varying the mobile-phase composition is by far the most common mode of gradient elution. Under these conditions, the mobile phase is very "liquid-like." Variations in pressure have

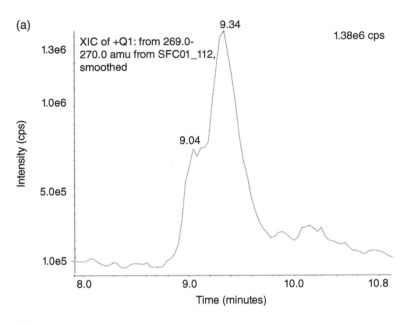

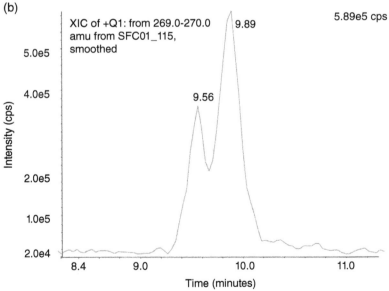

Figure 4.5 SFC/ESI-MS Separation of a pair of isomers in a product extract using identical conditions except (a) total flow introduction with mechanical BPR and (b) total flow introduction with pressure-regulating-fluid interface. The column was a 50-cm (2×25-cm) × 4.6-mm Inertsil Phenyl operated at a downstream pressure of 180 bar. The mobile phase and flow rate were 2% MeOH in CO_2 (isocratic) and 3 mL/min, respectively. The column temperature was held at 70 °C. *Source:* Reprinted from reference [44] with permission. Copyright 2005, Sage Publishing.

relatively little effect on the strength of the mobile phase and the selectivity of the separation. Also, the conditions on the column and in the transfer line are well above those at which phase transition (boiling) occurs.

Figure 4.6 illustrates a prime example of an application for which the passive BPR approach is well suited. Hoke et al. analyzed an entire 96-well plate containing dextromethorphan in less than 10 minutes using SFC/MS/MS [49]. The drug was extracted from human serum. The time scale is expanded in the upper traces, with the topmost trace showing two injection cycles. The injection-to-injection cycle time is 5 seconds, and the k' is ~1.5. This provides enough separation of the peak from the void volume to avoid ionization suppression effects. Intra-day accuracy was better than 9% over 3 days, and RSDs were less than 15%, even for the lowest quality-control standard of 0.554 ng/mL.

4.4 Postdecompression Detection

A wide variety of postdecompression detectors have been used in SFC over the years. In fact they have contributed to the well-earned reputation of SFC having perhaps the widest range of detector compatibility of any of the major chromatographic techniques [50]. Commercially available postdecompression detectors for SFC range from flame-based detectors, most commonly used in gas chromatography, to nebulization-based detectors commonly used in HPLC, such as the ELSD, the corona charged aerosol detector (Corona CAD), and atmospheric pressure ionization (API) mass spectrometry.

4.4.1 Flame-Based Detectors

Flame-based detectors are widely used in GC. The flame ionization detector (FID) is the flagship among these. GC mobile phases (H_2, He, and N_2) do not respond in the FID. It is sensitive, with LODs in the sub-ng range, has a wide dynamic range (four to five orders of magnitude), is nearly universal for volatile organics, and is reliable and simple to operate. Just as strikingly, the FID is rarely used in conventional condensed phase chromatography, with the exception of a few mobile-phase removal approaches in HPLC [51] and thin-layer chromatography [52], and online HPLC-FID when pure water is used as the mobile phase [53–55]. This is simply due to the fact that most HPLC mobile phases produce a huge background signal in the FID. Despite the advances in the ELSD and the corona CAD detectors, which share some degree of "universality," neither approaches the performance of the FID.

In contrast, pure CO_2 does not respond in the FID. Addition of a relatively low flow (a few mL/min, measured in the gas phase) of CO_2 to the hydrogen and air used in a conventional FID will not negatively affect its performance (though higher CO_2 flow rates can cause the flame to become unstable). It is

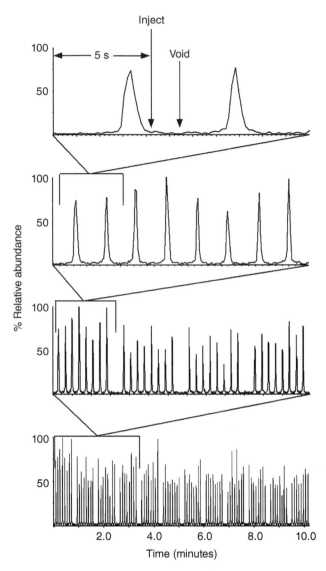

Figure 4.6 Ninety-six chromatograms of d3-dextromethorphan obtained by SFC-MS/MS analysis of repeated 10-µL injections of a methanol solution containing 4 ng/mL of d3-dextromethorphan (40 pg on column for each injection). *Source:* Reprinted with permission from reference [49]. Copyright 2001 American Chemical Society.

therefore not surprising that researchers eagerly coupled SFC, with its ability to solubilize and elute nonvolatile organics, with the FID. Most of the early work was performed with open-tubular capillary SFC [56]. The mobile phase was most often pure CO_2, and the total column flow was compatible with a conventional FID. The open-tubular column could be coupled to a flow restrictor [57] using a near-zero-dead-volume union, and the outlet of the restrictor was placed at the base of the FID flame. Alternatively, the end of the column could even be fashioned into an "integral" restrictor [58], which was placed at the base of the flame, avoiding the need for a union altogether.

Work in open-tubular SFC/FID produced impressive results, allowing the separation and FID detection of nonvolatile and thermally labile analytes, such as sucrose octaesters, which have molecular weights ranging to over 2000 [59], organometallics [60], oligosaccharides with DP (degree of polymerization) over 20 [61], small polymers, and peroxides [62]. Yet open-tubular SFC was remarkably non-user-friendly and had significant limitations in injection volume. Packed column SFC has gradually taken over the field, as described in Chapter 1 [50].

Most packed column SFC is performed with modified mobile phases, which are not compatible with the FID. But some important applications, notably in the petroleum [63], polymer [64, 65], and lipids [66] area, use SFC with pure CO_2 or with water-modified-CO_2 mobile phase. These can be coupled to an FID in a straightforward manner using the preBPR split approach, with a small fraction of the effluent directed to the FID. One critical component of this coupling is the passive flow restrictor through which the effluent passes into the base of the FID flame. Most of these are made of narrow-bore fused silica tubing. Early on, straight-walled, narrow-bore tubing was shown to be not suitable for higher molecular weight, nonvolatile analytes. These would fall out of solution as the pressure dropped across the restrictor, and small particles would cause FID "spiking" and baseline noise [67, 68]. So a variety of tapered and fritted restrictors have been used to delay pressure drop and to enhance heat transfer into the SFC effluent. Most notably, these include the "thin-walled tapered" [59, 69], the "integral" (incorporated into the end of the chromatographic column) [58], and the "frit" [57] restrictors. These are described in Chapter 3. The performance of the various designs was studied in some detail [70]. The integral restrictor, while challenging to prepare, was easily installed, easily cleared if it became partially plugged, and provided the best chromatographic performance for higher molecular weight analytes. While the integral and other restrictors were widely used in open-tubular SFC, the proper choice of restrictor is still important in packed column SFC/FID. At the time of this writing, SFC/FID systems and flow restrictors are commercially available from Selerity Technologies, Inc., and a channel partner of Agilent, Scientific Instruments Manufacturer GmbH.

The advantages of SFC/FID for group-type separations in the petroleum industry are especially clear. The nonpolar nature of petroleum constituents allows the use of pure CO_2 mobile phase, and the preBPR split is used to direct a portion of the effluent to an FID [71]. (The bulk flow can pass through a UV-absorbance detector on its way to the BPR to detect specific compound classes.) Various packed columns can be used to resolve functional group classes in very complex petroleum-derived mixtures. ASTM International has validated and promulgated two SFC/FID methods for petroleum products: D5186 for determining aromatics and polynuclear aromatics in diesel and aviation turbine fuels [72], and D6550, for measuring the total olefin content in gasoline. Figure 4.7 shows the reproducibility of the retention times and FID responses for a calibration standard used in D5186 [73].

The low viscosity of a CO_2 mobile phase allows the use of longer columns packed with small particles. This, in combination with the FID, provides a powerful tool for studying low molecular weight polymers, especially those which are not easily detected in HPLC. Campbell et al. described the use of 1.7-μm-particle columns for Triton-X100 analysis [74]. They showed more than a 15-fold reduction in analysis time relative to a 5-μm-particle column. The low viscosity and pressure drop of SFC allowed them to couple columns for the analysis of polymeric species, generating more than 100 000 plates for a 1.7-μm, 30-cm-long column (two coupled 15-cm columns).

There are a multitude of other flame-based and similar GC-style detectors that are compatible with SFC using pure CO_2 mobile phase [75]. Many of these are more selective than the FID, and can be used in much the same way as they are used in GC. These include the flame photometric detector (FPD) [76, 77], the thermionic ionization detector (TID), the photoionization

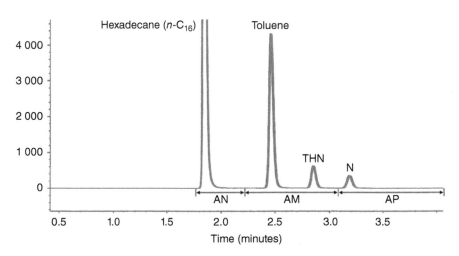

Figure 4.7 Calibration and reproducibility of the ASTM 5186 Method for aromatics and polynuclear aromatics in petroleum products: Overlay of five chromatograms of the calibration standard with the assignment to the fractions of a diesel sample (THN = Tetralin [1,2,3,4-tetrahydronaphthalene], N = Naphthalene, AN = Area of nonaromatics, AM = Area of monoaromatics, and AP = Area of polyaromatics). *Source:* Reprinted with permission from reference [73]. Copyright 2015, Agilent Technologies, Inc.

detector (PID) [78, 79], the nitrogen–phosphorus detector (NPD), the sulfur chemiluminescence detector (SCD) [80–85], and the electron capture detector (ECD) [82, 85].

A novel flame-based detector for SFC and HPLC, the "universal acoustic flame detector" (AFD) has been studied by Thurbide and colleagues [86–89]. One unique aspect of this detector is that it is compatible with modified CO_2 mobile phases. The detector is based upon changes in the frequency of acoustic emissions from an oscillating premixed hydrogen/oxygen flame. The frequency increases proportionately to the carbon content of organic analytes, and has a uniform response across analytes. While the sensitivity of the detector is reasonable (measured at 18 ng of carbon per second), a mobile-phase density or composition gradient introduces a baseline shift [86].

4.4.2 Evaporative Light Scattering Detection (ELSD) and Charged Aerosol Detection (Corona CAD)

These two detectors are the most common "universal" (for nonvolatile analytes) detectors in the HPLC world. In general, the two function in a similar manner. The effluent is nebulized and large droplets are trapped and drained away. Solvent evaporates from the smaller droplets, sometimes with the help of added heat. Once the solvent has evaporated, the analytes contained in the droplets remain as small particles. The particles are swept into a light scattering region (ELSD), or into a region where they acquire a charge (CAD). Scattered light is detected in the ELSD, while charge is detected in the CAD. The operation and HPLC applications of both the ELSD [90, 91] and the CAD [92] have been reviewed.

One of the challenges of the operation of these detectors in the HPLC world is efficient nebulization of the HPLC effluent. The explosive decompression of the SFC mobile phase assists greatly in the nebulization process when SFC is coupled to these detectors. However, this decompression also results in significant Joule–Thompson cooling, which can be detrimental to the nebulization and solvent-evaporation processes. At least one manufacturer of ELSD instruments (the Sedex line from SEDERE) has modified the ELSD for SFC operation. The vendor suggests using lower nebulizing gas flow (less is needed due to the expansion of CO_2), and provides a heated nebulization region to counteract the Joule–Thompson cooling. These features are summarized by Dreux et al. [93]. The other major ELSD and CAD systems incorporate a heated nebulization region. An optimum supply of heat is important: sufficient heat to evaporate the solvent and counteract Joule–Thompson cooling is required, but too much heat can result in evaporation of some of the more volatile analytes, and reduces sensitivity. Strode and Taylor extensively studied the performance characteristics of the ELSD for SFC detection [94]. In most of their work, they used the pre-BPR split interfacing approach. But, in general, most versions of the ELSD and CAD are also compatible with total flow introduction. Applications of SFC/ELSD were reviewed by Dreux and Lafosse [95].

Unlike the FID, the ELSD and CAD operate perfectly well with the introduction of SFC mobile-phase modifiers, which are usually small, polar organics, and are volatile. However, some mobile-phase additives, such as citric acid, are not volatile, and can introduce significant baseline noise if used with one of these detectors. The limit of detection of SFC/ELSD and SFC/CAD is on the order of 10 ng for most analytes. Though this is not as low as with detectors such as mass spectrometry, the ELSD and CAD allow detection of analytes not detectable using UV absorbance, at a reasonable cost. Figure 4.8 illustrates an LOD of ~25 ng for underivatized fatty acids using SFC/ELSD [96].

In general, the CAD has been shown to have better sensitivity and a greater dynamic range than the ELSD [92, 97, 98]. While the ELSD typically has ~2 orders-of-magnitude dynamic range, the CAD has up to 4. The improved sensitivity and dynamic range of the CAD was demonstrated in a comparison of separations of polyethylene glycol oligomers using SFC/ELSD and SFC/CAD [99].

4.4.3 Mass Spectrometric Detection

Most of the discussion in the following sections focuses on SFC/MS interfacing approaches and ionization methods. But it's important to note that there are no restrictions in SFC/MS with regard to mass analyzer. Any mass analyzer

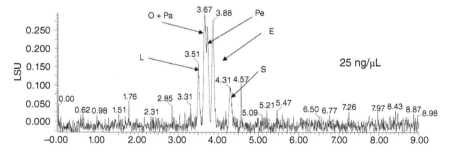

Figure 4.8 Detection limit using UHPSFC/ELSD – 1 μL injection of a solution of 25 μg/mL per compound. Chromatographic conditions: time (min) 0, 98/2 (A/B), time 0.5, 98/2, time 8, 96/4, time 9, 80/20, time 11, 80/20, time 11, 98/2, time 12, 98/2. Mobile phases: A = CO_2, B = MeOH with 0.1% (v/v) formic acid, flow: 1 mL/min, oven temp.: 25 °C, BPR pressure: 1500 psi, column: HSS C18, 150×3.0 mm, 1.8 μm, ELSD with makeup (IPA) flow rate: 0.2 mL/min, mixture: six free fatty acid standards: O = oleic, Pa = palmitic, L = linoleic, Pe = petroselinic, E = eliadic, S = stearic. *Source:* Reprinted with permission from reference [96]. Copyright 2015 Elsevier.

used for LC/MS (quadrupole, time-of-flight, ion-cyclotron-resonance, Orbitrap, and even magnetic sector) can and has been used for SFC/MS.

4.4.3.1 Interfacing and Ionization Approaches

Mass spectrometric detection for SFC was an important goal of much research from the very early stages of work in SFC, both for open-tubular/capillary and packed column SFC. Most open-tubular/capillary SFC/MS was performed with total-flow introduction into a conventional chemical ionization (CI) ion source [100–102]. The total effluent flow of CO_2 (nearly always pure CO_2) was low enough (a few mL/min of expanded gas), that the CI source operated with little interference with a standard differentially pumped mass spectrometer. Most open tubular/capillary SFC/MS was performed with pressure programming. Conventional CI spectra were delivered over the course of most of the pressure program, though a mixture of CI and CO_2 charge exchange ionization (providing more electron-ionization-like spectra) was sometimes observed at the very highest pressures and flow rates. Little open tubular/capillary SFC/MS has been performed since the late 1990s, so this area will not be discussed further.

Early packed column SFC/MS was performed using a number of novel interfacing approaches, all designed to make the high flow rate of the SFC system compatible with the high vacuum of the mass spectrometer. Some, like the thermospray source [103–107], used high temperature nebulization, ionization at relatively high pressures via either thermal processes or high voltage discharge, and electrostatic fields to enrich the flow of ions into the vacuum system. Niessen et al. described an interesting study of the use of the repeller voltage (directing ions into the high vacuum of the mass spectrometer) in the thermospray source to increase or decrease fragmentation via collisionally induced dissociation [108]. The thermospray interface proved useful for SFC/MS. For example, Perkins et al. used thermospray SFC/MS for the analysis of sulfonamide [105] and veterinary [109] drugs.

The particle–beam interface was another novel approach to couple HPLC and SFC to mass spectrometry [110]. The effluent is nebulized, and nonvolatile analytes form particles as the solvent is vaporized. The particles are transmitted to the mass spectrometer ion source via a momentum separator (one or more nozzle/skimmer combinations), while the vaporized solvent is pumped from the system in the momentum separator, and does not penetrate to the ion source. Typically, the particles impact a hot surface in the ion source and are vaporized. The analytes are ionized with either traditional electron ionization or chemical ionization. Jedrzejewski and Taylor explored the use of the particle–beam interface for SFC/MS [111]. They found the combination delivered library-quality electron ionization spectra with low-ng sensitivities. Both the particle–beam interface and the thermospray interface used total flow introduction of the SFC effluent.

These ionization methods hinted at the promise of packed column SFC/MS, but the thermospray and particle beam interfaces are no longer available commercially. The real explosion in the acceptance and more widespread use of SFC/MS came through the atmospheric pressure ionization approaches: atmospheric pressure chemical ionization, pneumatically assisted electrospray ionization, and atmospheric pressure photoionization. These atmospheric pressure ionization methods are all widely available. The respective sources are designed to be easily interchangeable, and some vendors provide "two-in-one" sources that can perform both atmospheric pressure chemical ionization and pneumatically assisted electrospray ionization.

4.4.3.2 Atmospheric Pressure Chemical Ionization (APCI)

In APCI, a high voltage discharge is used to create reagent ions from the mobile phase, which in turn ionize the analytes [112]. Huang et al. were the first to describe SFC/MS with ionization at atmospheric pressure with the use of APCI [113]. These researchers used methanol modifier to perform a double duty: both as a mobile phase modifier, and as the source of reagent ions (via corona discharge) to ionize analytes. Figure 4.9 shows an illustration of their impressive results, even by later standards: the detection of 20 ppb of trenbolone from a bovine liver homogenate, with a signal-to-noise of at least 10.

APCI has been widely used for SFC/MS since Huang et al.'s first publication [114–116], including application to polycyclic aromatic hydrocarbons [117,118], explosives [119], surfactants [32], petroleum products [120, 121], and many, many applications to pharmaceuticals [122–125]. Some have even advocated that APCI is the most widely applicable of the atmospheric pressure ionization methods. But many practitioners of pneumatically assisted electrospray ionization for SFC/MS are equally optimistic about their ionization method of choice.

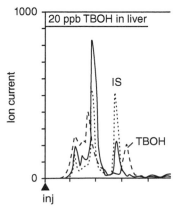

Figure 4.9 Early SFC/APCI-MS by Huang et al. [113] showing impressive results, even by later standards: 20 ppb of trenbolone (TBOH) in a bovine liver tissue homogenate. Internal standard (IS), 19-nortestosterone, present at 15 ppb. Acquisition was conducted in selected reaction monitoring (SRM) mode. *Source:* Reprinted with permission from reference [113]. Copyright 1990 Elsevier.

4.4.3.3 Pneumatically Assisted Electrospray Ionization (ESI)

ESI operates under a dramatically different ionization mechanism than does APCI [126,127]. The effluent passes through an aperture held at high voltage, with pneumatically assisted nebulization. The effluent droplets are charged, and the field strength about the droplets increases as solvent evaporates and the droplet size decreases. Eventually, field strength is so high that analyte and solvent ions are emitted, through a process called "coulombic explosion." Some gas-phase chemical ionization is believed to also occur, after ion emission. Electrospray generally provides lower energy ionization, with less fragmentation, than does APCI.

Sadoun, Virelizier, and Arpino (Arpino of early LC/MS fame [128]) were the first to couple SFC and MS using ESI [129]. Because ESI depends on the formation of charged solvent droplets, the authors found that SFC/ESI-MS required the addition of a polar organic modifier to the mobile phase, but this is the most reasonable approach for the elution of polar analytes using SFC, so is really not a significant disadvantage. Note that a droplet-forming solvent can also be added to the effluent postcolumn, if this is more appropriate for the desired separation. Baker and Pinkston described a "sheath flow interface" for SFC/ESI-MS which was specifically designed to allow the addition of solvents and additives to the effluent postcolumn for the optimization of ionization conditions [48]. Figure 4.10 provides a schematic diagram of their sheath-flow

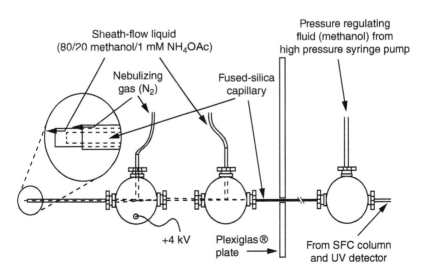

Figure 4.10 Sheath-flow interface designed to allow the postcolumn introduction of solvents and additives for optimizing the ionization process in SFC/ESI-MS [48]. This interface also shows the use of the pressure-regulating-fluid approach for backpressure regulation, rather than a mechanical device. *Source:* Reprinted with permission from reference [48]. Copyright 1998 Springer Nature.

interface. More common in recent commercial SFC/MS systems is the simple addition of a solvent or a solvent/buffer mixture to the effluent at or just after the pressure regulation point to ensure good mass transfer to the electrospray source and efficient ionization.

Despite concerns by some practitioners of SFC/APCI-MS, ESI has proven quite versatile for SFC/MS. Applications range from a wide array of pharmaceuticals [130], vitamins [131], polymers [132], surfactants [133], and small proteins [134]. Only the most nonpolar analytes, such as hydrocarbons, are not efficiently ionized by ESI. Figure 4.11 shows the base-peak chromatogram and a few representative ESI spectra from the SFC/MS analysis of a small, "di-capped" block copolymer [132]. The structure of the small polymer is shown in the figure. These results were acquired using a 1-m-long column and the pressure-regulating-fluid interface (see above).

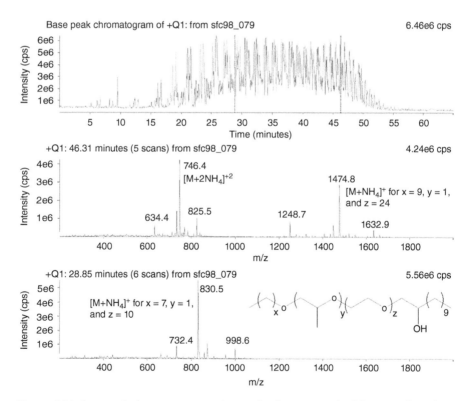

Figure 4.11 Base-peak chromatogram and a couple of representative ESI spectra from the SFC/MS analysis of a small, "di-capped" polyethoxylate-polypropoxylate block copolymer [132]. Base peak chromatogram (top) of the di-capped polymer, and mass spectra of two components of this mixture: The $x = 9$, $y = 1$, and $z = 24$ component elutes at 46.3 minutes (middle), while the $x = 7$, $y = 1$, and $z = 10$ component elutes at 28.8 minutes (bottom). *Source:* Reprinted with permission from reference [132]. Copyright 2002 Springer Nature.

4.4.3.4 Atmospheric Pressure Photoionization (APPI)

Photoionization has been shown in LC/MS and SFC/MS to be especially useful for nonpolar analytes that are not efficiently ionized by APCI or ESI [78, 79, 135]. The mobile phase modifier can act as the "dopant" which is ionized, and then in turn ionizes the analyte molecules, or another dopant can be added. APPI, though not as widely used as APCI or ESI, is quite versatile. Mejean et al., for example, examined the determination of tocopherols and tocotrienols (vitamin E congeners) in soybean oil using SFC/MS with APPI [136]. They found that APPI outperformed APCI and ESI.

4.4.4 Postdecompression Detection Using Less Common Approaches – Deposition IR

Rapid elimination of the primary mobile phase component in SFC allows the coupling of SFC with less common, novel methods of detection. For example, a number of researchers have explored SFC/deposition FTIR [33, 137–143], where the SFC effluent is deposited on a surface, the mobile phase is rapidly removed, and the chromatographic separation is preserved in space, rather than in time. An IR microscope is used to acquire IR spectra of the separated components. A distinct advantage of this approach is removal of the normal time constraints of online chromatographic detection, so that long acquisition times can be used to improve the signal-to-noise ratio for low concentration components, and interesting regions of the chromatographic separation can be re-examined at later times. Furthermore, potential interference from a mobile-phase modifier in the IR region is removed.

Examples of applications of SFC/deposition FTIR include polycyclic aromatic compounds in a coal tar pitch [144], pharmaceuticals (sulfanilamides [145]), polymer additives [143, 146], and complex mixtures of aliphatic and phenolic carboxylic acids [147]. Searchable spectra were acquired for sub-ng quantities of strong IR absorbers, such as caffeine [142].

4.5 Concluding Remarks

SFC is indeed a friend to many detectors as described in this chapter's introduction. In its various implementations, the SFC mobile phase is compatible with a wider range of detectors than either GC or HPLC alone, ranging from GC-style, flame-based detectors, to condensed phase detectors common in HPLC, to powerful mass spectrometric and spectroscopic detectors. Coupling SFC to detectors operating at atmospheric or lower pressures can be challenging. A variety of interfacing approaches have been described here, and all have advantages and limitations. But the compatibility of SFC with a wide array of detectors has been, and will continue to be, one of its chief advantages.

References

1 Yoshino, P.R. (2003). *Chemical Physics* 290: 251–256.
2 Jackson, H.B. Burdick & Jackson Solvent Guide. http://www51.honeywell. com/sm/rlss/bandj/common/documents/honeywell-burdick-jackson-product-guide.pdf.
3 Berger, T.A. and Fogelman, K. (2009). Improving signal to noise ratio and dynamic range in supercritical fluid chromatography with UV detection. The Peak. (17–33 September).
4 Sun, Y., Shekunov, B.Y., and York, P. (2003). Refractive index of supercritical CO_2-ethanol solvents. *Chemical Engineering Communications* 190 (1): 1–14.
5 Berger, T.A. and Berger, B.K. (2011). Minimizing UV noise in supercritical fluid chromatography. I. Improving back pressure regulator pressure noise. *Journal of Chromatography A* 1218 (16): 2320–2326.
6 Berger, T.A. (2015). Instrumentation for analytical scale supercritical fluid chromatography. *Journal of Chromatography A* 1421: 171–183.
7 Fjeldsted, J.C., Richter, B.E., Jackson, W.P., and Lee, M.L. (1983). Scanning fluorescence detection in capillary supercritical fluid chromatography. *Journal of Chromatography A* 279 (0): 423–430.
8 Smith, R.M., Chienthavorn, O., Danks, N., and Wilson, I.D. (1998). Fluorescence detection in packed-column supercritical fluid chromatographic separations. *Journal of Chromatography A* 798 (1–2): 203–206.
9 Nomura, A., Yamada, J., Yarita, T., and Maeda, T. (1995). Supercritical-fluid chromatograms of fuel oils on ods-silica gel column using fluorescence, UV-absorption, and flame-ionization detectors. *The Journal of Supercritical Fluids.* 8 (4): 329–333.
10 Kanomata, T., Horikawa, Y., and Kikuchi, S. (2013). High-pressure fluorescence flow cell, flow cell assembly, fluorescence detector, and supercritical fluid chromatograph. US Patents 9,267,887.
11 Toghill, K.E., Méndez, M.A., and Voyame, P. (2014). Electrochemistry in supercritical fluids: a mini review. *Electrochemistry Communications* 44: 27–30.
12 Abbott, A.P. and Harper, J.C. (1996). Electrochemical investigations in supercritical carbon dioxide. *Journal of the Chemical Society, Faraday Transactions* 92 (20): 3895–3898.
13 Di Maso, M., Purdy, W.C., and McClintock, S.A. (1990). Electrochemical detection system for supercritical fluid chromatography. *Journal of Chromatography A* 519 (1): 256–262.
14 Almquist, S.R., Nyholm, L., and Markides, K.E. (1994). Electrochemical detection in open tubular column supercritical fluid chromatography using a platinum microelectrode and CO2/water as mobile phase. *Journal of Microcolumn Separations* 6 (5): 495–501.

15 Dressman, S.F. and Michael, A.C. (1995). Online Electrochemical Detectors for Supercritical Fluid Chromatography. *Analytical Chemistry* 67 (8): 1339–1345.

16 Dressman, S.F., Simeone, A.M., and Michael, A.C. (1996). Supercritical Fluid Chromatography with Electrochemical Detection of Phenols and Polyaromatic Hydrocarbons. *Analytical Chemistry* 68 (18): 3121–3127.

17 Señoráns, F.J., Markides, K.E., and Nyholm, L. (1999). Determination of tocopherols and vitamin A in vegetable oils using packed capillary column supercritical fluid chromatography with electrochemical detection. *Journal of Microcolumn Separations* 11 (5): 385–391.

18 Toniolo, R., Comisso, N., Schiavon, G., and Bontempelli, G. (2004). Porous Electrodes Supported on Ion-Exchange Membranes as Electrochemical Detectors for Supercritical Fluid Chromatography. *Analytical Chemistry* 76 (7): 2133–2137.

19 Shafer, K.H. and Griffiths, P.R. (1983). On-line supercritical fluid chromatography/Fourier transform infrared spectrometry. *Analytical Chemistry* 55 (12): 1939–1942.

20 Olesik, S.V., French, S.B., and Novotny, M. Development of capillary supercritical fluid chromatography/Fourier transform infrared spectrometry. *Chromatographia* 18 (9): 489–495.

21 Johnson, C.C., Jordan, J.W., Taylor, L.T., and Vidrine, D.W. On-line supercritical fluid chromatography with Fourier transform infrared spectrometric detection employing packed columns and a high pressure lightpipe flow cell. *Chromatographia* 20 (12): 717–723.

22 Shah, S., Ashraf-Khorassani, M., and Taylor, L.T. Normal-phase liquid and supercritical-fluid chromatographic separations of steroids with Fourier-transform infrared spectrometric detection. *Chromatographia* 27 (9): 441–448.

23 Jordan, J.W. and Taylor, L.T. (1986). Mobile Phase and Flow Cell Comparisons in Packed Column Supercritical Fluid Chromatography/Fourier Transform Infrared Spectrometry. *Journal of Chromatographic Science* 24 (3): 82–88.

24 Calvey, E.M., Taylor, L.T., and Palmer, J.K. (1988). Supercritical fluid chromatography with Fourier transform infrared detection of peracetylated aldononitrile derivatives and byproducts from monosaccharides. *Journal of High Resolution Chromatography* 11 (10): 739–741.

25 Ashraf-Khorassani, M. and Taylor, L.T. (1989). Qualitative supercritical fluid chromatography/Fourier transform infrared spectroscopy study of methylene chloride and supercritical carbon dioxide extracts of double-base propellant. *Analytical Chemistry* 61 (2): 145–148.

26 Ashraf-Khorassani, M. and Taylor, L.T. (1989). Analysis of propellant stabilizer components via packed and capillary supercritical fluid chromatography/Fourier transform infrared spectrometry. *Journal of High Resolution Chromatography* 12 (1): 40–44.

27 Vidrine, D.W. and Allhands, D.R. (1986). Light-pipe flow cell for supercritical fluid chromatography. US Patents 4,588,893.

28 Morin, P., Caude, M., Richard, H., and Rosset, R. Carbon dioxide supercritical fluid chromatography-Fourier transform infrared spectrometry. *Chromatographia* 21 (9): 523–530.

29 Wieboldt, R.C., Adams, G.E., and Later, D.W. (1988). Sensitivity improvement in infrared detection for supercritical fluid chromatography. *Analytical Chemistry* 60 (21): 2422–2427.

30 Raynor, M.W., Clifford, A.A., Bartle, K.D. et al. (1989). Microcolumn supercritical fluid chromatography with on-line Fourier transform infrared detection. *Journal of Microcolumn Separations* 1 (2): 101–109.

31 Raynor, M.W., Shilstone, G.F., Bartle, K.D. et al. (1989). Use of xenon as a mobile phase for on–line capillary supercritical fluid chromatography–Fourier transform infrared spectrometry. *Journal of High Resolution Chromatography* 12 (5): 300–302.

32 Auerbach, R.H., Dost, K., Jones, D.C., and Davidson, G. (1999). Supercritical fluid extraction and chromatography of non-ionic surfactants combined with FTIR, APCI-MS and FID detection. *The Analyst* 124 (10): 1501–1505.

33 Jinno, K. Interfacing between supercritical fluid chromatography and infrared spectroscopy – a review. *Chromatographia* 23 (1): 55–62.

34 Taylor, L.T. and Calvey, E.M. (1989). Supercritical fluid chromatography with infrared spectrometric detection. *Chemical Reviews* 89 (2): 321–330.

35 Allen, L.A., Glass, T.E., and Dorn, H.C. (1988). Direct monitoring of supercritical fluids and supercritical chromatographic separations by proton nuclear magnetic resonance. *Analytical Chemistry* 60 (5): 390–394.

36 Albert, K., Braumann, U., Streck, R. et al. Application of direct on-line coupling of HPLC and SFC with 1HNMR spectroscopy for the investigation of monomeric acrylates. *Fresenius' Journal of Analytical Chemistry* 352 (5): 521–528.

37 Albert, K., Braumann, U., Tseng, L.-H. et al. (1994). Online Coupling of Supercritical Fluid Chromatography and Proton High-Field Nuclear Magnetic Resonance Spectroscopy. *Analytical Chemistry* 66 (19): 3042–3046.

38 Albert, K. (1995). On-line use of NMR detection in separation chemistry. *Journal of Chromatography A* 703 (1–2): 123–147.

39 Albert, K. (1997). Supercritical fluid chromatography-proton nuclear magnetic resonance spectroscopy coupling. *Journal of Chromatography A* 785 (1–2): 65–83.

40 Braumann, U., Händel, H., Strohschein, S. et al. (1997). Separation and identification of vitamin A acetate isomers by supercritical fluid chromatography – 1H NMR coupling. *Journal of Chromatography A* 761 (1–2): 336–340.

41 Fischer, H., Gyllenhaal, O., Vessman, J., and Albert, K. (2003). Reaction Monitoring of Aliphatic Amines in Supercritical Carbon Dioxide by Proton

Nuclear Magnetic Resonance Spectroscopy and Implications for Supercritical Fluid Chromatography. *Analytical Chemistry* 75 (3): 622–626.

42 Pinkston, J.D., Hentschel, R., and Smith, C.A. (1994). An Investigation of the Reactions of Amines and Amine Derivatives with Supercritical CO2 Using Open-Tubular Supercritical Fluid Chromatography/Mass Spectrometry. *The 5th International Symposium on Supercritical Fluid Chromatography and Extraction* (11–14 January 1994). Baltimore, Maryland, USA.

43 Hirata, Y., Kawaguchi, Y., and Funada, Y. (1996). Refractive Index Detection Using an Ultraviolet Detector with a Capillary Flow Cell in Preparative SFC. *Journal of Chromatographic Science* 34 (1): 58–62.

44 Pinkston, J.D. (2005). Advantages and drawbacks of popular supercritical fluid chromatography/mass spectrometry interfacing approaches - a user's perspective. *European Journal of Mass Spectrometry* 11 (2): 189–197.

45 Bolaños, B., Greig, M., Ventura, M. et al. (2004). SFC/MS in drug discovery at Pfizer, La Jolla. *International Journal of Mass Spectrometry* 238 (2): 85–97.

46 Berger, T.A. Simple correction for variable post column split ratios using pure carbon dioxide in packed column supercritical fluid chromatography with independent pressure and flow control. *Chromatographia* 54 (11): 783–788.

47 Chester, T.L. and Pinkston, J.D. (1998). Pressure-regulating fluid interface and phase behavior considerations in the coupling of packed-column supercritical fluid chromatography with low-pressure detectors. *Journal of Chromatography A* 807 (2): 265–273.

48 Baker, T.R. and Pinkston, J.D. (1998). Development and Application of Packed-Column Supercritical Fluid Chromatography/Pneumatically Assisted Electrospray Mass Spectrometry. *Journal of the American Society for Mass Spectrometry* 9 (5): 498–509.

49 Hoke, S.H., Tomlinson, J.A., Bolden, R.D. et al. (2001). Increasing Bioanalytical Throughput Using pcSFC-MS/MS: 10 Minutes per 96-Well Plate. *Analytical Chemistry* 73 (13): 3083–3088.

50 Saito, M. (2013). History of supercritical fluid chromatography: instrumental development. *Journal of Bioscience and Bioengineering* 115 (6): 590–599.

51 Pearson, C.D. and Gharfeh, S.G. (1986). Automated high-performance liquid chromatography determination of hydrocarbon types in crude oil residues using a flame ionization detector. *Analytical Chemistry* 58 (2): 307–311.

52 Shantha, N.C. (1992). Thin-layer chromatography-flame ionization detection Iatroscan system. *Journal of Chromatography A* 624 (1): 21–35.

53 Miller, D.J. and Hawthorne, S.B. (1997). Subcritical water chromatography with flame ionization detection. *Analytical Chemistry* 69 (4): 623–627.

54 Yang, Y., Kondo, T., and Kennedy, T.J. (2005). HPLC separations with micro-bore columns using high-temperature water and flame ionization detection. *Journal of Chromatographic Science* 43 (10): 518–521.

55 Coym, J.W. and Dorsey, J.G. (2004). Superheated water chromatography: a brief review of an emerging technique. *Analytical Letters* 37 (5): 1013–1023.

56 Fjeldsted, J.C., Kong, R.C., and Lee, M.L. (1983). Capillary supercritical-fluid chromatography with conventional flame detectors. *Journal of Chromatography A* 279 (0): 449–455.

57 Cortes, H., Pfeiffer, C.D., Richter, B.E., and Stevens, T.S. (1988). Chromatography columns with cast porous plugs and methods of fabricating same. US Patents 4,793,920.

58 Guthrie, E.J. and Schwartz, H.E. (1986). Integral pressure restrictor for capillary SFC. *Journal of Chromatographic Science* 24 (6): 236–241.

59 Chester, T.L., Innis, D.P., and Owens, G.D. (1985). Separation of sucrose polyesters by capillary supercritical-fluid chromatography/flame ionization detection with robot-pulled capillary restrictors. *Analytical Chemistry* 57 (12): 2243–2247.

60 Vela, N.P. and Caruso, J.A. (1993). Comparison of flame ionization and inductively coupled plasma mass spectrometry for the detection of organometallics separated by capillary supercritical fluid chromatography. *Journal of Chromatography A* 641 (2): 337–345.

61 Chester, T.L., Pinkston, J.D., and Owens, G.D. (1989). Separation of malto-oligosaccharide derivatives by capillary supercritical fluid chromatography and supercritical fluid chromatography-mass spectrometry. *Carbohydrate Research* 194: 273–279.

62 David Pinkston, J., Bowling, D.J., and Delaney, T.E. (1989). Industrial applications of supercritical-fluid chromatography – mass spectrometry involving oligometric materials of low volatility and thermally labile materials. *Journal of Chromatography A* 474 (1): 97–111.

63 Thiebaut, D. (2012). Separations of petroleum products involving supercritical fluid chromatography. *Journal of Chromatography A* 1252: 177–188.

64 Pyo, D. (2008). Supercritical fluid chromatographic separation of polyethylene glycol polymer. *Bulletin of the Korean Chemical Society* 29 (1): 231–233.

65 Pyo, D. and Lim, C. (2005). Supercritical fluid chromatographic separation of dimethylpolysiloxane polymer. *Bulletin of the Korean Chemical Society* 26 (2): 312–314.

66 France, J.E., Snyder, J.M., and King, J.W. (1991). Packed-microbe supercritical fluid chromatography with flame ionization detection of abused vegetable oils. *Journal of Chromatography A* 540 (0): 271–278.

67 Bally, R.W. and Cramers, C.A. (1986). Tapered versus constant diameter post-column restrictors in capillary SFC. *Journal of High Resolution Chromatography* 9 (11): 626–632.

68 Smith, R.D., Fulton, J.L., Petersen, R.C. et al. (1986). Performance of capillary restrictors in supercritical fluid chromatography. *Analytical Chemistry* 58 (9): 2057–2064.

69 Chester, T.L. (1984). Capillary supercritical-fluid chromatography with flame-ionization detection: reduction of detection artifacts and extension of detectable molecular weight range. *Journal of Chromatography. A* 299: 424–431.

70 Pinkston, J.D. and Hentschel, R.T. (1993). Evaluation of flow restrictors for open-tubular supercritical fluid chromatography at pressures up to 560 atm. *Journal of High Resolution Chromatography* 16 (5): 269–274.

71 Di Sanzo, F.P. and Yoder, R.E. (1991). Determination of aromatics in jet and diesel fuels by supercritical fluid chromatography with flame ionization detection (SFC – FID): a quantitative study. *Journal of Chromatographic Science* 29 (1): 4–7.

72 and ASfT, (ASTM) M, http://www.astm.org/Standards/D5186.htm (accessed April 1. ASTM D5186-03: Standard Test Method for Determination of the Aromatic Content and Polynuclear Aromatic Content of Diesel Fuels and Aviation Turbine Fuels by Supercritical Fluid Chromatography. http:// wwwastmorg/Standards/D5186htm (accessed April 1, 2015)2009.

73 Noll-Borchers, M. and Hölscher, T. (2015). Determination of Aromatic Content in Diesel Fuel According to ASTM D5186. Scientific Instruments Manufacturer, GmbH; Agilent Technologies, Inc. Agilent Pub. No. 5991-5682EN.

74 Robert, M., Campbell, J.S., Oravetz, L., et al. (2013). High Speed and High Efficiency SFC Separations of Complex Mixtures. *SFC 2013 7th International Conference on Packed Column SFC*, Boston, MA, USA.

75 Richter, B.E., Bornhop, D.J., Swanson, J.T. et al. (1989). Gas Chromatographic Detectors in SFC. *Journal of Chromatographic Science* 27 (6): 303–308.

76 Olesik, S.V., Pekay, L.A., and Paliwoda, E.A. (1989). Characterization and optimization of flame photometric detection in supercritical fluid chromatography. *Analytical Chemistry* 61 (1): 58–65.

77 Dachs, J. and Bayona, J.M. (1993). Optimization of a flame photometric detector for supercritical fluid chromatography of organotin compounds. *Journal of Chromatography A* 636 (2): 277–283.

78 Onuska, F.I. and Terry, K.A. (1991). Preliminary examination of the photoionization detector for microbore column supercritical fluid chromatography. *Journal of Microcolumn Separations* 3 (1): 33–37.

79 Zegers, B.N., Hessels, R., Jagesar, J. et al. (1995). Photoionization detection in packed-capillary liquid and supercritical-fluid chromatography. *Journal of Liquid Chromatography* 18 (3): 413–440.

80 Foreman, W.T., Shellum, C.L., Birks, J.W., and Sievers, R.E. (1989). Supercritical fluid chromatography with sulfur chemiluminescence detection. *Journal of Chromatography A* 465 (1): 23–33.

81 Bornhop, D.J., Murphy, B.J., and Krieger-Jones, L. (1989). Chemiluminescence sulfur detection in capillary supercritical fluid chromatography. *Analytical Chemistry* 61 (7): 797–800.

82 Chang, H.C.K. and Taylor, L.T. (1990). Use of sulfur chemiluminescence detection after supercritical fluid chromatography. *Journal of Chromatography A* 517 (0): 491–501.

83 Shi, H., Taylor, L.T., Fujinari, E.M., and Yan, X. (1997). Sulfur-selective chemiluminescence detection with packed column supercritical fluid chromatography. *Journal of Chromatography A* 779 (1–2): 307–313.

84 Jiménez, A.M. and Navas, M.J. (1999). Chemiluminescence detection systems in supercritical fluid methods. *TrAC Trends in Analytical Chemistry* 18 (5): 353–361.

85 Strode Iii, J.T.B. and Taylor, L.T. (1996). Optimization of electron-capture detector when using packed-column supercritical fluid chromatography with modified carbon dioxide. *Journal of Chromatography A* 723 (2): 361–369.

86 Xia, Z.P. and Thurbide, K.B. (2006). Universal acoustic flame detection for modified supercritial fluid chromatography. *Journal of Chromatography A* 1105 (1-2): 180–185.

87 Mah, C. and Thurbide, K.B. (2008). An improved interface for universal acoustic flame detection in modified supercritical fluid chromatography. *Journal of Separation Science* 31 (8): 1314–1321.

88 Mah, C. and Thurbide, K.B. (2011). Increased flow rate compatibility for universal acoustic flame detection in liquid chromatography. *Journal of Chromatography A* 1218 (2): 362–365.

89 Scott, A.F. and Thurbide, K.B. (2014). Comparative response characterization of a universal acoustic flame detector for chromatography. *Chromatographia* 77 (13–14): 865–872.

90 Megoulas, N.C. and Koupparis, M.A. (2005). Twenty years of evaporative light scattering detection. *Critical Reviews in Analytical Chemistry* 35 (4): 301–316.

91 Lucena, R., Cardenas, S., and Valcarcel, M. (2007). Evaporative light scattering detection: trends in its analytical uses. *Analytical and Bioanalytical Chemistry* 388 (8): 1663–1672.

92 Vehovec, T. and Obreza, A. (2010). Review of operating principle and applications of the charged aerosol detector. *Journal of Chromatography A* 1217 (10): 1549–1556.

93 Dreux, M., Lafosse, M., and Morin-Allory, L. (1996). The evaporative light scattering detector – a universal instrument for non-volatile solutes in LC and SFC. *LC-GC International.* 1996: 148–156.

94 Strode, J.T.B. and Taylor, L.T. (1996). Evaporative light scattering detection for supercritical fluid chromatography. *Journal of Chromatographic Science* 34 (6): 261–271.

95 Dreux, M. and Lafosse, M. (1997). Review of evaporative light scattering detection for packed-column SFC. *LC-GC International.* 1997: 382–390.

96 Ashraf-Khorassani, M., Isaac, G., Rainville, P. et al. (2015). Study of ultraHigh performance supercritical fluid chromatography to measure free fatty acids without fatty acid ester preparation. *Journal of Chromatography B, Analytical Technologies in the Biomedical and Life Sciences* 997: 45–55.

97 Magnusson, L.-E., Risley, D.S., and Koropchak, J.A. (2015). Aerosol-based detectors for liquid chromatography. *Journal of Chromatography A* 1421: 68–81.

98 Bu, X., Regalado, E.L., Cuff, J. et al. (2016). Chiral analysis of poor UV absorbing pharmaceuticals by supercritical fluid chromatography-charged aerosol detection. *The Journal of Supercritical Fluids.* 116: 20–25.

99 Takahashi, K., Kinugasa, S., Senda, M. et al. (2008). Quantitative comparison of a corona-charged aerosol detector and an evaporative light-scattering detector for the analysis of a synthetic polymer by supercritical fluid chromatography. *Journal of Chromatography A* 1193 (1-2): 151–155.

100 Smith, R.D., Fjeldsted, J.C., and Lee, M.L. (1982). Direct fluid injection interface for capillary supercritical fluid chromatography-mass spectrometry. *Journal of Chromatography A* 247 (2): 231–243.

101 Pinkston, J.D., Owens, G.D., Burkes, L.J. et al. (1988). Capillary supercritical fluid chromatography-mass spectrometry using a "high mass" quadrupole and splitless injection. *Analytical Chemistry* 60 (10): 962–966.

102 Kalinoski, H.T., Udseth, H.R., Chess, E.K., and Smith, R.D. (1987). Capillary supercritical fluid chromatography – mass spectrometry. *Journal of Chromatography A* 394 (1): 3–14.

103 Chapman, J.R. and Jennings, K.R. (1988). Coupled supercritical fluid chromatography/mass spectrometry using a thermospray source. *Rapid Communications in Mass Spectrometry* 2 (1): 6–7.

104 Niessen, W.M.A., Van Der Hoeven, R.A.M., De Kraa, M.A.G. et al. (1989). Repeller effects in discharge ionization in liquid and supercritical-fluid chromatography-mass spectrometry using a thermospray interface. *Journal of Chromatography A* 478: 325–338.

105 Perkins, J.R., Games, D.E., Startin, J.R., and Gilbert, J. (1991). Analysis of sulphonamides using supercritical fluid chromatography and supercritical fluid chromatography – mass spectrometry. *Journal of Chromatography A* 540 (0): 239–256.

106 Via, J. and Taylor, L.T. (1994). Packed-column supercritical fluid chromatography/chemical ionization mass spectrometry of energetic material extracts using a thermospray interface. *Analytical Chemistry* 66 (9): 1385–1395.

107 Combs, M.T., Ashraf-Khorassani, M., and Taylor, L.T. (1997). Packed column supercritical fluid chromatography-mass spectroscopy: a review. *Journal of Chromatography A* 785 (1–2): 85–100.

108 Niessen, W.M.A., Van Der Hoeven, R.A.M., De Kraa, M.A.G. et al. (1989). Repeller effects in discharge ionization in liquid and supercritical-fluid chromatography-mass spectrometry using a thermospray interface: II. Changes in some analyte spectra. *Journal of Chromatography A* 478 (0): 325–338.

109 Perkins, J.R., Games, D.E., Startin, J.R., and Gilbert, J. (1991). Analysis of veterinary drugs using supercritical fluid chromatography and supercritical fluid chromatography – mass spectrometry. *Journal of Chromatography A* 540 (0): 257–270.

110 Ligon, W.V. and Dorn, S.B. (1990). Particle beam interface for liquid chromatography/mass spectrometry. *Analytical Chemistry* 62 (23): 2573–2580.

111 Jedrzejewski, P.T. and Taylor, L.T. (1995). Packed column supercritical fluid chromatography-mass spectrometry with particle beam interface aided with particle forming solvent. *Journal of Chromatography A* 703 (1–2): 489–501.

112 French, J.B., Thomson, B.A., Davidson, W.R. et al. (1985). Atmospheric pressure chemical ionization mass spectrometry. In: *Mass Spectrometry in Environmental Sciences* (eds. F.W. Karasek, O. Hutzinger and S. Safe), 101–121. Boston, MA: Springer US.

113 Huang, E., Henion, J., and Covey, T.R. (1990). Packed-column supercritical fluid chromatography – mass spectrometry and supercritical fluid chromatography – tandem mass spectrometry with ionization at atmospheric pressure. *Journal of Chromatography A* 511 (0): 257–270.

114 Niessen, W.M.A. (1996). Chapter 1. Developments in interface technology for combined liquid chromatography, capillary electrophoresis and supercritical fluid chromatography-mass spectrometry. In: *Journal of Chromatography Library*, vol. 59 (ed. D. Barceló), 3–70. Elsevier.

115 Lazar, I.M., Lee, M.L., and Lee, E.D. (1996). Design and optimization of a corona discharge ion source for supercritical fluid chromatography time-of-flight mass spectrometry. *Analytical Chemistry* 68 (11): 1924–1932.

116 Morgan, D.G., Harbol, K.L., and Kitrinos, N.P. Jr. (1998). Optimization of a supercritical fluid chromatograph–atmospheric pressure chemical ionization mass spectrometer interface using an ion trap and two quadrupole mass spectrometers. *Journal of Chromatography A* 800 (1): 39–49.

117 Anacleto, J.F., Ramaley, L., Boyd, R.K. et al. (1991). Analysis of polycyclic aromatic compounds by supercritical fluid chromatography/mass spectrometry using atmospheric-pressure chemical ionization. *Rapid Communications in Mass Spectrometry* 5 (4): 149–155.

118 Thomas, D., Greig Sim, P., and Benoit, F.M. (1994). Capillary column supercritical fluid chromatography/mass spectrometry of polycyclic aromatic compounds using atmospheric pressure chemical ionization. *Rapid Communications in Mass Spectrometry* 8 (1): 105–110.

119 McAvoy, Y., Dost, K., Jones, D.C. et al. (1999). A preliminary study of the analysis of explosives using packed-column supercritical fluid chromatography with atmospheric pressure chemical ionisation mass spectrometric detection. *Forensic Science International* 99 (2): 123–141.

120 Lavison-Bompard, G., Thiébaut, D., Beziau, J.-F. et al. (2009). Hyphenation of atmospheric pressure chemical ionisation mass spectrometry to supercritical fluid chromatography for polar car lubricant additives analysis. *Journal of Chromatography A* 1216 (5): 837–844.

121 Pereira, A.S. and Martin, J.W. (2015). Exploring the complexity of oil sands process-affected water by high efficiency supercritical fluid chromatography/ orbitrap mass spectrometry. *Rapid Communications in Mass Spectrometry* 29 (8): 735–744.

122 Dost, K. and Davidson, G. (2000). Development of a packed-column supercritical fluid chromatography/atmospheric pressure chemical-ionisation mass spectrometric technique for the analysis of atropine. *Journal of Biochemical and Biophysical Methods* 43 (1–3): 125–134.

123 Hsieh, Y., Favreau, L., Schwerdt, J., and Cheng, K.C. (2006). Supercritical fluid chromatography/tandem mass spectrometric method for analysis of pharmaceutical compounds in metabolic stability samples. *Journal of Pharmaceutical and Biomedical Analysis* 40 (3): 799–804.

124 Chen, J.W., Hsieh, Y.S., Cook, J. et al. (2006). Supercritical fluid chromatography-tandem mass spectrometry for the enantioselective determination of propranolol and pindolol in mouse blood by serial sampling. *Analytical Chemistry* 78 (4): 1212–1217.

125 Coe, R.A., Rathe, J.O., and Lee, J.W. (2006). Supercritical fluid chromatography–tandem mass spectrometry for fast bioanalysis of R/S-warfarin in human plasma. *Journal of Pharmaceutical and Biomedical Analysis* 42 (5): 573–580.

126 Fenn, J.B., Mann, M., Meng, C.K. et al. (1990). Electrospray ionization–principles and practice. *Mass Spectrometry Reviews* 9 (1): 37–70.

127 Kebarle, P. and Tang, L. (1993). From ions in solution to ions in the gas phase - the mechanism of electrospray mass spectrometry. *Analytical Chemistry* 65 (22): 972A–986A.

128 Arpino, P.J., Dawkins, B.G., and McLafferty, F.W. (1974). A liquid chromatography/mass spectrometry system providing continuous monitoring with nanogram sensitivity. *Journal of Chromatographic Science* 12 (10): 574–578.

129 Sadoun, F., Virelizier, H., and Arpino, P.J. (1993). Packed-column supercritical fluid chromatography coupled with electrospray ionization mass spectrometry. *Journal of Chromatography A* 647 (2): 351–359.

130 Pinkston, J.D., Wen, D., Morand, K.L. et al. (2006). Comparison of LC/MS and SFC/MS for screening of a large and diverse library of pharmaceutically relevant compounds. *Analytical Chemistry* 78 (21): 7467–7472.

131 Matsubara, A., Uchikata, T., Shinohara, M. et al. (2012). Highly sensitive and rapid profiling method for carotenoids and their epoxidized products using supercritical fluid chromatography coupled with electrospray ionization-triple quadrupole mass spectrometry. *Journal of Bioscience and Bioengineering* 113 (6): 782–787.

132 Pinkston, J.D., Marapane, S.B., Jordan, G.T., and Clair, B.D. (2002). Characterization of low molecular weight alkoxylated polymers using long column SFC/MS and an image analysis based quantitation approach. *Journal of the American Society for Mass Spectrometry* 13 (10): 1195–1208.

133 Hoffman, B.J., Taylor, L.T., Rumbelow, S. et al. (2004). Determination of alcohol polyether average molar oligomer value/distribution via supercritical

fluid chromatography coupled with UV and MS detection. *Journal of Chromatography A* 1043 (2): 285–290.

134 Zheng, J., Pinkston, J.G., Zoutendam, P.H., and Taylor, L.T. (2006). Feasibility of supercritical fluid chromatography/mass spectrometry of polypeptides with up to 40-mers. *Analytical Chemistry* 78 (5): 1535–1545.

135 Robb, D.B., Covey, T.R., and Bruins, A.P. (2000). Atmospheric pressure photoionization: an ionization method for liquid chromatography-mass spectrometry. *Analytical Chemistry* 72 (15): 3653–3659.

136 Mejean, M., Brunelle, A., and Touboul, D. (2015). Quantification of tocopherols and tocotrienols in soybean oil by supercritical-fluid chromatography coupled to high-resolution mass spectrometry. *Analytical and Bioanalytical Chemistry* 407 (17): 5133–5142.

137 Fujimoto, C., Hirata, Y., and Jinno, K. (1985). Supercritical fluid chromatography-infrared spectroscopy of oligomers: use of buffer-memory technique. *Journal of Chromatography A* 332 (0): 47–56.

138 Pentoney, S.L., Shafer, K.H., and Griffiths, P.R. (1986). A solvent elimination interface for capillary supercritical fluid chromatography/Fourier transform infrared spectrometry using an infrared microscope. *Journal of Chromatographic Science* 24 (6): 230–235.

139 Raymer, J.H., Moseley, M.A., Pellizzari, E.D., and Velez, G.R. (1988). Capillary supercritical fluid chromatography coupled with matrix isolation Fourier transform infrared spectroscopy (SFC-MI-FTIR). *Journal of High Resolution Chromatography* 11 (2): 209–211.

140 Fuoco, R., Pentoney, S.L., and Griffiths, P.R. (1989). Comparison of sampling techniques for combined supercritical fluid chromatography and Fourier-transform infrared spectrometry with mobile phase elimination. *Analytical Chemistry* 61 (19): 2212–2218.

141 Gurka, D.F., Pyle, S., Titus, R., and Shafter, E. (1994). Direct-deposition infrared spectrometry with gas and supercritical fluid chromatography. *Analytical Chemistry* 66 (15): 2521–2528.

142 Norton, K.L. and Griffiths, P.R. (1995). Performance characteristics of a real-time direct deposition supercritical fluid chromatography-Fourier transform infrared spectrophotometry system. *Journal of Chromatography A* 703 (1–2): 503–522.

143 Smith, S.H., Jordan, S.L., Taylor, L.T. et al. (1997). Packed column supercritical fluid chromatography–mobile phase elimination Fourier transform infrared spectrometry employing modified fluids. *Journal of Chromatography A* 764 (2): 295–300.

144 Raynor, M.W., Davies, I.L., Bartle, K.D. et al. (1988). Supercritical fluid extraction/capillary supercritical fluid chromatography/Fourier transform infrared microspectrometry of polycyclic aromatic compounds in a coal tar pitch. *Journal of High Resolution Chromatography* 11 (11): 766–775.

145 Yang, J. and Griffiths, P.R. (1997). Separation and identification of sulfanilamides by capillary supercritical fluid chromatography-Fourier transform infrared spectroscopy. *Journal of Chromatography A* 785 (1–2): 111–119.

146 Raynor, M.W., Bartle, K.D., Davies, I.L. et al. (1988). Polymer additive characterization by capillary supercritical fluid chromatography/Fourier transform infrared microspectrometry. *Analytical Chemistry* 60 (5): 427–433.

147 Raynor, M.W., Bartle, K.D., Clifford, A.A. et al. (1990). Analysis of aliphatic and phenolic carboxylic acids by capillary supercritical fluid chromatography-Fourier-transform infrared microspectrometry. *Journal of Chromatography A* 505 (1): 179–190.

5

Chiral Analytical Scale SFC – Method Development, Stationary Phases, and Mobile Phases

5.1 Introduction

Chromatographic enantioseparations are critical to many areas of pharmaceutical discovery and development. They are used to analyze synthetic starting materials, evaluate asymmetric reactions or crystallizations, study drug metabolism, and determine enantiomeric purity of reaction intermediates and final products. The technique is also used extensively for the preparative resolution of enantiomers.

Resolution of enantiomers can be achieved indirectly or directly. Indirect chromatographic resolution involves derivatization of the enantiomer/racemate to form a pair of diastereomers, followed by separation on an achiral stationary phase. This approach requires high enantiomeric purity of the

Modern Supercritical Fluid Chromatography: Carbon Dioxide Containing Mobile Phases,
First Edition. Larry M. Miller, J. David Pinkston, and Larry T. Taylor.
© 2020 John Wiley & Sons, Inc. Published 2020 by John Wiley & Sons, Inc.

derivatizing agent and time for diastereomer formation. To achieve direct enantiomeric resolution, a source of chirality must be introduced into the chromatographic system through either the addition of a chiral agent to the mobile phase, or preferably the use of a chiral stationary phase (CSP). To achieve a chiral separation, three points of interactions are required, one of which must depend on stereochemistry [1]. The interaction between the analyte and the CSP creates a transient diastereomer complex. Ideally the diastereomer complex formed by one enantiomer will have different free energies relative to the complex formed by the other enantiomer, thereby resulting in differences in retention, and affording a separation. This is illustrated in Figure 5.1. The enantiomer depicted on the top has a spatial orientation that allows three points of interactions (π–π stacking and two H-bonding) with the chiral stationary phase (left). The opposite enantiomer, depicted on the bottom, has a spatial orientation that allows only two points of interactions with the phase (second H-bonding interaction is not possible). The top enantiomer has stronger interactions with the phase, and will be more retained relative to the bottom enantiomer. Advances in CSP design and stability as well as chromatography equipment have made the use of chiral stationary phases the method of choice for chromatographic resolution.

Chiral chromatographic resolution using HPLC has been used in the pharmaceutical industry since the first separations were reported in the 1980s

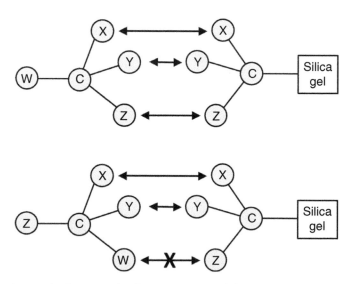

Figure 5.1 Interaction between chiral stationary phase (left) and both enantiomers of a biphenyl derivative (right).

[2–4]. Recently supercritical fluid chromatography (SFC) has become the predominant technique for chiral analysis and resolution, especially in pharmaceutical research [5–8]. While SFC was developed over 50 years ago, it is only during the past 15–20 years that it has become routine. Mourier and coworkers first reported an example of enantiomer separations using packed column SFC in 1985 [9]. Numerous authors have published reviews of enantioseparations using SFC [5, 10–13].

Throughout this chapter, the term "racemate" is used to describe the compounds being resolved. The techniques described in this chapter are relevant for any mixture of enantiomers, not just racemates.

5.2 Chiral Stationary Phases for SFC

There are currently hundreds of chiral stationary phases (CSPs) on the market. The main types are (i) polysaccharides [14–16], (ii) "Pirkle" type [17, 18], (iii) proteins [19], (iv) cyclodextrin [20, 21], and (v) macrocyclic glycopeptide [22]. Each of these phases have different characteristics, are designed to operate under specific operating conditions, and are able to resolve a different range of racemates. Characteristics for the five most popular types of CSP are summarized in Table 5.1 and structures of some of them are shown in Tables 5.2–5.4. While each type of CSPs (excluding protein based) has been evaluated with SFC, the main types currently used in SFC operation are polysaccharides and Pirkle phases. These two types of CSP are also widely used for HPLC enantioseparations. Their wide use is due to their ability to resolve a wide range of

Table 5.1 Characteristics of chiral stationary phases (CSP).

	Polysaccharides	Immobilized polysaccharides	Pirkle type	Protein based	Cyclodextrin	Macrocyclic Glycopeptides
Solvent limitations	Severe	Very few	Very few	Reversed phase only	Very few	Very few
Loadability	High	Medium	Medium	Very low	Low	Medium
Range of resolution	High	High	Low	Medium	Low	Medium
Large column sizes/bulk availability	Yes	Yes	Yes	No	2 cm i.d. and less	2 cm i.d. and less

Table 5.2 Polysaccharide based, coated chiral stationary phases.

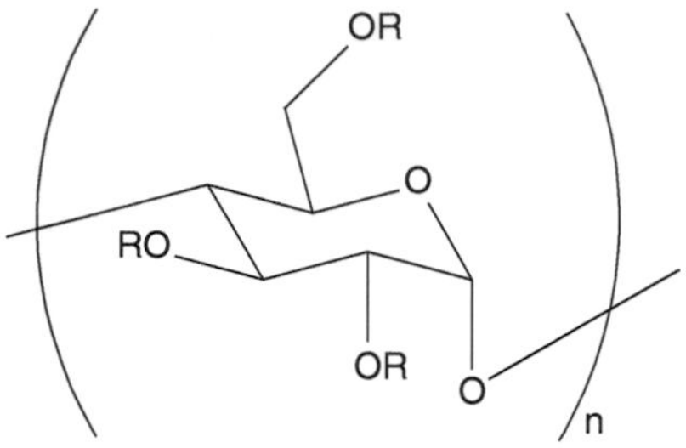

Amylose

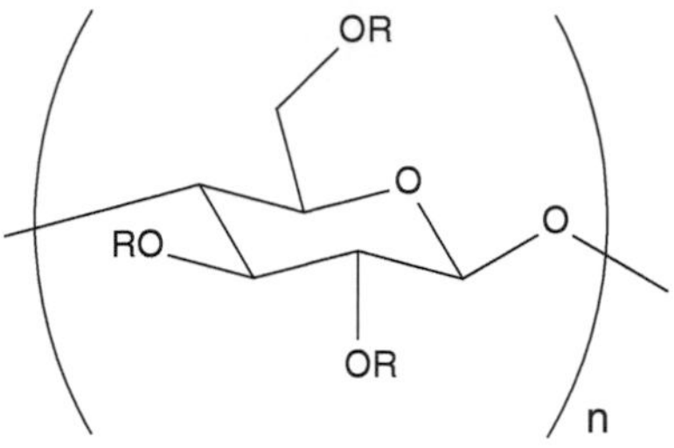

Cellulose

R Group	Amylose based	Cellulose based
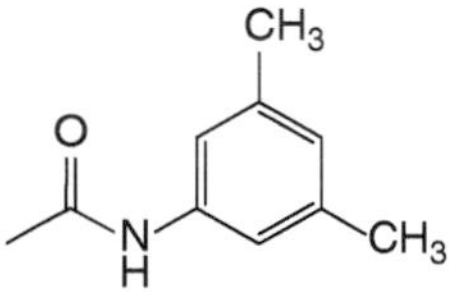 *tris* (**3,5-dimethylphenylcarbamate**)	Chiralpak AD (CT) Amylose-1 (YMC) ChromegaChiral CCA (ES) RegisPack (Regis)	Chiralcel OD (CT) Cellulose-1 (YMC) ChromegaChiral CCO (ES) RegisCell (Regis) Lux Cellulose-1 (Phenomenex)
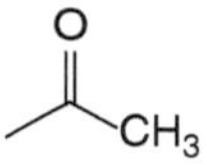 **Cellulose triacetate**		Chiralcel OA (CT)
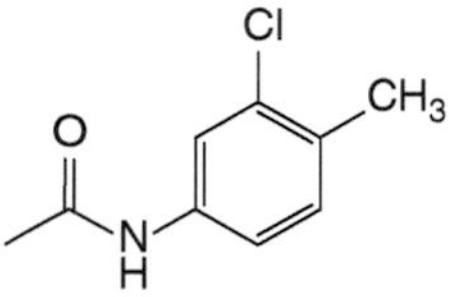 *tris* (**3-chloro-4-methylphenylcarbamate**)	Chiralpak AZ (CT)	ChromegaChiral CC4 (ES)
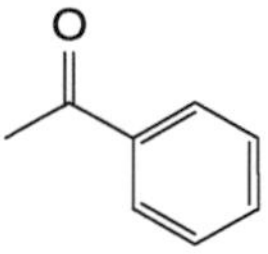 **Cellulose tribenzoate**		Chiralcel OB (CT)

Table 5.2 (Continued)

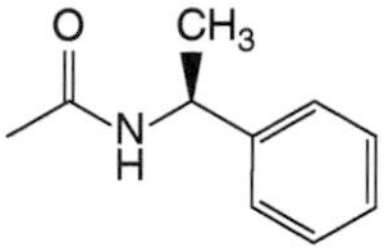 **tris (S)-α-methylbenzylcarbamate**	Chiralpak AS (CT) ChromegaChiral CCS (ES)
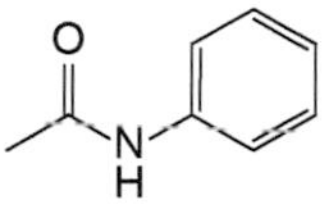 *tris* **(phenylcarbamate)**	Chiralcel OC (CT)
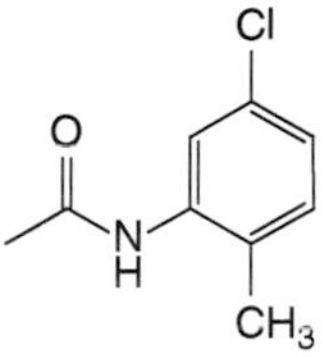 *tris* **(5-chloro-2-methylphenylcarbamate)**	Chiralpak AY (CT) ChromegaChiral CC3 (ES) Lux Amylose-2 (Phenomenex) RegisPack CLA-4 (Regis)
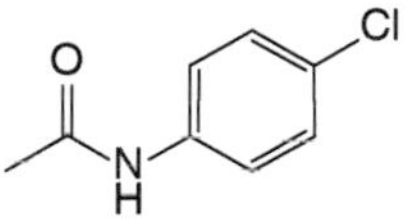 *tris* **(4-chlorophenylcarbamate)**	Chiralcel OF (CT)
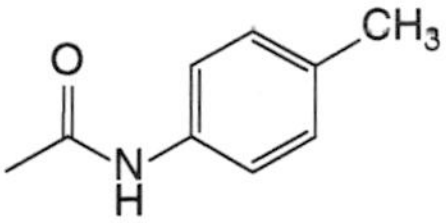 *tris* **(4-methylphenylcarbamate)**	Chiralcel OG (CT)
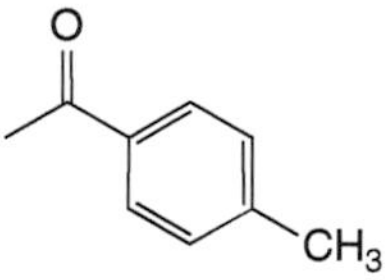 *tris* **(4-methylbenzoate)**	Chiralcel OJ (CT) ChromegaChiral CCJ (ES) Lux Cellulose-3 (Phenomenex)
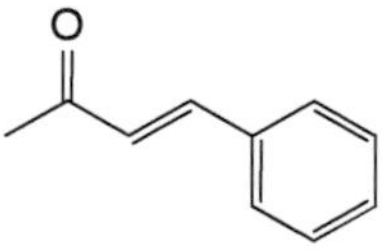 **Cellulose tricinnamate**	Chiralcel OK (CT)

(Continued)

Table 5.2 (Continued)

Structure		
tris (4-chloro-3-methylphenylcarbamate)	Chiralcel OX (CT) Lux Cellulose-4 (Phenomenex)	
tris (3-chloro-4-methylphenylcarbamate)	Chiralcel OZ (CT) ChromegaChiral CC2 (ES) Lux Cellulose-2 (Phenomenex)	
tris (2-fluoro-5-methylphenylcarbamate)	ChromegaChiral CCO-F2 (ES)	
tris (4-fluoro-3-methylphenylcarbamate)	ChromegaChiral CCO-F4 (ES)	ChromegaChiral CCA-F4 (ES)
tris (4-fluoro-3-triflourophenylcarbamate)	ChromegaChiral CCO-F4T3 (ES)	

CT = Chiral Technologies.
ES = ES Industries.
Regis = Regis Technologies.

Table 5.3 Pirkle type chiral stationary phases.

Name	Structure
DACH-DNB	
Leucine	
Phenylglycine	

(*Continued*)

Table 5.3 (Continued)

Pirkle-1 J

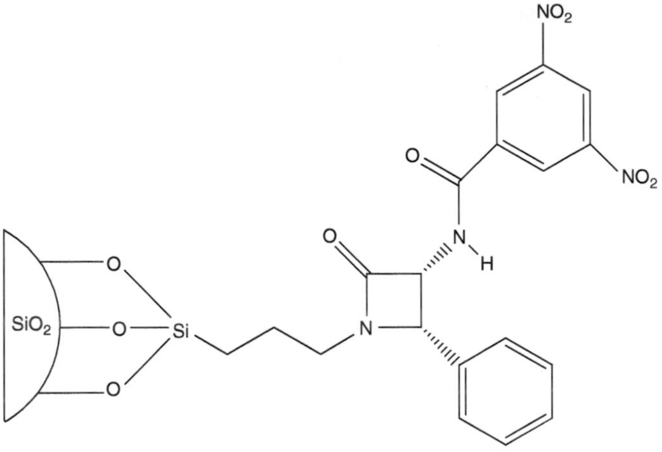

ULMO

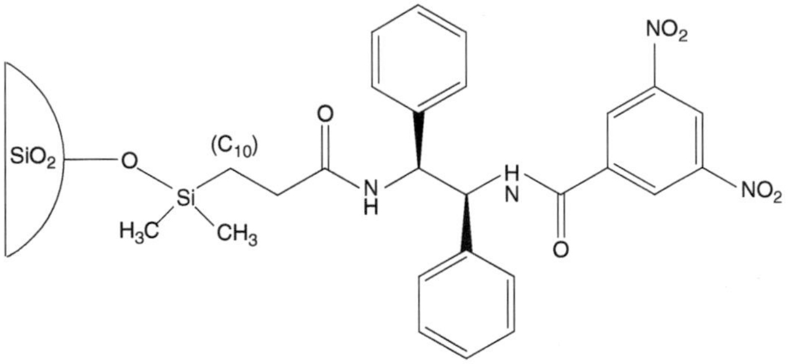

Whelk-O1

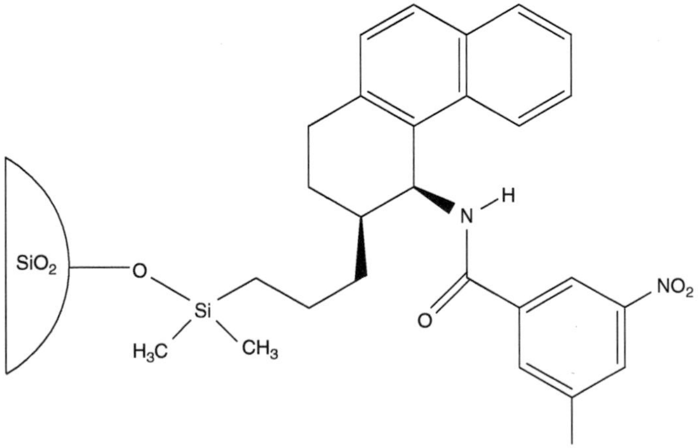

Table 5.3 (Continued)

Whelk-O2

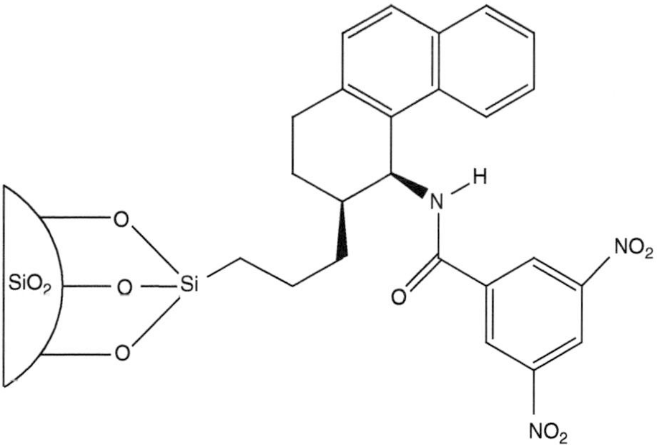

α-Burke 2

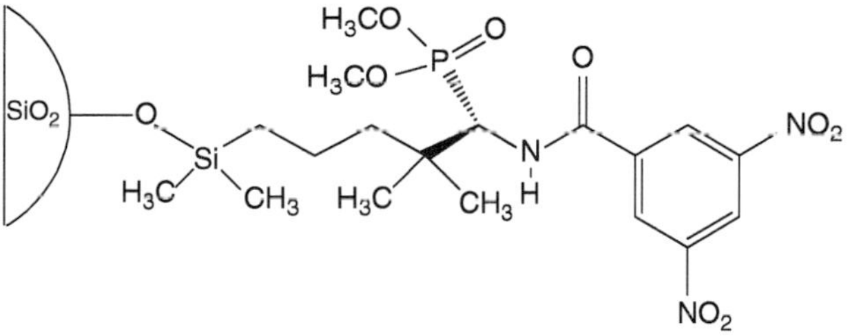

β-Gem 1

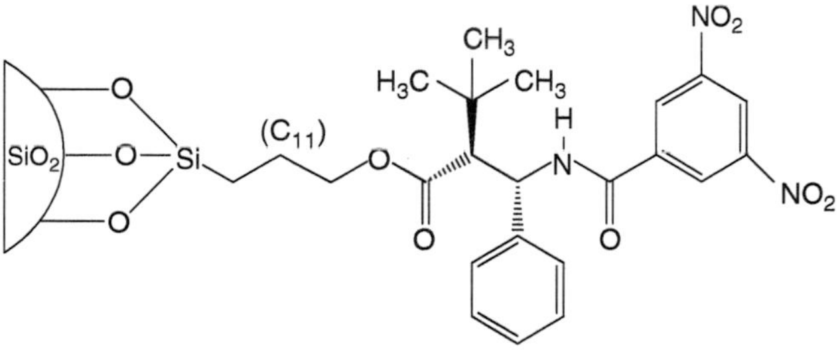

Table 5.4 Polysaccharide based, immobilized chiral stationary phases.

R Group	Amylose based	Cellulose based
tris (3,5-dimethylphenylcarbamate)	Chiralpak IA (CT) Amylose-SA (YMC) i-Amylose-1 (Phenomenex)	Chiralpak IB (CT) Cellulose-SB (YMC)
tris (3,5-dichlorophenylcarbamate)	Chiralpak IE (CT) Amylose-SE (YMC)	Chiralpak IC (CT) Cellulose-SC (YMC) i-Cellulose-5 (Phenomenex)
tris (3-chlorophenylcarbamate)	Chiralpak ID (CT)	
tris (3-chloro-4-methylphenylcarbamate)	Chiralpak IF (CT)	
tris (3-chloro-5-methylphenylcarbamate)	Chiralpak IG (CT)	
tris (S)-α-methylbenzylcarbamate	Chiralpak IH (CT)	
tris (4-methylbenzoate)		Cellulose-SJ (YMC)

CT = Chiral technologies.

racemates. In addition, many of the chiral SFC methods are developed for ultimate scale-up to preparative separations. Polysaccharide and Pirkle type CSPs have high-loading capacities which are critical for preparative resolutions. Additional details can be found in Chapter 7. The most commonly used polysaccharide based CSP are shown in Table 5.2 while the most commonly used Pirkle type CSP are shown in Table 5.3. Additional CSP such as cyclofructan [23] and cation exchange type [24, 25] have been recently introduced to the market. While these have been evaluated, they are not routinely used for SFC separations.

While polysaccharide-based based CSP have the ability to resolve a wide range of racemate structures, they have limited solvent compatibility. Most polysaccharide based chiral selectors are adsorbed on silica rather than bonded to the silica. This means that the phase can be dissolved and washed off the silica if a noncompatible solvent is used. Noncompatible solvents include, but are not limited to, ethyl acetate, toluene, dichloromethane, and tetrahydrofuran. This limitation has been eliminated through the introduction of immobilized polysaccharide-based CSPs (Table 5.4) which have no solvent restrictions and have proven useful when selectivity cannot be obtained using typical SFC modifiers. As one might expect, immobilized CSPs are not only useful for analytical-scale chiral separations but are also useful for preparative separations in which the racemate has poor solubility in traditional SFC modifiers [26]. Additional information on immobilized CSPs is provided later in this chapter.

Currently the vast majority of chiral SFC separations are performed using 3 and 5 µm particles. The past 10 years have seen tremendous advances in particle design for LC applications. These advances include sub-2 µm and superficially porous silica (SPS) stationary phases. The advantages of SPS over fully porous silica (FPS) are well documented and include significantly higher efficiency and shift to higher optimal flow rates [27–29]. The use of SPS for LC chiral separations has also been investigated [30]. Recently the use of SPS for SFC chiral separations has also been investigated [31]. Columns made with sub-2 µm particles can be shorter and have efficiencies comparable to longer columns with larger particles. Use of smaller particle phases results in higher plate number and increased resolution, or in faster analysis times and reduced solvent use with the same resolution. Recent publications have discussed the use of sub-2 µm particles for chiral SFC separations [32]. Both SPS and sub-2 µm particles offer the potential of ultrafast chiral SFC separations (on the order of seconds). While their use is currently limited to niche applications and research for a number of reasons (including SFC system limitations and limited stationary phase offerings), usage should increase once these are addressed.

5.3   Chiral SFC vs. Chiral HPLC

Over the past decade, SFC has become the technique of choice for chromatographic resolution of enantiomers. Despite some statements to the contrary, SFC does not routinely offer improved selectivity relative to HPLC. Nevertheless, SFC offers a number of important advantages over HPLC for chiral separations. These include the following:

- Increased flow rate shortens time for analysis and method development
- Reduced organic solvent use
- Lower separation costs
- Miscibility of methanol and acetonitrile with carbon dioxide (in comparison to heptane) allows increased method development options relative to normal phase HPLC
- Acidic nature of carbon dioxide based eluent mixtures eliminates need for acidic additives during separation of slightly acidic racemates

As discussed earlier, the low viscosity of carbon dioxide results in a lower pressure drop, which allows higher flow rates before system pressure limitations are reached. Higher solute diffusivity in supercritical fluid mobile phases results in higher mass transfer rates. The resulting improved C-term in the Van Deemter equation allows increased flow rates without sacrificing efficiency. Typical chiral separations using SFC are three to five times faster than HPLC. Several examples of rapid separations are shown in Figure 5.1. It is not unusual to routinely achieve separations in less than 60 seconds. All separations shown in Figure 5.2 use 4.6 mm i.d. ×100 mm length columns packed with 5 μm particles. Use of shorter columns and smaller particles, including superficially porous particles, has allowed separations in under 30 seconds [31, 33, 34].

The interactions required for chiral resolution are complex, making prediction of conditions that afford enantioseparation difficult. Standard practice involves screening a number of stationary and mobile phases during the method development process. Additional information on method development can be found later in this chapter. The high flow rates and reduced equilibration times characteristic of SFC allow faster method development. The operator therefore has the ability to perform more analyses per instrument; reducing equipment, operation and maintenance costs as well as laboratory space requirements.

While the use of carbon dioxide offers chromatographic advantages of speed and reduced solvent consumption, its presence can also impact enantioselectivity. Figure 5.3a shows the HPLC and SFC separation of propranolol. HPLC analysis using Chiralcel OD and a mobile phase of methanol w/0.2% diethylamine does not yield an enantioseparation. Keeping all conditions the same and adding 75% CO_2 to the mobile phase (i.e. SFC conditions) we see a

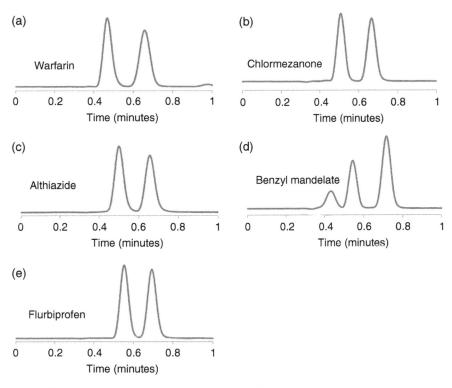

Figure 5.2 Examples of rapid chiral resolutions. Conditions for all analyses: column dimensions 4.6 × 100 mm, 5 μm, flow rate of 5 mL/min, 100 bar back pressure. 10 μl of 1 mg/mL methanol solution is injected. (a) Warfarin, Chiralpak AD-H 55% ethanol, (b) Chlormezanone, Chiralpak AD-H 50% IPA, (c) Althiazide, Chiralpak AD-H 50% IPA, (d) Benzyl mandelate, Chiralpak AD-H 45% methanol, and (e) Flurbiprofen, Chiralpak AD-H 30 methanol.

baseline separation of the enantiomers. The opposite effect is seen in Figure 5.3b for the separation of alpha-methyl-alpha-phenylsuccinimide; addition of CO_2 to the mobile phase results in a reduction of selectivity. Based on polarity, one would expect retention to increase when nonpolar CO_2 is added to the mobile phase. The separation of chlormexanone in Figure 5.3c, however, shows this to not always be the case. Here, the addition of CO_2 to the mobile phase results in a decrease in retention. These examples show that carbon dioxide does not act as an uninvolved component of the mobile phase (as one might consider the carrier gas in GC) for SFC enantioseparations. It can increase or decrease selectivity and/or retention. The examples in Figure 5.3 all use polysaccharide based CSPs. It appears that the introduction of CO_2 results in changes to the tertiary structure of the stationary phase, which is evident by the changing retention and selectivity.

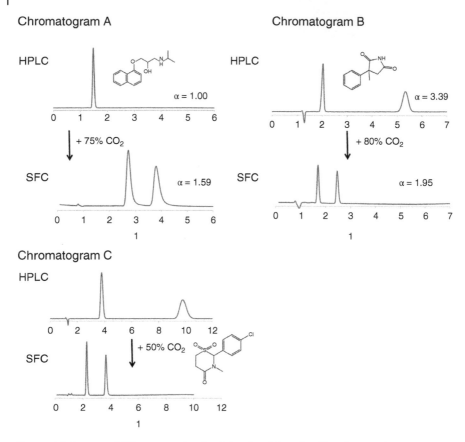

Figure 5.3 Impact of CO_2 on enantioselectivity. Conditions for all analyses: column dimensions 4.6×100 mm, 5 μm, 2 mL/min. 10 μl of 1 mg/ml methanol solution is injected. Chromatogram A: Resolution of propranolol, Chiralpak OD-H, methanol w/0.2% diethylamine. Chromatogram B: Resolution of α-methyl-α-phenylsuccinimide, Chiralcel OJ-H, ethanol w/0.2% diethylamine. Chromatogram C: Resolution of chloromexanone, Chiralpak AD-H, ethanol w/0.2% diethylamine.

5.4 Method Development Approaches

The complexity of interactions between analyte and the mobile and stationary phases make it difficult to predict conditions that will resolve a racemate, especially for polysaccharide-based phases. Spectroscopic studies have been performed as part of mechanistic studies on chiral discrimination on polysaccharide phases [35]. While some reports of successful separation prediction have been reported for polysaccharide CSP, it is difficult, if not impossible, to predict the best separation conditions. Given the limited success of making useful predictions about selectivity on most CSPs, the standard practice is to screen a set of chiral phases along with multiple mobile phases. Successful predictions, on the other hand, have been accomplished with Pirkle type CSP, which have less complex structures and interactions relative to polysaccharide CSP [36, 37].

It is known that minor structural changes can have a large impact on enantioselectivity. This is illustrated in Table 5.5 that lists the resolution of a series

Table 5.5 Effect of structural changes on enantioselectivity.

R =				
Chiralpak IC – MeOH/DEA	0.00	0.973	3.45	4.00
Chiralcel OZ – MeOH/DEA	4.579	2.145	3.448	4.406
Chiralpak AD – MeOH/DEA	1.157	2.367	0.00	2.344
Chiralcel OZ – IPA/DEA	4.142	2.92	4.084	4.372
Chiralpak AD – IPA/DEA	1.813	3.715	1.942	1.18

of closely related racemates under identical SFC conditions. The racemates are identical except for minor changes in the left side of the molecule. The changes include location of a fluorine on the pyridine ring as well as replacement of the fluorine with a chlorine. Even though the changes are five to six carbons removed from the chiral center, they greatly impact enantioselectivity. This is best illustrated in the first row for separation using a Chiralpak IC CSP and methanol-diethylamine modifier. When the fluorine is para to the nitrogen, no separation is observed. Moving the fluorine to the ortho position results in an increase in resolution, but baseline separation is not obtained. With the fluorine in the meta position, resolution increases drastically. When analyzed using a Chiralpak AD CSP and methanol-diethylamine modifier, no separation is seen for the meta analog while the best resolution is observed with the ortho analog. Similar changes are observed for the meta-compound when the fluorine is replaced with a chlorine. This is a good illustration of the need to perform method development for all chiral SFC separations.

The choice of CSP can also greatly impact enantioselectivity; at times the impact can be as large as to result in reversal of elution order. This is shown in Figure 5.4 for the separation of an ~10:1 mixture of R:S benzyl mandelate. Using a Chiralpak AD and Chiralcel OJ CSP (chromatograms A and C) the minor enantiomer elutes first. With a Chiralpak AS CSP (chromatogram B), the minor enantiomer elutes second. Elution order reversal is common when using different CSPs, and has also been observed when using the same CSP and different mobile-phase modifiers.

5.4.1 Modifiers for Chiral SFC

Most pharmaceutically applicable compounds analyzed by SFC are moderately polar, and CO_2 alone as a mobile phase is insufficient for elution from a chromatographic column. In most cases, the addition of a polar modifier is required. For chiral separations, the modifier is most often a low molecular weight alcohol such as methanol, ethanol, or isopropanol. Acetonitrile has also been shown to be useful for chiral resolution. In addition to the solvents listed above, nontraditional modifiers can be used. The use of nontraditional modifiers will be discussed later in this chapter. When working with nonpolar racemates, traditional alcohol modifiers can be too polar, resulting in compounds having no retention. Some researchers have even evaluated using mixtures of heptane or hexane with isopropanol as an SFC modifier to reduce mobile phase polarity and achieve retention [38].

The choice of modifier can have a drastic impact on chiral recognition. For example the modifier can impact the tertiary structure of the polysaccharide based CSPs, resulting in changes in enantiomer/CSP interactions,

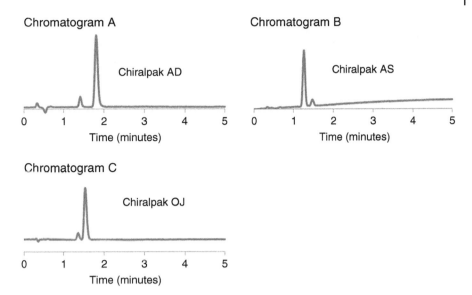

Figure 5.4 Example of elution reversal (~10/90 mix of S/R Benzyl Mandelate). Conditions for all analyses: Column dimensions 4.6 × 100 mm, 5 μm, flow rate of 5 mL/min, 100 bar back pressure. Chromatogram A: Chiralpak AD-H, 10% methanol w/0.2% diethylamine. Chromatogram B: Chiralpak AS-H, 5% methanol w/0.2% diethylamine. Chromatogram C: Chiralcel OJ, 10% methanol w/0.2% diethylamine.

which may result in enantioselectivity changes. Figure 5.5 shows the impact of modifier on the chiral resolution of α-(2,4-dichlorophenyl)-1H-imidazole-1-ethanol. While separation is observed with all four modifiers, the degree of separation varies. For the alcohols, the best separation is seen with methanol, whereas enantioselectivity decreases with ethanol and the poorest separation is seen with isopropanol. Higher enantioselectivity is seen with acetonitrile as the modifier; but the amount of modifier required for elution (45%) is higher than that required with alcohol modifiers (20%). Note that acetonitrile, while polar, differs from methanol rather drastically as acetonitrile is not a hydrogen-bond donor (although it can be an H-bond acceptor).

5.4.2 Additives for Chiral SFC

SFC analysis of strongly acidic or basic racemates using only carbon dioxide and typical organic mobile phase modifiers may result in poor peak shape due to strong interactions between the CSP and the analyte. Addition of a basic (for basic racemates) or acidic (for acidic racemates) additives to the mobile phase

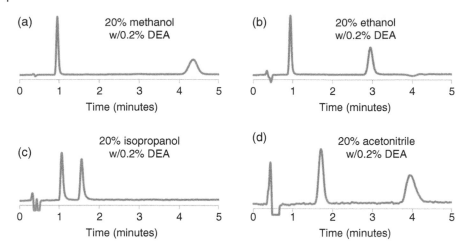

Figure 5.5 Impact of modifier on enantioselectivity of α-(2,4-Dichlorophenyl)-1H-imidazole-1-ethanol. Conditions for all analyses: Chiralpak AD-H, 4.6 × 100 mm, 5 μm, flow rate of 5 mL/min, 100 bar back pressure. Modifier: (a) 20% methanol w/0.2% diethylamine, (b) 20% ethanol w/0.2% diethylamine, (c) 20% isopropanol w/0.2% diethylamine, and (d) 45% acetonitrile w/0.2% diethylamine.

can result in improved peak shapes. For basic racemates typical additives include diethylamine, trimethylamine, isopropylamine, ammonium hydroxide, or others [39]. Under SFC conditions, it has been demonstrated that carbon dioxide can interact with aliphatic alcohols such as methanol, forming carbonate species. These species have pKa values less than four [40]. Due to the presence of the acidic carbonate species in the SFC mobile phase, an acidic additive may not be needed for moderately acidic racemates. This is illustrated in Figure 5.6 for the analysis of a racemate containing an acidic sulfonamide. During HPLC method development, tailing peaks were observed for a neutral mobile phase. Trifluoroacetic acid was added to the mobile phase to improve peak shape (chromatogram B). Under SFC conditions, an acidic additive was not necessary (chromatogram A) to achieve good peak shape. However, for more acidic compounds, it may be necessary to add acid at low levels (<0.5%) to the SFC modifier. Typical acidic additives include trifluoroacetic acid, acetic acid, or formic acid [41, 42].

De Klerck et al. have reported on the simultaneous use of acidic and basic additives in chiral SFC [43]. Their work showed that compared to individual additives, an increase in enantioselectivity can occur when combining trifluoroacetic acid and isopropylamine in the SFC mobile phase. It is important to keep in mind that the combination of acidic and basic additives can also lead to the formation of salts between the two additives. This is

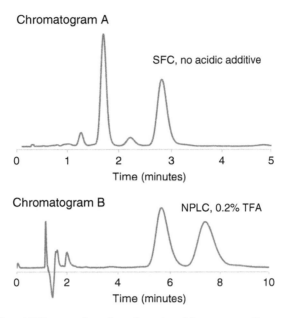

Figure 5.6 HPLC and SFC separation of moderately acidic racemate. Chromatogram A: Analytical SFC Separation: Chiralpak AD-H, 4.6 × 100 mm, 20% methanol/80% CO_2, 5 mL/min. Chromatogram B: Analytical HPLC separation: Chiralcel OD-H, 4.6 × 100 mm, 20% ethanol w/0.2% trifluoroacetic acid/80% heptane, 1 mL/min.

especially important if a method for purification is being developed, as these salts are nonvolatile and remain mixed with the isolated enantiomers after purification.

5.4.3 Nontraditional Modifiers

Silica-coated polysaccharide CSPs are the most widely used phases for SFC resolution of enantiomers. However, as noted earlier, these coated CSPs have severe solvent restrictions due to the derivatized cellulose or amylose being simply adsorbed onto the silica gel. Contact with certain solvents (i.e. ethyl acetate, tetrahydrofuran, and methylene chloride) results in dissolution of the cellulose/amylose polymer and loss of resolution and/or column destruction. This limitation reduces mobile phase modifier choices. The introduction of immobilized cellulose/amylose CSPs increases the range of solvents that can be explored during the method development process [44]. The increased choices for solvents are especially important in preparative purifications in which compound solubility in the mobile phase has a major impact on purification productivity.

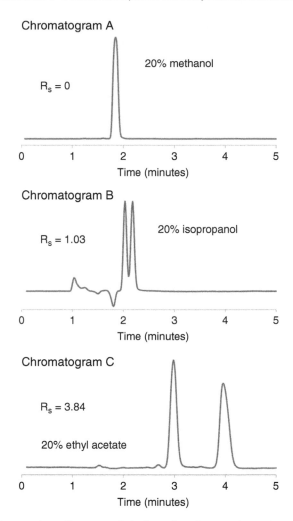

Figure 5.7 SFC separation of benzoin ethyl ether. All analyses performed on Chiralpak IA (4.6 × 100 mm), 5 mL/min, 100 bar BPR. The following modifiers were utilized: Chromatogram A, 20% methanol; Chromatogram B, 20% isopropanol; and Chromatogram C, 20% ethyl acetate.

Nontraditional solvents have been used for HPLC analysis and purification for greater than 15 years [26]. Only recently have nontraditional modifiers been explored for SFC separations [45–48]. The advantage of using nontraditional modifiers is illustrated in Figure 5.7 for the resolution of benzoin ethyl ether. Using a Chiralpak IA CSP and methanol modifier, enantioselectivity is not achieved. Switching to isopropanol, minor enantioseparation is obtained ($R_s = 1.08$). When ethyl acetate, a nontradition modifier is used, a large increase in enantioseparation is observed ($R_s = 3.84$).

5.4.4 Method Development Approaches

The goal of chiral method development is to determine chromatographic conditions that provide a suitable separation as quickly as possible. There are currently more than 100 commercially available chiral columns. Evaluation of all available CSPs, along with a number of modifiers, is time consuming and expensive, and is usually not necessary. Numerous laboratories have evaluated optimization of method development approaches [7, 8, 49–52]. The majority of these approaches were optimized for analytical scale SFC enantioseparations. Additional information on preparative method development approaches is discussed in Chapter 7. The majority of published approaches use polysaccharide-based CSP as they have been shown to resolve the largest range of racemates. Most screening strategies use gradient elution; this allows one strategy to be used for the analysis of a large number of compounds with a range of polarities. In addition, screening strategies are not designed to achieve baseline resolution, but to identify conditions that serve as a starting point for further optimization.

Early method development strategies often used four to six CSPs (5 μm) and two or three modifiers. Examples of this approach include Miller and Potter [8] as well as White [7]. The results of one approach, for a proprietary structure, are shown in Figure 5.8. Gradient screens were performed on five CSPs. The best gradient separation was then converted to isocratic

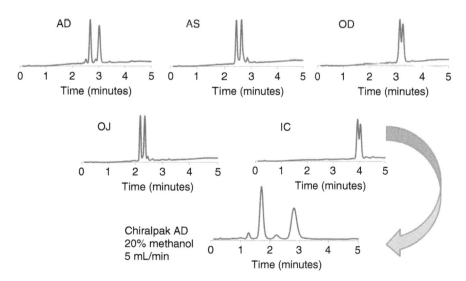

Figure 5.8 Method development example. Gradient conditions: 5–55% methanol over 210 seconds, hold at 55% for 60 seconds. Flow rate of 5 mL/min, column dimensions: 4.6 × 100 mm, 5 μm.

conditions: Chiralpak AD, 20% methanol. These screening strategies use SFC advantages of higher flow rates as well as reduced equilibration times to allow screening times of 60–90 minutes or less. More recent reports of SFC method development optimization have identified a smaller number of CSP and modifier combinations (four to eight) that give a high probability of achieving enantioselectivity. Some of these approaches are summarized in Table 5.6. De Klerck [52] was able to baseline resolve 19 racemates from a 20 compound library. Hamman [51] demonstrated that the use of six CSPs with one modifier was able to resolve 78 out of the 80 racemates with 51 being baseline resolved. More recent method optimization takes advantages of second generation SFC equipment that allows the use of smaller particle (3 μm) with smaller i.d. columns [51, 53] and/or shorter columns [49]. While these approaches may result in shorter method development times, their greatest advantage is an increase in sensitivity combined with reduced solvent requirements.

Similar method optimization has been performed for immobilized CSP, including the work of De Klerck [54], DaSilva [47], and Lee [48]. As the range of solvent possibilities using immobilized phase is significantly larger than with coated phases, the number of methods to be explored is higher, and method development time is longer. Additional method development strategies have been developed and published by the column manufacturer [55].

Table 5.6 Method development strategies.

Reference	CSP	Modifier
De Klerck et al.[a]	Chiralpak OZ-H or Lux Cellulose-2	20% methanol[b]
	Chiralpak AD-H	20% isopropanol[b]
	Chiralcel OD-H or Lux Cellulose-1	20% methanol[b]
	Lux Cellulose-4	20% isopropanol[b]
Hamman et al[c]	Chiralpak AD	Ethanol w/0.1% NH$_4$OH
	Chiralpak AS	
	Chiralpak AY	
	Chiralpak ID	
	CC4	
	Whelk-O	

[a] De Klerck et al. [52].
[b] With 0.1% isopropylamine and trifluoroacetic acid.
[c] Hamman et al. [51].

One advantage of SFC over HPLC is the lower mobile phase viscosity, which results in lower column pressure drops. The lower pressures allow the use of longer columns, or coupling of columns to achieve separations. Column coupling can use columns of different chiral phases [56] or an achiral phase coupled to a chiral phase [57, 58]. Zhang and coworkers [59] developed a unique application of column coupling. The technique entitled "simulating moving columns" uses two or three short chiral column in series. Once both enantiomers elute from the first column, a series of valves allow the first column to be placed at the end of the second or third column. This process is repeated until sufficient resolution is achieved. Using this technique a simulated column length of 490 cm achieved an efficiency of 320 000 theoretical plates.

5.5 High Throughput Method Development

Initial screening of packed column SFC for chiral method development can be extremely rapid; often completed in less than one hour. In laboratories that develop a large number of methods, higher throughput approaches must be used. One such high throughput method is parallel SFC [60, 61]. A parallel SFC system is designed to divide mobile phase flow equally across a number of columns. Each of these columns is connected to an individual detector, most often UV based. Existing commercial parallel SFC systems use either five (Waters) or eight (Sepiatec) columns. An example of an eight column, eight modifier (64 separate methods) parallel screen of trans stilbene oxide is shown in Figure 5.9. Method development times can be as rapid as 1 minute/method using parallel SFC equipment.

Another high throughput method development approach is sample pooling [62]. With this technique multiple racemates are combined into one vial, submitted to SFC method development, and a MS detector used in addition to UV-visible detection to add a layer of analyte-specific information. An example of this technique is shown in Figure 5.10 for the chiral separation of clenbuterol, pindolol, athiazide, and propranolol. As seen in Figure 5.10, the cumulative UV and MS traces for this separation are difficult to interpret, but when the specific mass of each racemate is extracted, the separation is easily visualized. This technique has the advantage of reducing method development time, but requires an SFC system equipped with a mass spectrometer. Another limitation of sample pooling is that it is difficult to determine which racemate contains impurities that are observed (e.g. the peak near 1.00 minutes in the total ion chromatogram in (B) in Figure 5.10). In these situations each racemate must be injected individually, essentially negating the time-saving advantage of simultaneous racemate screening.

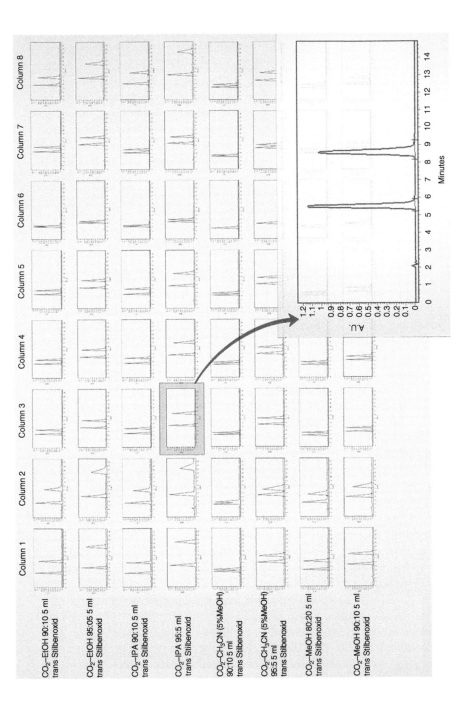

Figure 5.9 Parallel SFC screening example of chiral resolution of trans stilbene oxide. Columns are a mixture of amylose and cellulose chiral phases, 4.6×250 m, 3 and 5 μm. Modifiers: (1) 10% ethanol, (2) 5% ethanol, (3) 10% isopropanol, (4) 5% isopropanol, (5) 10% acetonitrile with 5% (v/v) methanol, (6) 5% acetonitrile with 5% (v/v) methanol, (7) 20% methanol, and (8) 10% methanol. *Source:* Data courtesy of Sepiatec.

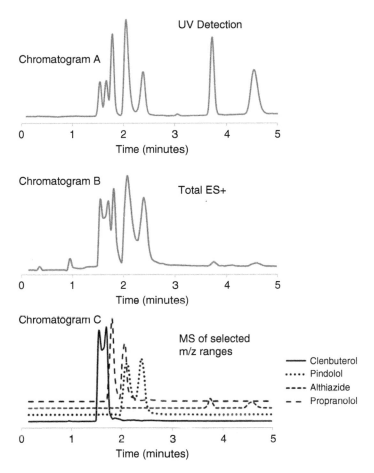

Figure 5.10 Sample pooling separation of clenbuterol, pindolol, althiazide, and propranolol. Chiralpak AD-H, 4.6 × 100 mm, 5 mL/min, 100 bar BPR. Modifier: methanol w/0.2% diethylamine. Gradient conditions: 5% for 30 seconds, from 5 to 50% over 3.5 minutes, held at 50% for 1 minute. Chromatogram A: UV trace (total response from 210 to 400 nm). Chromatogram B: ES+ total ion current trace. Chromatogram C: Extracted from the mass spectra for selected m/z ranges.

5.6 Summary

SFC is the most efficient and effective approach for developing chromatographic methods for enantiomer resolution. The speed advantages of SFC relative to HPLC allow routine development of SFC methods in less than one hour. For the past 15 years SFC has been the preferred method for enantioseparations in support of pharmaceutical discovery. With advances offered by the newest generation of SFC equipment and columns, the use of SFC will also soon become the preferred method for enantioseparations in pharmaceutical development and manufacturing.

References

1 Dalgliesh, C.E. The optical resolution of aromatic amino-acids on paper chromatograms. *Journal of the Chemical Society (Resumed)* 1952 (137): 3940–3942.

2 Miller, L. and Bush, H. (1989). Preparative resolution of enantiomers of prostaglandin precursors by liquid chromatography on a chiral stationary phase. *Journal of Chromatography* 484: 337–345.

3 Berger, C. and Perrut, M. (1990). Preparative supercritical fluid chromatography. *Journal of Chromatography* 505: 3743–3749.

4 Miller, L. and Weyker, C. (1990). Analytical and preparative resolution of enantiomers of prostaglandin precursors and prostaglandins by liquid chromatography on derivatized cellulose chiral stationary phases. *Journal of Chromatography* 511: 97–107.

5 Tefloth, G. (2001). Enantioseparations in super- and subcritical fluid chromatography. *Journal of Chromatography A* 906: 301–307.

6 Welch, C.J., Leonard, W.R. Jr., Dasilva, J.O. et al. (2005). Preparative chiral SFC as a green technology for rapid access to enantiopurity in pharmaceutical process research. *LCGC North America* 23 (1).

7 White, C. (2005). Integration of supercritical fluid chromatography into drug discovery as a routine support tool. *Journal of Chromatography A* 1074 (1–2): 163–173.

8 Miller, L. and Potter, M. (2008). Preparative chromatographic resolution of racemates using HPLC and SFC in a pharmaceutical discovery environment. *Journal of Chromatography B, Analytical Technologies in the Biomedical and Life Sciences* 875 (1): 230–236.

9 Mourier, P.A., Eliot, E., Caude, M.H. et al. (1985). Supercritical and subcritical fluid chromatography on a chiral stationary phase for the resolution of phosphine oxide enantiomers. *Analytical Chemistry* 57 (14): 2819–2823.

10 Anton, K., E, J., Fredericksen, L. et al. (1994). Chiral separations by packed-column super- and subcritical fluid chromatography. *Journal of Chromatography A* 666: 495–401.

11 Mangelings, D. and Vander Heyden, Y. (2008). Chiral separations in sub- and supercritical fluid chromatography. *Journal of Separation Science* 31 (8): 1252–1273.

12 Kalikova, K., Slechtova, T., Vozka, J., and Tesarova, E. (2014). Supercritical fluid chromatography as a tool for enantioselective separation; a review. *Analytica Chimica Acta* 821: 1–33.

13 West, C. (2014). Enantioselective separations with supercritical fluids – review. *Current Analytical Chemistry* 10: 99–120.

14 Ali, I. and Aboul-Enein, H.Y. (2008). Role of polysaccharides in chiral separations by liquid chromatography and capillary electrophoresis. In: *Chiral Separation Techniques: A Practical Approach* (ed. G. Subramanian), 29–98. Wiley-VCH Verlag GmbH & Co. KGaA.

15 Okamoto, Y. and Yashima, E. (1998). Polysaccharide derivatives for chromatographic separation of enantiomers. *Angewandte Chemie, International Edition* 37 (8): 1020–1043.

16 Yashima, E. (2001). Polysaccharide-based chiral stationary phases for high-performance liquid chromatographic enantioseparation. *Journal of Chromatography A* 906 (1): 105–125.

17 Gasparrini, F., Misiti, D., and Villani, C. (2001). High-performance liquid chromatography chiral stationary phases based on low-molecular-mass selectors. *Journal of Chromatography A* 906 (1–2): 35–50.

18 Welch, C.J. (1994). Evolution of chiral stationary phase design in the Pirkle laboratories. *Journal of Chromatography A* 666 (1): 3–26.

19 Haginaka, J. (2001). Protein-based chiral stationary phases for high-performance liquid chromatography enantioseparations. *Journal of Chromatography A* 906 (1–2): 253–273.

20 Cabrera, K. and Ludbda, D. (1992). Chemically-bonded β-cyclodextrin as chiral stationary phase for the separation of enantiomers of pharmaceutical drugs. *GIT Spezial, Chromatographie* 12 (2): 77–79.

21 Menges, R.A. and Armstrong, D.W. (1991). Chiral separations using native and functionalized cyclodextrin-bonded stationary phases in high-pressure liquid chromatography. In: *Chiral Separations by Liquid Chromatography* (ed. S. Ahuja), 67–100. American Chemical Society.

22 Ward, T.J. and Farris, A.B. III (2001). Chiral separations using the macrocyclic antibiotics: a review. *Journal of Chromatography A* 906 (1): 73–89.

23 Vozka, J., Kalíková, K., Roussel, C. et al. (2013). An insight into the use of dimethylphenyl carbamate cyclofructan 7 chiral stationary phase in supercritical fluid chromatography: the basic comparison with HPLC. *Journal of Separation Science* 36 (11): 1711–1719.

24 Pell, R. and Lindner, W. (2012). Potential of chiral anion-exchangers operated in various subcritical fluid chromatography modes for resolution of chiral acids. *Journal of Chromatography A* 1245: 175–182.

25 Wolrab, D., Kohout, M., Boras, M., and Lindner, W. (2013). Strong cation exchange-type chiral stationary phase for enantioseparation of chiral amines in subcritical fluid chromatography. *Journal of Chromatography A* 1289: 94–104.

26 Zhang, T. and Franco, P. (2007). Analytical and preparative potential of immobilized polysaccharide-derived chiral stationary phases. In: *Chiral Separation Techniques: A Practical Approach* (ed. G. Subramanian), 99–134. Wiley-VCH Verlag GmbH & Co. KGaA.

27 Perrenoud, A.G., Farrell, W.P., Aurigemma, C.M. et al. (2014). Evaluation of stationary phases packed with superficially porous particles for the analysis of pharmaceutical compounds using supercritical fluid chromatography. *Journal of Chromatography A* 1360: 275–287.

28 DeStefano, J.J., Schuster, S.A., Lawhorn, J.M., and Kirkland, J.J. (2012). Performance characteristics of new superficially porous particles. *Journal of Chromatography A* 1258: 76–83.

29 Hayes, R., Ahmed, A., Edge, T., and Zhang, H. (2014). Core-shell particles: preparation, fundamentals and applications in high performance liquid chromatography. *Journal of Chromatography A* 1357C: 36–52.

30 Kharaishvili, Q., Jibuti, G., Farkas, T., and Chankvetadze, B. (2016) Further proof to the utility of polysaccharide-based chiral selectors in combination with superficially porous silica particles as effective chiral stationary phases for separation of enantiomers in high-performance liquid chromatography. *Journal of Chromatography A* 1467: 163–168.

31 Patel, D.C., Breitbach, Z.S., Wahab, M.F. et al. (2015). Gone in Seconds: Praxis, Performance, and Peculiarities of Ultrafast Chiral Liquid Chromatography with Superficially Porous Particles. *Analytical Chemistry* 87 (18): 9137–9148.

32 Biba, M., Regalado, E.L., Wu, N., and Welch, C.J. (2014). Effect of particle size on the speed and resolution of chiral separations using supercritical fluid chromatography. *Journal of Chromatography A* 1363: 250–256.

33 Regalado, E.L. and Welch, C.J. (2015). Pushing the speed limit in enantioselective supercritical fluid chromatography. *Journal of Separation Science* 38 (16): 2826–2832.

34 Kotoni, D., Ciogli, A., Molinaro, C. et al. (2012). Introducing enantioselective ultrahigh-pressure liquid chromatography (eUHPLC): theoretical inspections and ultrafast separations on a new sub-2-mum Whelk-O1 stationary phase. *Analytical Chemistry* 84 (15): 6805–6813.

35 Ma, S., Shen, S., Lee, H. et al. (2009). Mechanistic studies on the chiral recognition of polysaccharide-based chiral stationary phases using liquid chromatography and vibrational circular dichroism: Reversal of elution order of N-substituted alpha-methyl phenylalanine esters. *Journal of Chromatography A* 1216 (18): 3784–3793.

36 Norinder, U. and Sundholm, E.G. (1987). The use of computer aided chemistry to predict chiral separation in liquid chromatography. *Journal of Liquid Chromatography* 10 (13): 2825–2844.

37 Del Rio, A. and Gasteiger, J. (2008). Simple method for the prediction of the separation of racemates with high-performance liquid chromatography on Whelk-O1 chiral stationary phase. *Journal of Chromatography A* 1185 (1): 49–58.

38 Wu, H., Yu, S., and Zeng, L. (2016). Effects of Hexane in Supercritical Fluid Chromatography for the Separation of Enantiomers. *Chirality* 28 (3): 192–198.

39 Yun, K.Y., Lynam, K.G., and Stringham, R.W. (2004). Effect of amine mobile phase additives on chiral subcritical fluid chromatography using polysaccharide stationary phases. *Journal of Chromatography A* 1041 (1): 211–217.

40 Vidal, D.T.R., Nogueira, T., Saito, R.M., and do Lago, C.L. (2011). Investigating the formation and the properties of monoalkyl carbonates in aqueous medium

using capillary electrophoresis with capacitively coupled contactless conductivity detection. *Electrophoresis* 32 (8): 850–856.

41 Blackwell, J.A. (1999). Effect of acidic mobile phase additives on chiral selectivity for phenylalanine analogs using subcritical fluid chromatography. *Chirality* 11 (2): 91–97.

42 Stringham, R.W. (2005). Chiral separation of amines in subcritical fluid chromatography using polysaccharide stationary phases and acidic additives. *Journal of Chromatography A* 1070 (1–2): 163–170.

43 De Klerck, K., Mangelings, D., Clicq, D. et al. (2012). Combined use of isopropylamine and trifluoroacetic acid in methanol-containing mobile phases for chiral supercritical fluid chromatography. *Journal of Chromatography A* 1234: 72–79.

44 Shen, J., Ikai, T., and Okamoto, Y. (2014). Synthesis and application of immobilized polysaccharide-based chiral stationary phases for enantioseparation by high-performance liquid chromatography. *Journal of Chromatography A* 1363C: 51–61.

45 Miller, L. (2012). Evaluation of non-traditional modifiers for analytical and preparative enantioseparations using supercritical fluid chromatography. *Journal of Chromatography A* 1256: 261–266.

46 Miller, L. (2014). Use of dichloromethane for preparative supercritical fluid chromatographic enantioseparations. *Journal of Chromatography A* 1363: 323–330.

47 Dasilva, J.O., Coes, B., Frey, L. et al. (2014). Evaluation of non-conventional polar modifiers on immobilized chiral stationary phases for improved resolution of enantiomers by supercritical fluid chromatography. *Journal of Chromatography A* 1328: 98–103.

48 Lee, J., Lee, J.T., Watts, W.L. et al. (2014). On the method development of immobilized polysaccharide chiral stationary phases in supercritical fluid chromatography using an extended range of modifiers. *Journal of Chromatography A* 1374: 238–246.

49 Hamman, C., Wong, M., Hayes, M., and Gibbons, P. (2011). A high throughput approach to purifying chiral molecules using 3mum analytical chiral stationary phases via supercritical fluid chromatography. *Journal of Chromatography A* 1218 (22): 3529–3536.

50 De Klerck, K., Parewyck, G., Mangelings, D., and Vander Heyden, Y. (2012). Enantioselectivity of polysaccharide-based chiral stationary phases in supercritical fluid chromatography using methanol-containing carbon dioxide mobile phases. *Journal of Chromatography A* 1269: 336–345.

51 Hamman, C., Wong, M., Aliagas, I. et al. (2013). The evaluation of 25 chiral stationary phases and the utilization of sub-2.0mum coated polysaccharide chiral stationary phases via supercritical fluid chromatography. *Journal of Chromatography A* 1305: 310–319.

52 De Klerck, K., Vander Heyden, Y., and Mangelings, D. (2014). Generic chiral method development in supercritical fluid chromatography and

ultra-performance supercritical fluid chromatography. *Journal of Chromatography A* 1363: 311–322.

53 Schafer, W., Chandrasekaran, T., Pirzada, Z. et al. (2013). Improved chiral SFC screening for analytical method development. *Chirality* 25 (11): 799–804.

54 De Klerck, K., Vander Heyden, Y., and Mangelings, D. (2014). Pharmaceutical-enantiomers resolution using immobilized polysaccharide-based chiral stationary phases in supercritical fluid chromatography. *Journal of Chromatography A* 1328: 85–97.

55 Technologies C (2016). Method Development Strategy. http://chiraltech.com/method-development-strategies.

56 Welch, C.J., Biba, M., Gouker, J.R. et al. (2007). Solving multicomponent chiral separation challenges using a new SFC tandem column screening tool. *Chirality* 19 (3): 184–189.

57 Barnhart, W.W., Gahm, K.H., Thomas, S. et al. (2005). Supercritical fluid chromatography tandem-column method development in pharmaceutical sciences for a mixture of four stereoisomers. *Journal of Separation Science* 28 (7): 619–626.

58 Phinney, K.W., Sander, L.C., and Wise, S.A. (1998). Coupled achiral/chiral column techniques in subcritical fluid chromatography for the separation of chiral and nonchiral compounds. *Analytical Chemistry* 70 (11): 2331–2335.

59 Zhang, Y., Dai, J., Wang-Iverson, D.B., and Tymiak, A.A. (2007). Simulated moving columns technique for enantioselective supercritical fluid chromatography. *Chirality* 19 (9): 683–692.

60 Zeng, L., Xu, R., Laskar, D.B., and Kassel, D.B. (2007). Parallel supercritical fluid chromatography/mass spectrometry system for high-throughput enantioselective optimization and separation. *Journal of Chromatography A* 1169 (1–2): 193–204.

61 Laskar, D.B., Zeng, L., Xu, R., and Kassel, D.B. (2008). Parallel SFC/MS-MUX screening to assess enantiomeric purity. *Chirality* 20 (8): 885–895.

62 Zhao, Y., Woo, G., Thomas, S. et al. (2003). Rapid method development for chiral separation in drug discovery using sample pooling and supercritical fluid chromatography–mass spectrometry. *Journal of Chromatography A* 1003 (1–2): 157–166.

6

Achiral Analytical Scale SFC – Method Development, Stationary Phases, and Mobile Phases

6.1   Introduction

Most new practitioners of SFC will enter the field with some previous experience in chromatography. While some will have practiced gas chromatography, it is likely that the majority have worked with high performance liquid chromatography (HPLC). Method development in HPLC can be complex. It involves choices from myriad stationary phases, mobile phase compositions, gradient designs, not to mention multiple methods of detection. There are hundreds of choices among C18 stationary phases alone, each with slightly different retention characteristics. Mobile phases can be aqueous or nonaqueous, and include a wide range of acidic, neutral, and basic additives. Despite these complexities,

Modern Supercritical Fluid Chromatography: Carbon Dioxide Containing Mobile Phases,
First Edition. Larry M. Miller, J. David Pinkston, and Larry T. Taylor.
© 2020 John Wiley & Sons, Inc. Published 2020 by John Wiley & Sons, Inc.

many practitioners of HPLC develop great expertise in method development, and true optimization of methods is facilitated by software programs, which are widely used.

Method development and optimization in HPLC are indeed complex, yet SFC presents even greater complexity because of the added importance of column temperature and system pressure, and the impact of these parameters on mobile-phase density. These two variables provide the opportunity for more easily adjusting selectivity in SFC than in HPLC, but they also make method development and optimization even more challenging in SFC than in HPLC. This chapter provides systematic guidance for method development in SFC, starting with the nature of the mixture to be separated, to choices in stationary phases and in mobile phases. Finally, a method development flow chart (a so-called "decision tree") provides reasonable starting-point conditions for SFC separations of a variety of analytes.

6.2   The Mixture To Be Separated

The first step in developing any separation should be a careful consideration of the mixture to be separated. This may appear to be obvious, but practitioners sometimes try to force-fit a generic method that has worked in the past, without considering the range of compound polarities and intermolecular interactions present in the mixture to be separated. Let's review how relatively simple principles can be used to make intelligent choices in the method development process.

6.2.1   Molecular Interactions

Molecules exhibit polarity due to unequal sharing of electrons by their constituent atoms. It's a fundamental property of molecules, of mobile phases, of stationary phases, and of chromatographic supports. Polarity can, at first glance, explain many of the retention mechanisms of chromatographic separations. In broad strokes, most SFC separations are normal phase separations, and operate within a polarity "window": The stationary phase is a bit more polar than the mixture to be separated, and the mobile phase is a bit less polar. In fact, the mobile phase in ANY chromatographic method should be a mediocre solvent for the analytes to be separated. If the mobile phase is too strong a solvent, partitioning between the mobile phase and stationary phase is prevented, and the solute elutes as an unretained peak. If the mobile phase is too weak of a solvent, the solute take far too long to elute from the column.

But beyond polarity, there are many more fundamental intermolecular interactions that the chromatographer should consider when choosing a mobile phase and a stationary phase. In order of increasing strength, these include dispersion interactions (also called London dispersion forces or van der Waals forces) between nonpolar groups, induction forces, dipole-induced dipole, dipole–dipole attraction, forces involved in hydrogen bond donation and acceptance, and ionic interactions [1]. Stearic interactions are special cases where one or more of these interactions occur between stationary phase and solute because of specific spacing and arrangement in space. A clear example of the latter type of interaction is chiral chromatography, described in Chapter 5.

6.2.2   Molecular "Handles"

With these potential interactions in mind, consider the solutes within the mixture to be separated. Are they nonpolar, polar, ionic? As a first indication, in what solvents are they soluble? Are dispersion forces likely to be an important aspect of the potential interactions the solutes will exhibit? What about dipole interactions, or even hydrogen bonding? What are the molecular weights or sizes of the solutes? Are chiral or stearic differences important? Finally, but just as importantly, are the solutes reactive? At what temperature will they begin to decompose?

In summary, what are the molecular "handles" one may take advantage of to accomplish the desired separation? These features dictate reasonable choices in terms of stationary phases, mobile-phase modifiers, additives, and operating temperatures.

Regarding the range of molecular "handles" that are accessible and useful in SFC, a slow and quiet revolution has taken place over the past few decades. In the early days of SFC, especially when pure CO_2 was chosen as the mobile phase, only nonpolar solutes [2], or solutes that were derivatized to render them nonpolar [3], had sufficient solubility in CO_2 to be eluted, even with highly deactivated open-tubular columns. As mobile-phase modifiers, such as small alcohols, were introduced to provide greater polarity and hydrogen bonding capabilities to the mobile phase [4], much more polar solutes were separated and eluted, such as, for example, sugars [5]. In recent years, practitioners of SFC have become adept in choosing low concentrations of additives, designed to provide specific molecular interactions with the stationary phase or the analyte, though present at only low levels. Examples include small acids and small amines [6], volatile salts [7], ammonium hydroxide [8], and low concentrations of water [9]. This has opened up SFC to applications never imagined by early practitioners, even including small proteins with molecular masses up to 4000–5000 Da [10, 11]. Once alcohol modifiers were introduced, a long-held rule of thumb was that only solutes soluble in methanol or less polar solvents were amenable to SFC. But this adage has not held in recent years. When additives are chosen with strategic attention to the molecular interactions present in the solute/stationary phase/mobile phase system, this limitation no longer applies.

6.3   Achiral SFC Stationary Phases

At first glance, choosing a column for an SFC method can be a daunting task, even when only considering "traditional" spherical, silica based, and ligand-bound stationary phases. Columns come with a variety of particle sizes and pore sizes, in various lengths and diameters, and with a withering array of bound ligand structures. But systematic consideration of the important requirements for a separation can help narrow the choices.

A good SFC method requires a stationary phase to perform well in three areas: efficiency (plate count, N), retention (k), and selectivity (α) [12]. With regard to efficiency, in general, a large N (or, conversely, a small height equivalent to a theoretical plate [HETP]) is desirable for a separation of similar structures. Column packing efficiency, particle size, and column length are primary drivers of HETP. Column manufacturers have developed sophisticated methods for ensuring good packing efficiency, and one can expect nearly all commercially available columns to be well packed.

6.3.1   Column Safety and Compatibility

Before considering these requirements, here is an important safety note: much early packed column SFC was performed with columns designed for HPLC. But not all HPLC columns are compatible with the range of pressures, temperatures, and mobile phase mixtures used in SFC. A first requirement in choosing a column is therefore to ensure that the column construction materials are fully compatible with the SFC conditions that may be chosen. Column temperature is an example. Many HPLC columns have a maximum temperature limit of 60 °C or 80 °C. While much SFC is performed below these temperatures, some SFC separations benefit from higher temperatures. For example, separations of thermally stable, oligomeric mixtures improved as column temperature was raised to as high as 200 °C [13]. If the maximum column temperature is exceeded, the stationary phase may be damaged, or the column seals may fail, followed by system decompression.

6.3.2   Efficiency

The influence of particle size on efficiency is illustrated in Figure 6.1, a plot of the Van Deemter equation for various particle structures and sizes used in analytical-scale SFC [14]. Not only does the smaller particle provide a smaller HETP, but also the rise in HETP is not as steep as the mobile phase linear velocity increases beyond the optimum velocity with smaller particles. So loss in efficiency is not as great as one increases the mobile-phase velocity using a column packed with a smaller particle. (Note that the superficially porous particles also perform well, though their loading capacity is typically not as great

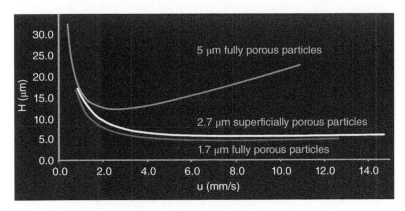

Figure 6.1 Van Deemter plot (HETP vs. mobile phase linear velocity, u) for various stationary phase particle sizes and structures in SFC. *Source:* Reprinted from reference [14] with permission. Copyright 2014, Elsevier.

as a fully porous particle.) Surprisingly, anecdotal early results from SFC experiments with columns packed with sub-2-μm particles showed little advantage over 5-μm-particle columns. But, as described in Chapter 3, the most recent generation of SFC instruments has been designed with careful attention to reduce extra column volume and to optimize detector response time [15]. Experiments using these instruments have clearly shown the advantages of reduced particle size [16–20]. It's therefore no surprise that small particle sizes (sub-2-μm) have taken SFC by storm [19, 21], much as in HPLC.

A longer column will also provide more plates in a simple linear manner. Therefore, longer columns and smaller particles provide high efficiency separations, at the expense of column backpressure and analysis time. The advantage of SFC over HPLC in taking advantage of these two approaches in improving efficiency is obvious when considering Figure 6.2, showing Van Deemter (HETP vs. mobile phase velocity, u) and pressure drop vs. mobile phase velocity plots for typical separations in HPLC, SFC, UHPLC, and UHPSFC ("ultra-high performance liquid chromatography" and "ultra-high performance SFC," respectively) [22]. The HETP for the SFC and HPLC columns is dictated by the particle size, and is therefore approximately the same, but the SFC system achieves this optimum at far higher mobile-phase velocity, and the HETP for the SFC system rises more gradually with increase in mobile-phase velocity. The situation is the same for the UHPSFC and UHPLC columns. Because most condensed phase chromatography is performed at greater than optimum velocity to save time and money, the advantage of SFC is obvious. Figure 6.2b shows why many more columns can be coupled serially in SFC than in HPLC [13, 23], providing higher efficiency while maintaining a reasonable flow rate and analysis time, before the system hits its pressure limit.

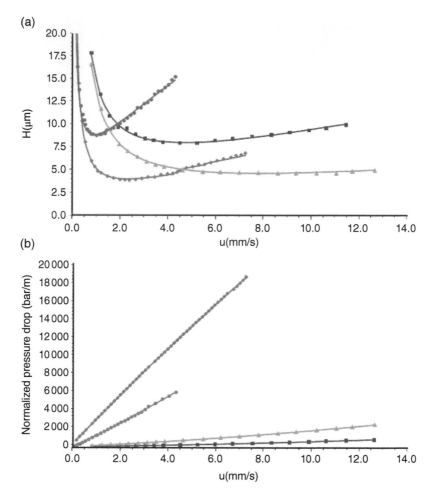

Figure 6.2 Kinetic performance and pressure drop for various HPLC and SFC configurations. (a) Van Deemter curves on four different chromatographic systems equipped with the most suitable column dimensions: HPLC column 150×4.6 mm, 5-μm (next to darkest grey points), SFC column 150×4.6 mm, 5-μm (darkest grey points), UHPLC column 50×2.1 mm, 1.7-μm (next to lightest grey points – top line in graph (b)), UHPSFC column 100×3.0 mm,1.7-μm (lightest grey points). (b) Corresponding system pressure drops. *Source:* Reprinted from reference [22] with permission. Copyright 2012, Elsevier.

Therefore, in regard to efficiency alone, one should choose the smallest particle and the longest column compatible with the flow rate (i.e. analysis time) and system pressure limit. Because the former considerations revolve around dimensions, it is appropriate to discuss column diameter choices. When a variety of column diameters are available, the first considerations when choosing a diameter are the desired injection volume and the flow rate compatibility of the detector. For example, a 4.6-mm-i.d. column is compatible with relatively large injection volumes (up to 5–10 µL of a "chromatographically strong" injection solvent), while much smaller volumes (0.5–1 µL) may be required when using a 2.1-mm-i.d. column [24]. Ultraviolet absorbance detectors are compatible with the "standard" SFC flow of 2–4 mL/min from a 4.6-mm-i.d. column, while this flow might require splitting to one quarter or less of this value when the SFC separation is coupled with electrospray ionization mass spectrometric detection. Alternatively, the analyst could choose a 2.1-mm-i.d. column and use it at the "standard" SFC flow rate of 0.4–0.8 mL/min.

6.3.3 Retention

The second requirement, retention, is a function of the molecular interactions of the solute with the stationary phase and the mobile phase. Regarding the stationary phase, a large number of columns now exist specifically designed, both in terms of stationary phase and column construction, for SFC. The development of new and improved stationary phases for SFC is an area of active research and development [25]. Much of this work involves silica-based, bonded stationary phases in which the pendant group contains one or more molecular structures that can interact both with active sites on the underlying silica and with the solutes to be separated [26]. The interactions with the underlying silica can "hide" surface active sites from the solutes. Stationary phase manufactures go to great efforts to reduce metal impurities and other sources of nonspecific interactions on the silica surface. Despite these efforts, surface "hot spots," in addition to the expected silanols, may be present even after ligand bonding and end-capping. Another common theme is that many new stationary phases are generally polar, providing more normal-phase separations via a variety of molecular interactions with analytes, far beyond the dispersion interactions provided by nonpolar C_{18} stationary phases encountered in reversed-phase HPLC. Another interesting finding is that residual silanols on silica particles undergo reversible silyl ether formation with alcohol mobile-phase modifiers (see Section 6.4.2) under SFC conditions [27]. This reaction reduces the hydrophilicity of the stationary phase, and is reversible upon exposure to water.

Useful guidance in choosing an appropriate stationary phase for an SFC separation comes from considering the structure of the ligand bound to the silica, and the various molecular interactions that may be provided by a particular ligand. Figure 6.3 illustrates the structures of a number of common and a few less common stationary phase ligands. Table 6.1 (T.L. Chester, Stationary phase interactions, personal communication. 2008) lists the molecular interactions of these and other ligands can provide.

Of course, solutes must possess the capacity to interact in a particular mode for that molecular interaction to be relevant. For example, the cyano stationary phase has the ability to accept hydrogen bonds from analytes, a very strong

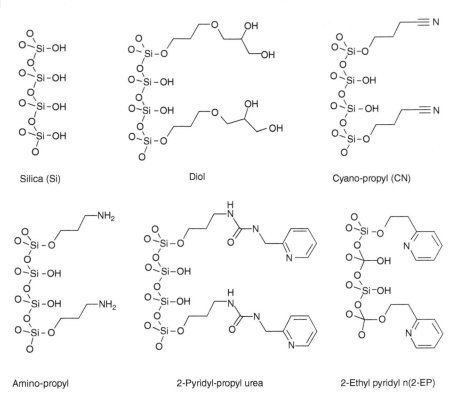

Figure 6.3 Structures of various stationary phase ligands.

Table 6.1 Common stationary phases in SFC and molecular interactions each can provide.

Phase	Intermolecular force (s)
C_{18}, C_8, etc.	London dispersion (L)
Phenyl	L + induction (i) + small dipole (d)
Cyano/cyano-propyl	L + d + H-bond acceptor
2-Ethylpyridine	L + i + d + H-bond acceptor
Silica (bare)	d + H-acceptor + H-donor
Amino/amino-propyl	L + d + H-acceptor + weak H-donor
2-Pyridyl-propyl urea	L + d + i + H-acceptor + weak H-donor
Diol	L + d + H-acceptor + H-donor
Sulfonic acid (SCX)	L + d + weak H-acceptor + strong H-donor

Source: T.L. Chester, Stationary phase interactions, personal communication. 2008.

type of interaction, while a traditional, well-deactivated C_{18} phase does not. However, in a separation of alkanes, the solutes do not have the capacity to interact with the stationary phase via hydrogen bonding, so the C_{18} phase may provide the strongest molecular interactions with the solutes, and the best separation. In contrast, a separation of alcohols (or amines), which can interact with the stationary phase by hydrogen-bond donation, might benefit from the cyano stationary phase rather than the C_{18} phase, if greater retention is desired.

In considering Figure 6.3, one sees, for example, the structural similarity and differences between bare silica and the diol phase. This similarity is reflected in Table 6.1, with the primary difference between the two being the capacity of the diol phase to interact with solutes via dispersion interactions. Thus, solutes with similar hydrogen-bond donating and accepting capacities might be better separated by the diol phase than by silica if they possess nonpolar structures that differ from one another. Figure 6.3 also illustrates how the ligand of one of the first phases designed specifically for SFC, the 2-ethylpyridine (2-EP) phase, may interact with surface silanols and other active sites to provide "self-deactivation" [26]. This self-deactivation may reduce the need for mobile-phase additives (see later in this chapter), if the additive is acting on surface active sites. Perrenoud et al. [28] compared the peak shape asymmetry of 92 basic pharmaceutical analytes on five different commercially available 2-EP phases with a simple CO_2/methanol mobile phase (i.e. without the use of mobile phase additive). Note that these basic pharmaceuticals were chosen because they typically required a mobile-phase additive, such as ammonium acetate or trimethylamine, in the mobile phase in order to exhibit symmetrical, Gaussian peak shape with traditional stationary phases. Performance of the 2-EP stationary phases differed dramatically, with the percentage of solutes exhibiting symmetric, Gaussian peak shapes ranging from 77 to 22%. So, the presence of the 2-EP ligand alone was not sufficient for satisfactory performance without additive. The underlying silica of the phases also played a major role. Yet the performance of the best 2-EP phases without mobile-phase additive was, in general, impressive.

Another way to use Table 6.1 is to see how, for example, the amino and the diol stationary phases should behave similarly. And this is, in fact, the case in general. Many practitioners know this from experience and intuition. They also are aware that the diol phase provides stronger retention for hydrogen-bond acceptors, like amines. Considering the structures in Figure 6.3 and the interactions listed in Table 6.1 help explain this behavior.

These general considerations about molecular interactions can certainly provide guidance in selecting a stationary phase during method development. But these general considerations can only go so far. The nature of each stationary phase is not only influenced by its bound ligand, but also by the density of this ligand binding, the nature and porosity of the underlying support, the extent of silanol "endcapping," etc. This leads to a great diversity of stationary phases.

This diversity of stationary phases has led to much effort and research to classify and organize the available choices [29–38]. Most of these efforts involve semi-empirical modeling, with probe analytes, to compare and characterize stationary phases, and to ultimately model retention in SFC.

The groups led by Lesellier and West have been very active in the area of stationary phase characterization and modeling. In 2016, they described the characterization of 31 "ultra-high-performance" stationary phases using 109 probe analytes, including neutral, acidic, and basic molecules [37]. The stationary phases consisted of both fully porous and superficially porous particles with sizes ranging from 1.7 to 2.7 µm, modified with a wide variety of commonly used, as well as many less-commonly used, ligands. The probe analytes were used to estimate the linear solvation energy relationships (LSER) with Abraham descriptors (for neutral species) as well as with descriptors added to assess the interactions of the stationary phases with ionizable species. Figure 6.4 shows a hierarchical cluster analysis (HCA) for 31 SFC stationary phases. Phases which are more similar are closer together. The shading is added to indicate the larger groupings of similar stationary phases. Figure 6.5 shows a spider diagram based on the same type of LSER analysis, calculated with seven molecular interaction descriptors (a (H-bonding with proton donors), b (H-bonding with proton acceptors), e (π–π interactions), s (dipole–dipole interactions), v (dispersive interactions)), including descriptors for ionizable species (d^+ (interactions with positively charged species), d^- (interactions with negatively charged species)). The most polar phases lie in the center right and lower right areas, where a and b dominate. Conversely, the nonpolar C_{18} phases lie in the center left area, where e and v dominate. During method development, one can use this type of analysis to choose stationary phases which are similar in retention characteristics, or, perhaps more importantly, to choose stationary phases which are different and may offer different selectivities. Note that the ligand alone may not dictate the grouping into which a stationary phase may best fit. For example, the Nucleoshell HILIC displays retention characteristics that place it closer to the grouping of polar phases like the BEH 2-ethyl pyridine and the Torus diol phases, rather than with the other HILIC phases.

6.3.4 Selectivity

The third consideration in choosing a column, selectivity, is really a function of the difference in the interactions between the solutes to be separated and the stationary and mobile phases. With regard to the stationary phase, there are general considerations that are useful in selecting a stationary phase for good selectivity. The analyst should consider the differences that exist between the solutes to be separated, and choose a stationary phase accordingly. For example, mixtures containing species that differ primarily in their nonpolar moieties (their "hydrocarbon volume" – such as a series of surfactants with differing

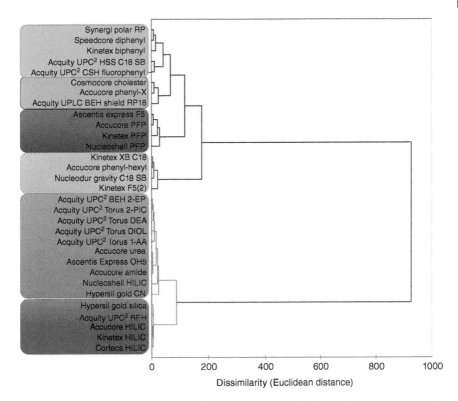

Figure 6.4 Hierarchical cluster analysis (HCA) on the normalized retention data (centered and reduced log k values) measured for 109 model analytes for 31 SFC stationary phases. Retention dissimilarity in the abscissa reflects Euclidean distance between the stationary phases. Chromatographic conditions: CO_2/MeOH 90:10 (v/v), column temperature: 25 °C, postcolumn ("downstream") pressure: 150 bar, flow rate: 1 or 3 mL/min depending on column dimensions. *Source:* Reprinted with permission from reference [37]. Copyright 2016 Elsevier.

nonpolar "tails") might best be separated using a C_8 or C_{18} stationary phase. Similarly, species which differ in more polar structures, some of which are capable of hydrogen-bonding interactions, might best be separated using an amino, a cyano, or a 2-ethylpyridine stationary phase. Mixtures that differ by a variety of structures might best be separated using an aromatic phase.

6.4 Mobile-Phase Choices

Contemporary SFC is almost always performed with a mobile phase consisting of a mixture. The mixture typically contains a primary component, a secondary component, generally called the modifier, and, sometimes, one

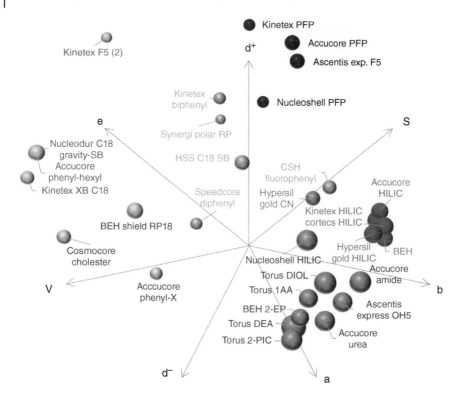

Figure 6.5 Spider diagram based on LSER models calculated with 7 molecular interaction descriptors (a, b, d$^+$, d$^-$, e, s, v), with the retention data measured for 85 neutral probe analytes and 24 ionizable probe analytes, for 31 SFC stationary phases. Stationary phase groupings are identified with colors. Chromatographic conditions are the same as identified in Figure 6.4. *Source:* Reprinted with permission from reference [37]. Copyright 2016 Elsevier.

or more minor components, typically added to the modifier at levels below 1%, called the additive(s).

6.4.1 Primary Mobile-Phase Component

As described in Chapter 2, there are a variety of fluids that form supercritical fluids at fairly reasonable temperatures and pressures, and a large number of these were explored as the primary (or the sole) mobile phase component during the early days of SFC. Examples include small hydrocarbons [39–41], Freons [42, 43], sulfur hexafluoride [44], nitrous oxide [45], ammonia [46], and even xenon [47]. But all of these suffered from one or more disadvantage when compared to carbon dioxide, including higher cost, environmental or human safety concerns, corrosiveness, flammable or oxidizing nature, low solvating power, etc. These experiments with alternate primary mobile phase components have largely disappeared, leaving CO_2 as practically the only primary mobile phase component used in SFC. This slow displacement of alternate mobile phases has been so complete that some advocate replacing the term "SFC" with "chromatography with CO_2–based mobile phases." This suggestion has been, understandably, controversial – the two terms are not synonymous when considering the strict definition of "supercritical fluid chromatography." But this suggestion is understandable, given the contemporary situation. Indeed, as described in Chapter 2, carbon dioxide has a great many advantages as a chromatographic mobile phase. It is inexpensive, widely available in high purity, readily recycled or recoverable from natural processes, relatively inert, polarizable, miscible with many polar cosolvents, and has few health concerns. The remainder of this discussion of SFC mobile phases will therefore assume that the primary mobile phase component is CO_2.

The solvating power of pure CO_2 is largely related to its density, and the density is a function of temperature and pressure. Separations using pure CO_2 as the mobile phase therefore often rely upon a gradient from low to high pressure (controlled by the backpressure regulator, postcolumn, often referred to as the "downstream" or "outlet" pressure), and sometimes changes in both pressure and temperature, to raise the density, and thereby provide a gradient from low to high solvating power of the mobile phase. This is, for example, a method often used for nonpolar or polarizable hydrocarbons [48] or species that have been rendered nonpolar through chemical derivatization. Derivatization was more common in earlier work with open-tubular capillary SFC of polar analytes, such as carbohydrates, and flame ionization detection [3, 49].

6.4.2 Secondary Mobile-Phase Component – The "Modifier"

It was immediately clear to early researchers in SFC that the realm of applications would be limited to nonpolar solutes, where the mobile phase limited to a single (nonpolar) component such as one of the small alkanes or CO_2. Therefore "cosolvents," now usually referred to as mobile-phase modifiers, were used in some of the earliest SFC separations. For example, in 1978 both Klesper and Hartmann [40] and Conaway et al. [50] described the addition of methanol or 2-propanol, respectively, to the nonpolar SFC mobile phase (the primary component was pentane in both cases) for the separation of polystyrene oligomers. The concept and application of mobile phase gradients in SFC were introduced by Klesper and coworkers in 1983 [51]. In this case, the primary mobile-phase component was alkanes or diethyl ether, and the modifier was an alcohol, cyclohexane, or dioxane. The beneficial effects of the modifier on reduced retention and improved peak shape were immediately obvious to these pioneers. The modifiers were hypothesized to both act on the surface of the stationary phase, and to provide an improvement in the solvating power of the mobile phase. In 1985, Blilie and Greibrokk stated, "the modifiers functioned as deactivation agents by direct interactions with residual silanol groups and also as modifiers of the eluting power of the mobile phase" [52]. The accuracy of these hypotheses has been supported by much research over the ensuing years [53–55].

With regard to method development, the most widely used modifier, by far, is methanol. It is one of the most polar small organics, and is miscible with CO_2 over wide ranges of temperature and pressure (more about this later). It is therefore a wise first choice when considering modifiers. Other small alcohols, such as ethanol and 2-propanol, are also widely used as modifiers. They are less polar than methanol, and this often results in distinctly different retention effects when compared to methanol [6]. Of course, methanol is also widely used as a mobile-phase component in HPLC.

Acetonitrile, also widely used in HPLC, is another story altogether. Some practitioners of HPLC have treated acetonitrile and methanol as nearly interchangeable, but this is in the presence of copious amounts of water, a strong acceptor and donator of hydrogen bond interactions. Like water, methanol is a potent hydrogen bond acceptor and donator. In contrast, acetonitrile can accept hydrogen bonds, but does not donate hydrogen bonds. This has limited the utility of acetonitrile as a modifier in SFC. Note that the increasingly popular use of low levels of water as a mobile-phase "additive" [56] (see below) may open the door to the more widespread use of acetonitrile as a modifier in combination with water as additive.

While methanol has shown great utility as a modifier, and is a reasonable first choice in method development, it's important to say a few words about the "allowed" combinations of mobile-phase pressure, temperature, and methanol percentage. This may come as a bit of a surprise to some practitioners of SFC, but methanol is not universally miscible with CO_2 under conditions which are easily chosen using SFC instruments. As an aside, it's important in almost all partition chromatography separations that the mobile-phase consist of a single phase. Both the partition process itself (and the subsequent peak shape) and many modes of detection are negatively affected if the mobile phase consists of a mixture of two immiscible phases. And, in fact, most SFC mobile-phase modifiers are not miscible with CO_2 over the full range of accessible temperatures and pressures. Early practitioners of SFC saw the negative effects of this behavior manifested in broad peak shapes and extremely noisy UV absorbance signals under the "wrong" conditions [57]. These effects are illustrated in Figure 6.6, where the methanol content in CO_2 is raised at constant temperature and pressure, such that the mixture moves from a single phase to a two-phase system. The negative effects of operating in the two-phase region are obvious. The separations shown in Figure 6.6 were performed with open-tubular SFC, but similar effects would be observed in packed-column SFC.

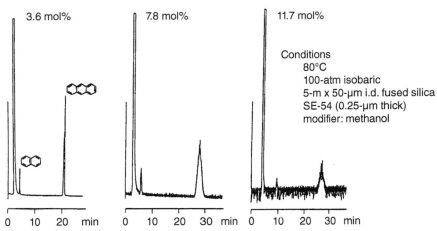

Figure 6.6 SFC separation of polynuclear aromatic hydrocarbons with CO_2–methanol mobile phase at constant temperature and pressure with increasing methanol content. The mobile phase moves from a single phase (far left) into a two-phase region (far right). In this case, the separation was performed with an open-tubular column and UV detection [57]. *Source:* Reprinted with permission from reference [57]. Copyright 2001 Elsevier.

So how does the SFC practitioner know which pressure–temperature–composition regions to avoid during method development? Luckily, some fairly straightforward guidelines have been established. Chester and coworkers published pressure–temperature plots of the locus of points above which the modifier–CO_2 mixture is always a single, fully miscible phase [58, 59]. Figure 6.7 shows an example of these plots for a variety of alcohols, including methanol. (In fact, these are compressions of three-dimensional pressure–temperature–composition plots, compressed along the composition axis. Not all combinations of pressure–temperature–composition below the locus of points results in two-phases, but some do. All combinations above the locus result in a single phase.) How does one use a plot such as this in method development? The simplest way is to choose a column temperature, and set the system pressure above the locus of points at that temperature. For example, for methanol at 40 °C, inspection of Figure 6.7 shows that the minimum pressure to ensure a one-phase mixture is ~80 atm. If the column temperature is raised to 80 °C, the minimum is ~125 atm. If 2-propanol is used as a modifier at 100 °C, the minimum pressure is ~150 atm.

In revisiting Figure 6.6, it's clear that the conditions chosen for the analysis are below the locus of points for CO_2–methanol in Figure 6.7. In hindsight, given the discussion in the previous paragraph, 80 °C and 100 atm may not have been the best choice. But why are the analytical results in the leftmost chromatogram (at 3.6 mole% methanol) entirely acceptable, while those at the same

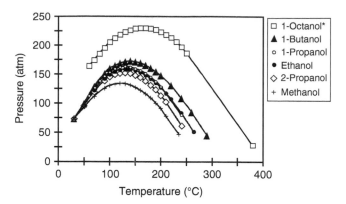

Figure 6.7 Pressure–temperature plots of critical-mixture loci for small alcohols in CO_2. The important point is that the two-component (i.e. alcohol and CO_2) mixtures always form a single phase at pressures and temperatures above the curve, regardless of the mixture composition. For example, methanol–CO_2 above ~130 atm is always a single-phase mixture, regardless of the temperature or composition. Pressure–temperature–composition combinations below the curve exist such that two phases are possible, one alcohol-rich, the other CO_2-rich. Another way to use the curve: If one sets the column temperature at 50 °C, the critical-mixture point is just below 100 atm. Therefore, regulating the pressure at 100 atm or above will ensure that any combination of CO_2 and methanol will form a single phase. *Octanol does not form a Type I mixture with CO_2, and there is therefore a miscibility gap below 50 °C. *Source:* Reprinted with permission from reference [58]. Copyright 1995 American Chemical Society.

temperature and pressure in the rightmost plot (at 11.7 mole% methanol) not? Recall that Figure 6.7 is a compression of a three-dimensional pressure–temperature–composition plot. The compressed plot doesn't tell the whole story. Conditions below the locus *may* result in two phases, but, depending on the composition of the mixture, may not. At 3.6 mole% methanol, the mobile phase is clearly a single phase. At 11.7 mole%, it is not. Had the investigators chosen a downstream (post column) pressure of 125 atm (see Figure 6.7), they would not have encountered the poor chromatographic results they observed at 11.7 mole% methanol.

Another important point should be made about the use of methanol as a modifier. CO_2 is often referred to as "inert" or "nonreactive." And this is, in general, true. However, there is evidence that the pH values of CO_2–methanol mixtures are acidic. Researchers have used solvatochromic methods to study such phenomena since the early days of SFC [53, 60]. As one might expect, the acidic nature of CO_2–methanol–water mixtures is even more pronounced [61]. While the acidic nature of CO_2–methanol undoubtedly shifts the equilibrium of small acid solutes, the shift is not sufficient to drive these solutes to full protonation, which would be expected to lead to improved peak shape in SFC. Stronger acidic additives, such as formic, acetic, or even trifluoroacetic acids, are often used as additives to improve the peak shape of acidic analytes [55, 62], as discussed in Section 6.4.3.

While methanol is by far the most common modifier, and is a good choice in initial method development, a host of other modifiers have been used over the years. In method development, a useful initial guide in the choice of a modifier is a solvent or mixture in which the solutes of interest are well soluble. Beyond methanol, the most common modifiers include acetonitrile [62] (as discussed above) and small alcohols [9] such as ethanol and 2-propanol. Less common, yet quite useful modifiers include dichloromethane [63, 64], tetrahydrofuran [65], and even hexane [66]. DaSilva et al., working in the field of chiral SFC, have explored other uncommon modifiers including chloroform, 2-methyl tetrahydrofuran, methyl tert-butyl ether, cyclopentyl methyl ether, acetone, ethyl acetate, toluene, 2,2,2-trifluoroethanol, and N,N-dimethylformamide [67]. These uncommon modifiers provided a range of useful selectivities. Though the latter work was in chiral SFC, there is no barrier for the use of these uncommon modifiers in achiral SFC.

While the introduction of modifiers provided a step change in the ability to apply SFC to polar analytes, researchers continued to encounter even more polar acidic, basic, or ionic species that could not be eluted, or could only be eluted with poor peak shape, in SFC. This called for the exploration of new and specialized mobile-phase components, the "additive." In fact, Berger and Deye stated, "additives provide a key to the separation of more polar solutes by SFC" [68], and we discuss this key next.

6.4.3 Tertiary Mobile-Phase Component – "Additives"

Additives are typically only poorly soluble, at best, in pure CO_2. They are usually dissolved in the modifier at low concentrations, with low mM levels being common. The modifier/additive solution is then added to the CO_2 to form the mobile phase. Additives are by far the most diverse of the three types of mobile phase components in SFC. While used at low concentrations, they can have dramatic effects on the quality of the SFC separation.

Pioneers in the exploration of additives in SFC were the groups of Ashraf-Khorassani, Taylor, and coworkers [69], Berger and Deye [68, 70], and Steuer, Erni, and coworkers [71, 72]. By the early 1990s, these and other groups had laid down the principles still operating today in the use and function of additives.

The logic behind the choice of the first additives is clear – acidic and basic solutes either could not be eluted at all in SFC using modified mobile phases (with CO_2 or other primary mobile-phase components) and packed columns which existed in the late 1980s or early 1990s, or could only be eluted with very poor peak shape. It stood to reason that adding a small amount of an acidic or basic additive to the mobile phase might help in the elution of acids or bases, respectively. Berger and Deye were some of the first to study the effects and mechanisms of low levels of acidic additives on the elution of acidic analytes [68]. They found that the improvement of peak shape provided by the additives was primarily due to suppression of solute ionization, and only secondarily due to coverage of active sites on the chromatographic stationary phase. They watched the loading and removal of additives from the surface of various stationary phases, and they studied the effect of the pK_a of the additive on the peak shape of the probe acidic analytes. Figure 6.8 shows an example of their results on three stationary phases, a nonpolar C8, a more polar (or polarizable) phenyl phase, and an even more polar cyanopropyl phase. It is clear that the stronger acidic additives provide better peak shape for the acidic analytes, in all three cases. In method development, choosing an acidic additive is still a reasonable choice when separating acidic analytes. Many reasonable choices exist, such as formic, acetic, trifluoroacetic, and citric acids. A typical starting concentration of the additive in the modifier lies between 0.05 and 0.5%.

Similarly, small amines were, and are still, frequently used as additives in analytical SFC [73, 74]. Similar to their results with acidic additives, Berger and Deye concluded that strongly basic additives improved the peak shape of basic analytes due to ionization suppression [74]. In their experiments, neither the use of acidic additives, which would form ion pairs with the basic analytes, nor deactivation of the stationary phase, provided improved peak shape. Examples of amine additives that are most commonly used today are the small alkyl amines, methyl- and ethylamines (i.e. methyl-, dimethyl-, trimethyl-, and so on). These small alkyl amines are as strong as, or stronger than most organic amines, so would indeed inhibit ionization of organic amine analytes. They would also interact with acidic active sites on the stationary phase.

Because CO_2 is nonpolar, the idea of adding an ionic salt to a mobile phase primarily composed of CO_2 made little sense to many practitioners. However, the use of small, volatile salts as mobile-phase additives has become popular. The use of salts as SFC mobile-phase additives had, in fact, been described as early as 1988 [69, 71, 72]. Steuer et al. used organic-soluble salts, such as sodium heptanesulphonate monohydrate, tetrabutylammonium bromide, and equimolar mixtures of tributylamine and acetic acid, as ion pairing agents to allow the elution of ionizable and ionic analytes, including pharmaceuticals and their degradation products [72]. Taylor and coworkers have expanded the use and understanding of ion-pairing in SFC [75–77].

But the more widespread use of small, volatile salts, such as ammonium acetate and ammonium formate, as SFC mobile phase additives was introduced in 2004 by Pinkston and coworkers [7]. Note that these additives were in widespread use at the time in LC/MS to help with both peak shape and ionization. The move toward volatile ammonium salts was originally driven from the perspective of compatibility with mass spectrometric detection. Small acids and small amines often result in ionization suppression, while no suppression is observed with volatile ammonium salts. Yet the improved chromatographic performance

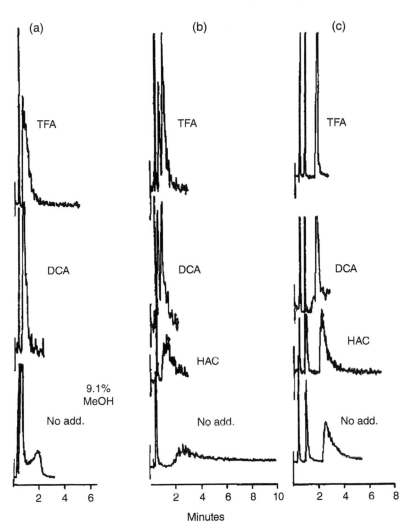

Figure 6.8 (a) C$_8$ column; (b) phenyl column; and (c) cyanopropyl column. Solute elution order: benzoic acid, phthalic acid, and trimellitic acid. Additives: HAC = acetic acid; DCA = dichloroacetic acid; TFA = trifluoroacetic acid. The methanol modifier contains 0.2% of each additive. Column dimension: 100 mm × 2 mm i.d.; flow-rate = 3 mL/min, 40 °C; 130 bar outlet pressure. *Source:* Reprinted with permission from reference [68]. Copyright 1991, Elsevier.

provided by the ammonium salts was immediately obvious. Figure 6.9 illustrates the effect of the introduction of low levels of ammonium acetate (in this case only 1.1 mM) in the methanol modifier on the chromatography of a polar analyte, reserpine. No peak is observed with pure methanol as modifier, but the addition of ammonium acetate provides a sharp, well-shaped peak. In contrast to

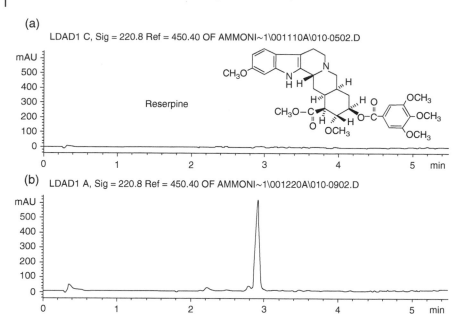

Figure 6.9 SFC/UV chromatograms at 220 nm of reserpine standard with pure methanol (a) and with 1.1 mM ammonium acetate in methanol (b) as mobile phase modifier. Chromatographic column: DeltaBond Cyano (Thermo Hypersil, Bellefonte, PA), 50 mm × 4.6 mm, particle size of 5 μm and pore size of 200 Å. Conditions: flow rate of 2 mL/min, postcolumn pressure of 160 bar, column oven temperature of 37 °C, and injection volume of 5 μL. The mobile phase composition was held at 1% modifier in CO_2 for 0.5 min after injection, then was raised to 50% methanol at a rate of 10%/min, where it was held until the end of the separation. *Source:* Reprinted with permission from reference [7]. Copyright 2004 Wiley.

the use of acidic and basic additives, the use of salts as additives provides improved peak shapes for both acidic and basic analytes [7]. While an ion pairing mechanism may be at play, further work using silicon-29 solid state NMR by Zheng, working with Taylor, Glass, and Pinkston, showed that deactivation of surface active sites was also an important factor in providing the improved peak shapes observed with low levels of small, volatile salts as additives [75].

As the use of small volatile salts as additives has grown, others have systematically explored their effects and utility. For example, Cazenave-Gassiot et al. studied the effects of increasing the concentration of ammonium acetate in a methanol modifier on three quite different stationary phases (2-ethylpyridine (2-EP), endcapped 2-EP, and bare silica) [78]. The concentration of the additive in methanol ranged from 0 to 30 mM on the 2-EP and endcapped 2-EP phases, and up to 60 mM on the silica phase. They also studied the effect of adding ammonium acetate to the sample injection solvent, where the ammonium acetate would form ion pairs with the ionic test solutes. Figure 6.10 shows

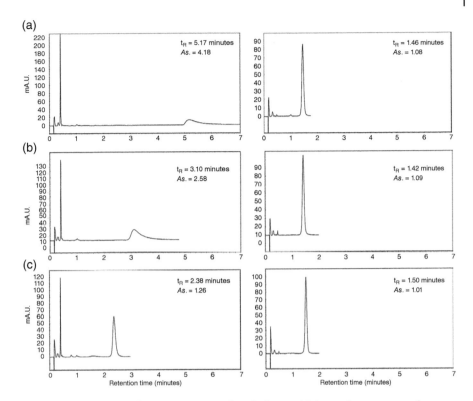

Figure 6.10 Evolution of retention time and peak shape with increasing amounts of NH$_4$OAc in the modifier, 2-EP stationary phase, 10% (v/v) modifier in CO$_2$, 100 bar downstream pressure, 35 °C oven temperature, and 4 mL/min flow rate. Left: compound 3, a sulfonamide synthesized by the authors of reference [78]; right: naproxen. Modifier: (a) 0.3 mM NH$_4$OAc in MeOH; (b) 5 mM NH$_4$OAc in MeOH; and (c) 30 mM NH$_4$OAc in MeOH. *Source:* Reprinted with permission from reference [78]. Copyright 2009 Elsevier.

some of their results. The addition of the additive has a dramatic effect on the retention and peak shape of the sulfonamide, but little effect on naproxen. The acidic nature of the CO$_2$–methanol mobile phase (as discussed above) is believed to protonate and neutralize the acidic naproxen. The ethyl-pyridine of the stationary phase and the two nitrogen atoms held by the sulfonamide solute are believed to be protonated, forming positive ions which do not interact well. The primary retention mechanism of the sulfonamide would therefore be strong retention by free silanols. The ammonium acetate is believed to deactivate the free silanols and other active sites on the stationary phases, dramatically decreasing the retention of the sulfonamide. This hypothesis agrees with the effect of ammonium acetate added to the injection solution. It has a relatively small effect on retention and peak shape, demonstrating that ion pairing is not a dominant mechanism. In summary, ammonium acetate is a versatile additive, providing positive effects for many analyte-stationary phase combinations, but rarely, if ever, resulting in negative effects. Some scientists add it as a matter of course to the modifier, especially when performing SFC/MS.

As mentioned earlier, the range of allowable additives used for a particular separation can be limited by the mode of detection. For example, UV absorbance detection is not compatible with additives that absorb strongly in the UV. Additives used with mass spectrometric, evaporative light scattering, and corona charged aerosol detection must be volatile under the conditions of the nebulization process associated with these detectors. Despite these limitations, a wide range of additives is nevertheless useful with these modes of detection. These include small volatile acids (e.g. formic, acetic, and trifluoroacetic), small volatile bases (e.g. trimethylamine and trimethylamine), and even many small volatile salts (e.g. ammonium formate and ammonium acetate). The limitations on the choice of additive in analytical scale SFC are small when compared to the limitations in preparative scale SFC, as discussed in Chapter 8. Additives used in preparative scale SFC must be easily removed from the desired product, after its isolation, without degrading or altering the product.

Some additives that have become popular for preparative scale SFC because they are indeed easily removed from the product have also become more and more widely used in analytical scale SFC. Ammonia, in the form of ammonium hydroxide, is one of these. As described in Section 6.3.3, Grand-Guillaume Perrenoud et al. provided a comprehensive comparison of five 2-ethylpyridine (2-EP) stationary phases and a bare silica phase using ammonium hydroxide as an additive [28]. Basic analytes have traditionally been difficult to analyze by SFC, and the 2-EP phases were designed to specifically address this challenging class of analytes. As discussed in Section 6.3.3, Grand-Guillaume Perrenoud et al. found that various commercial 2-EP phases provided a wide range of performance with 92 basic pharmaceutical analytes. The percentage of analytes for which the stationary phases provided Gaussian peak shapes ranged from 77 to 22% using CO_2–methanol and no additive, with the most successful phase being the PrincetonSFC 2-EP column. But the most striking result from this study, with regard to the use of additives, was the influence of ammonium hydroxide for the basic analytes on a hybrid silica stationary phase. Without additive in the CO_2–methanol mobile phase, only 17% of the analytes exhibited Gaussian peak shape. Adding 20 mM formic acid to the methanol modifier, which protonated the basic analytes, resulted in a drop in this value to 10%. But the addition of 20 mM ammonium hydroxide provided a dramatic improvement, with the percentage of analytes exhibiting Gaussian peak shapes rising to 81%. These kinds of results, coupled with the compatibility of ammonium hydroxide with preparative scale SFC and a wide variety of SFC detectors, explain the growing popularity of ammonium hydroxide as an additive.

Another rising star in the world of additives is simply water. The solubility of water in pure CO_2 is quite limited, and the ability to reproducibly and accurately add small amounts of water to CO_2 is challenging, so it is infrequently used as a binary mixture of CO_2 and water. But applications in which water is added to an alcohol modifier at concentrations ranging from 0.5 to 5% are becoming more widespread. In fact, water was frequently advocated as a cosolvent or additive in the early years of packed column SFC. Not only was it shown to improve chromatographic performance, but it was compatible with flame ionization detection. Geiser used water to improve the separation of free fatty acids [79]. Pyo used water-modified CO_2 for the separation of vitamins [80]. Camel et al. used water as part of an unusual modifier mixture (pyridine (or ethylene glycol)–methanol–water–trimethylamine) and evaporative light scattering detection for the SFC separation of underivatized amino acids [81]. And Salvador et al. used water as an additive in the SFC separation of polar monosaccharides and polyols [82] and of cyclodextrins [83].

Despite these clear successes, the use of water as an additive fell out of favor until recently. But, as reviewed by Taylor, researchers seem to have rediscovered the beneficial impact of low concentrations of water on chromatographic performance, and its compatibility with both preparative separations and a wide range of detectors [56]. Ashraf-Khorassani and Taylor evaluated the addition of low levels of water (up to 5%) in four small alcohol modifiers on the elution of water-soluble nucleobases from three stationary phases [9]. They found that water dramatically improved the chromatographic behavior of adenine and cytosine. The water additive provided similar results to those obtained with ammonium acetate. But both of these far surpassed the performance obtained using formic acid as an additive. Ashraf-Khorassani, Taylor, and Seest made a careful investigation of the effects of water as an additive, in conjunction with trifluoroacetic acid, 2-propylamine, or ammonium acetate with silica stationary phases and methanol modifier [84]. The beneficial effects of water persisted even after the water was removed from the additive. This, and other evidence, supported the authors' hypothesis that the primary effect of the water was adsorption on the surface of the silica stationary phase, providing partitioning of the analytes between the aqueous mobile phase and the surface-adsorbed water, much as in HILIC HPLC. dos Santos also explored the incorporation of CO_2 in HILIC mobile phases [85]. A CO_2–ethanol–water mobile phase provided similar results to a traditional HILIC mobile phase containing high proportions of acetonitrile.

Since this revival, the application of water as a mobile-phase additive has grown substantially [86–91]. For method development, it should be noted that some practitioners now consider a mixture of water and ammonium acetate (2–5% and 5–20 mM, respectively, added to the modifier) as a "universal" additive for a CO_2–methanol mobile phase [88]. This might be wise advice, because these additives rarely cause deleterious effects, and often provide improved performance.

We've reviewed the three types of mobile-phase components (primary component, modifier, and additive) in contemporary SFC. Because of the nature of compressed fluids, it can be challenging to measure the solvent properties of these mixed fluids using conventional methods. Therefore, as mentioned earlier, efforts have been made to peer into the polarity, the acid/base properties, and solvent strength of SFC mobile phase mixtures under realistic SFC conditions using solvatochromic probes since the early days of SFC [60, 92–94]. A masterful example of this type of work was provided by West et al. as they studied the influence of additives on SFC mobile phases [55]. Figure 6.11 shows an instructive example of their results. Figure 6.11a shows that additives – in this case the common water and ammonium acetate – do not modify the polarity in the immediate environment of the probe. Figure 6.11b shows the strong inverse correlation between mobile phase polarity and chromatographic retention of the probe analyte. The authors used a number of probes to estimate the pH of the mobile phase. They found that the pH of CO_2–methanol mixtures are near 5, with the pH dropping as a greater proportion of methanol is added during a gradient. This is presumably due to the formation of methoxycarbonic acid. Most acidic additives did not greatly decrease the pH, but trifluoroacetic acid, the strongest acidic additive explored, did have an effect, providing an apparent pH below 1.7. Basic and salt additives did not alter the pH. It was presumed that the bases are titrated by methoxycarbonic acid, and form ion pairs. The bases and salts could stabilize the pH during gradient runs. And all the additives are believed to exert influence on active sites on the stationary phase.

6.5 Influence of Column Temperature on Efficiency and Selectivity

As in any form of chromatography, diffusion is important in SFC. The Van Deemter equation teaches that high diffusion coefficients lead to higher chromatographic efficiencies. This is an important reason for the higher efficiencies observed in GC compared to HPLC. And higher temperatures lead to higher diffusion coefficients. Higher temperatures also provide lower viscosities, allowing the use of longer columns and/or higher flow rates. So some practitioners of SFC have advocated for the use of the highest temperatures possible in SFC, keeping in mind the temperature limits of the hardware and of the analytes. This is reasonable advice, especially in separations of hydrocarbons or small polymers where efficiency, and not selectivity, drives the quality of the separation [13]. But temperature affects more than just diffusion coefficients, viscosity, and chromatographic efficiency.

As described in Chapter 2, the changes in temperature can have more substantial effects on the density and solvating power of supercritical fluids than on traditional liquids. In SFC, these effects are most pronounced with mobile

(a)

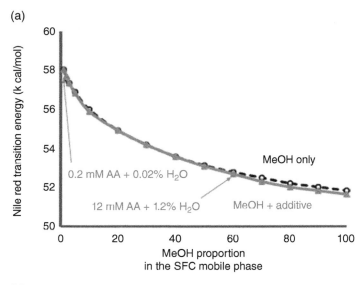

(b)

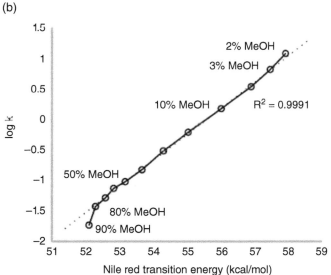

Figure 6.11 (a) Variation of mobile-phase polarity evaluated with Nile Red solvatochromic probe when CO_2–MeOH proportions vary. (b) Relation between mobile phase polarity and retention on the stationary phase. Nile Red injected on BEH silica (100 mm × 3.0 mm, 1.7 μm), 25 °C, 15 MPa, 0.8 mL/min. Open circles: no additive. Grey triangles: Methanol containing 2% water and 20 mM ammonium acetate (AA). *Source:* Reprinted with permission from reference [55]. Copyright 2017 Elsevier.

phases of pure CO_2 or at low modifier concentrations, and at low downstream (outlet) pressures. Mobile phases such as these are more "gas-like." Examining the effect of temperature on selectivity is a wise practice in SFC method development. Relatively small changes in temperature can provide significant differences in selectivity [95].

6.6 Where Do I Go from Here? Method Development Decision Tree and Summary

We've talked about the nature of the sample to be separated, described choices in stationary phase, the three constituents of mobile phases, and the influence of column temperature on efficiency and selectivity. But a new practitioner of SFC must begin somewhere in developing a new method. Figure 6.12 provides a "decision tree" which will help in developing a new SFC method. The recommendations and steps in the tree are based upon personal experience, careful inspection of the literature, both recent and older, and with conversations and exchanges with some of the leading experts in the field (P. Sandra, personal communication. 2006; T.L. Chester, Stationary phase interactions, personal communication. 2008; L.T. Taylor, personal communication. 2012).

The tree assumes the analyte(s) is/are soluble in a fluid phase of some sort. If this is not true, it is very unlikely that an SFC separation (or, in fact, any partitioning-based separation) will be successful. The tree starts at point "1," "Is/are the analyte(s) soluble in methanol?" If "no" move to point "A." At point A, the question is whether the sample is soluble in water. If the answer is "no," then the sample is not soluble in water or methanol, and is quite nonpolar. A good place to begin with such a sample is with a mobile phase of pure CO_2 with a pressure gradient. If the answer to question A is yes, then the analyte is very polar, so polar, in fact, that SFC may not be a wise choice. The decision tree suggests two possibilities in this case (point "C"): either hydrophilic interaction HPLC (HILIC), or chemical derivatization (methylation, acetylation, silylation, for example) and separation with pure CO_2 using a pressure gradient.

If the answer to question 1 (solubility in methanol) is "yes," then move to point "B," "Is the analyte ionic or ionizable?" If "no," then a separation with either a 2-ethylpyridine (2-EP) stationary phase with methanol modifier, or with a silica phase with methanol containing 5% water and 20 mM ammonium acetate would likely be successful. If the analyte(s) is/are ionic or ionizable, then three common cases are considered: peptides or small proteins, acids, or bases. So far, most successful separations of peptides have used a 2-EP or similar phase, with trifluoroacetic acid and water additives in a methanol modifier. As discussed earlier, an acidic additive (trifluoroacetic, formic, or citric, for example) with water in a methanol modifier works well for acids. An octadecyl or a diol stationary phase is suggested for acids under

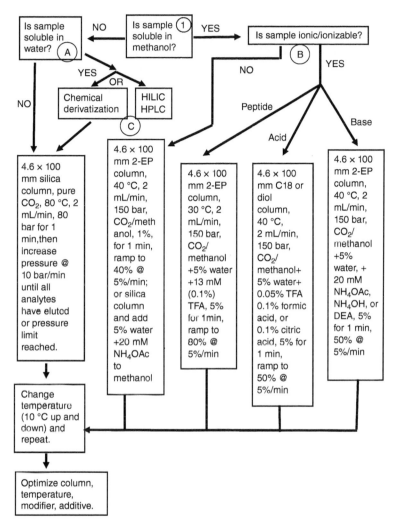

Figure 6.12 Decision tree for the method development process in SFC. See text for a description of the steps in the tree. The tree assumes the analyte(s) has(have) some solubility in a fluid phase of some sort.

these conditions. Ammonium acetate, ammonium hydroxide, or diethylamine additives in a methanol + 5% water modifier on a 2-EP column work well for most bases.

After an initial separation is achieved, repeating the separation with a higher, and a lower, column temperature is suggested to examine changes in selectivity for critical analyte pairs. A change of 10 °C in both directions is suggested. At this point, an acceptable separation may have been achieved. If not, the results of these early experiments may suggest changes in the stationary phase, the column temperature, the modifier, or the additive that may improve the separation. The column outlet (downstream) pressure can affect selectivity, but its influence is relatively small compared to these other factors for SFC separations where the modifier concentration is greater than a few percent. While univariate changes as described above can certainly improve a separation, true optimization requires a multivariate optimization design-of-experiment (DOE) approach [96].

For those accustomed to method development in HPLC, the added variables of temperature and pressure can initially make method development in SFC appear daunting. But knowledge of the analytes to be separated will lead to systematic, logical choices in SFC columns, mobile phases, and conditions leading to a successful separation, as described in this chapter. The added importance of temperature, modifiers, and additives in SFC can actually open the door to multiple avenues leading to excellent separations.

References

1 Morokuma, K. (1977). Why do molecules interact? The origin of electron donor-acceptor complexes, hydrogen bonding, and proton affinity. *Accounts of Chemical Research* 10 (8): 294–300.

2 Hawthorne, S.B. and Miller, D.J. (1989). Supercritical-fluid chromatography-mass spectrometry of polycyclic aromatic hydrocarbons with a simple capillary direct interface. *Journal of Chromatography A* 468 (0): 115–125.

3 Chester, T.L., Pinkston, J.D., and Owens, G.D. (1989). Separation of malto-oligosaccharide derivatives by capillary supercritical fluid chromatography and supercritical fluid chromatography-mass spectrometry. *Carbohydrate Research* 194: 273–279.

4 Blilie, A.L. and Greibrokk, T. (1985). Gradient programming and combined gradient-pressure programming in supercritical fluid chromatography. *Journal of Chromatography A* 349 (2): 317–322.

5 Herbreteau, B., Lafosse, M., Morin-Allory, L., and Dreux, M. (1990). Analysis of sugars by supercritical fluid chromatography using polar packed columns and light-scattering detection. *Journal of Chromatography A* 505 (1): 299–305.

6 Aranyi, A., Ilisz, I., Péter, A. et al. (2015). Exploring the enantioseparation of amino-naphthol analogues by supercritical fluid chromatography. *Journal of Chromatography A* 1387 (0): 123–133.

7 Pinkston, J.D., Stanton, D.T., and Wen, D. (2004). Elution and preliminary structure-retention modeling of polar and ionic substances in supercritical fluid chromatography using volatile ammonium salts as mobile phase additives. *Journal of Separation Science* 27 (1–2): 115–123.

8 Hamman, C., Schmidt, D.E. Jr., Wong, M., and Hayes, M. (2011). The use of ammonium hydroxide as an additive in supercritical fluid chromatography for achiral and chiral separations and purifications of small, basic medicinal molecules. *Journal of Chromatography. A* 1218 (43): 7886–7894.

9 Ashraf-Khorassani, M. and Taylor, L.T. (2010). Subcritical fluid chromatography of water soluble nucleobases on various polar stationary phases facilitated with alcohol-modified CO2 and water as the polar additive. *Journal of Separation Science* 33 (11): 1682–1691.

10 Zheng, J., Pinkston, J.D., Zoutendam, P.H., and Taylor, L.T. (2006). Feasibility of supercritical fluid chromatography/mass spectrometry of polypeptides with up to 40-mers. *Analytical Chemistry* 78 (5): 1535–1545.

11 Zhang, K., Scalf, M., Westphall, M.S., and Smith, L.M. (2008). Membrane protein separation and analysis by supercritical fluid chromatography-mass spectrometry. *Analytical Chemistry* 80 (7): 2590–2598.

12 West, C. and Lesellier, E. (2014). Strategies to choose a good column for achiral SFC. *SFC 2014: 8th International Conference on Packed-Column SFC* Basel, Switzerland (8–10 October 2014).

13 Pinkston, J.D., Marapane, S.B., Jordan, G.T., and Clair, B.D. (2002). Characterization of low molecular weight alkoxylated polymers using long column SFC/MS and an image analysis based quantitation approach. *Journal of the American Society for Mass Spectrometry* 13 (10): 1195–1208.

14 Perrenoud, A.G.G., Farrell, W.P., Aurigemma, C.M. et al. (2014). Evaluation of stationary phases packed with superficially porous particles for the analysis of pharmaceutical compounds using supercritical fluid chromatography. *Journal of Chromatography A* 1360: 275–287.

15 Berger, T.A. (2016). Instrument modifications that produced reduced plate heights <2 with sub-2 μm particles and 95% of theoretical efficiency at k = 2 in supercritical fluid chromatography. *Journal of Chromatography A* 1444: 129–144.

16 Berger, T.A. (2010). Demonstration of High Speeds with Low Pressure Drops Using 1.8 mu m Particles in SFC. *Chromatographia* 72 (7-8): 597–602.

17 Delahaye, S., Broeckhoven, K., Desmet, G., and Lynen, F. (2013). Application of the isopycnic kinetic plot method for elucidating the potential of sub-2 mu m and core-shell particles in SFC. *Talanta* 116: 1105–1112.

18 Khater, S., West, C., and Lesellier, E. (2013). Characterization of five chemistries and three particle sizes of stationary phases used in supercritical fluid chromatography. *Journal of Chromatography A* 1319: 148–159.

19 Grand-Guillaume Perrenoud, A., Veuthey, J.-L., and Guillarme, D. (2014). The use of columns packed with sub-2 μm particles in supercritical fluid chromatography. *TrAC Trends in Analytical Chemistry* 63 (0): 44–54.

20 Jones, M.D., Avula, B., Wang, Y.H. et al. (2014). Investigating sub-2 mu m particle stationary phase supercritical fluid chromatography coupled to mass spectrometry for chemical profiling of chamomile extracts. *Analytica Chimica Acta* 847: 61–72.

21 Nováková, L., Grand-Guillaume Perrenoud, A., Francois, I. et al. (2014). Modern analytical supercritical fluid chromatography using columns packed with sub-2 μm particles: a tutorial. *Analytica Chimica Acta* 824 (0): 18–35.

22 Grand-Guillaume Perrenoud, A., Veuthey, J.L., and Guillaume, D. (2012). Comparison of ultra-high performance supercritical fluid chromatography and ultra-high performance liquid chromatography for the analysis of pharmaceutical compounds. *Journal of Chromatography A* 1266: 158–167.

23 Brunelli, C., Zhao, Y., Brown, M.H., and Sandra, P. (2008). Development of a supercritical fluid chromatography high-resolution separation method suitable for pharmaceuticals using cyanopropyl silica. *Journal of Chromatography A* 1185 (2): 263–272.

24 De Pauw, R., Shoykhet Choikhet, K., Desmet, G., and Broeckhoven, K. (2015). Understanding and diminishing the extra-column band broadening effects in supercritical fluid chromatography. *Journal of Chromatography A* 1403: 132–137.

25 Desfontaine, V., Veuthey, J.-L., and Guillarme, D. (2016). Evaluation of innovative stationary phase ligand chemistries and analytical conditions for the analysis of basic drugs by supercritical fluid chromatography. *Journal of Chromatography A* 1438: 244–253.

26 Caldwell, W.B. and Caldwell, J. (2004). Selectivity of polar bonded phases for SFC and new developments. 11th International Symposium on Supercritical Fluid Chromatography, Extraction, and Processing, Pittsburgh, PA.

27 Fairchild, J.N., Brousmiche, D.W., Hill, J.F. et al. (2015). Chromatographic evidence of silyl ether formation (SEF) in supercritical fluid chromatography. *Analytical Chemistry* 87 (3): 1735–1742.

28 Perrenoud, A.G.G., Boccard, J., Veuthey, J.L., and Guillarme, D. (2012). Analysis of basic compounds by supercritical fluid chromatography: attempts to improve peak shape and maintain mass spectrometry compatibility. *Journal of Chromatography A* 1262: 205–213.

29 Lesellier, E. and Tchapla, A. (2005). A simple subcritical chromatographic test for an extended ODS high performance liquid chromatography column classification. *Journal of Chromatography A* 1100 (1): 45–59.

30 West, C. and Lesellier, E. (2006). Characterisation of stationary phases in subcritical fluid chromatography with the solvation parameter model IV. Aromatic stationary phases. *Journal of Chromatography A* 1115 (1-2): 233–245.

31 West, C. and Lesellier, E. (2007). Characterisation of stationary phases in supercritical fluid chromatography with the solvation parameter model – V. Elaboration of a reduced set of test solutes for rapid evaluation. *Journal of Chromatography A* 1169 (1-2): 205–219.

32 Lesellier, E. and West, C. (2007). Description and comparison of chromatographic tests and chemometric methods for packed column classification. *Journal of Chromatography A* 1158 (1–2): 329–360.

33 West, C. and Lesellier, E. (2008). A unified classification of stationary phases for packed column supercritical fluid chromatography. *Journal of Chromatography A* 1191 (1–2): 21–39.

34 West, C. and Lesellier, E. (2008). Orthogonal screening system of columns for supercritical fluid chromatography. *Journal of Chromatography A* 1203 (1): 105–113.

35 Lesellier, E. (2010). Extension of the C18 stationary phase knowledge by using the carotenoid test. *Journal of Separation Science* 33 (19): 3097–3105.

36 West, C. and Lesellier, E. (2012). Chemometric methods to classify stationary phases for achiral packed column supercritical fluid chromatography. *Journal of Chemometrics* 26 (3–4): 52–65.

37 West, C., Lemasson, E., Bertin, S. et al. (2016). An improved classification of stationary phases for ultra-high performance supercritical fluid chromatography. *Journal of Chromatography A* 1440: 212–228.

38 Galea, C., Mangelings, D., and Vander, H.Y. (2015). Characterization and classification of stationary phases in HPLC and SFC – a review. *Analytica Chimica Acta* 886 (0): 1–15.

39 Leyendecker, D., Leyendecker, D., Schmitz, F.P. et al. (1987). Supercritical fluid chromatography using mixtures of carbon dioxide or ethane with 1,4-dioxane as eluents. *Journal of Chromatography A* 398 (0): 105–123.

40 Klesper, E. and Hartmann, W. (1978). Apparatus and separations in supercritical fluid chromatography. *European Polymer Journal* 14 (2): 77–88.

41 Smith, R.D., Fjeldsted, J.C., and Lee, M.L. (1982). Direct fluid injection interface for capillary supercritical fluid chromatography-mass spectrometry. *Journal of Chromatography A* 247 (2): 231–243.

42 Jordan, J.W. and Taylor, L.T. (1986). Mobile phase and flow cell comparisons in packed column supercritical fluid chromatography/Fourier transform infrared spectrometry. *Journal of Chromatographic Science* 24 (3): 82–88.

43 Cantrell, G.O. and Blackwell, J.A. (1997). Comparison of 1,1,1,2-tetrafluoroethane and carbon dioxide-based mobile phases for packed column supercritical fluid chromatography. *Journal of Chromatography A* 782 (2): 237–246.

44 Kopner, A., Hamm, A., Ellert, J. et al. (1987). Determination of binary diffusion coefficients in supercritical chlorotrifluoromethane and sulphurhexafluoride with supercritical fluid chromatography (SFC). *Chemical Engineering Science* 42 (9): 2213–2218.

45 Kalinoski, H.T. and Hargiss, L.O. (1990). Supercritical fluid chromatography-mass spectrometry of non-ionic surfactant materials using chloride-attachment negative ion chemical ionization. *Journal of Chromatography A* 505 (1): 199–213.

46 Kuei, J.C., Markides, K.E., and Lee, M.L. (1987). Supercritical Ammonia as mobile phase in capillary chromatography. *Journal of Separation Science* 10 (5): 257–262.

47 Raynor, M.W., Shilstone, G.F., Bartle, K.D. et al. (1989). Use of xenon as a mobile phase for on-line capillary supercritical fluid chromatography – Fourier transform infrared spectrometry. *Journal of High Resolution Chromatography* 12 (5): 300–302.

48 Thiébaut, D. (2012). Separations of petroleum products involving supercritical fluid chromatography. *Journal of Chromatography A* 1252 (0): 177–188.

49 Lafosse, M., Herbreteau, B., and Morin-Allory, L. (1996). Supercritical fluid chromatography of carbohydrates. *Journal of Chromatography A* 720 (1–2): 61–73.

50 Conaway, J.E., Graham, J.A., and Rogers, L.B. (1978). Effects of pressure, temperature, adsorbent surface, and mobile phase composition on the supercritical fluid chromatographic fractionation of monodisperse polystyrenes. *Journal of Chromatographic Science* 16 (3): 102–110.

51 Schmitz, F.P., Hilgers, H., and Klesper, E. (1983). Gradient elution in supercritical fluid chromatography. *Journal of Chromatography A* 267 (0): 267–275.

52 Blilie, A.L. and Greibrokk, T. (1985). Modifier effects on retention and peak shape in supercritical fluid chromatography. *Analytical Chemistry* 57 (12): 2239–2242.

53 Pyo, D., Li, W., Lee, M.L. et al. (1996). Addition of methanol to the mobile phase in packed capillary column supercritical fluid chromatography retention mechanisms from linear solvation energy relationships. *Journal of Chromatography A* 753 (2): 291–298.

54 Vajda, P. and Guiochon, G. (2013). Modifier adsorption in supercritical fluid chromatography onto silica surface. *Journal of Chromatography A* 1305 (0): 293–299.

55 West, C., Melin, J., Ansouri, H., and Mengue, M.M. (2017). Unravelling the effects of mobile phase additives in supercritical fluid chromatography. Part I: polarity and acidity of the mobile phase. *Journal of Chromatography A* 1492: 136–143.

56 Taylor, L.T. (2012). Packed column supercritical fluid chromatography of hydrophilic analytes via water-rich modifiers. *Journal of Chromatography A* 1250 (0): 196–204.

57 Page, S.H., Goates, S.R., and Lee, M.L. (1991). Methanol/CO2 phase behavior in supercritical fluid chromatography and extraction. *The Journal of Supercritical Fluids* 4 (2): 109–117.

58 Ziegler, J.W., Dorsey, J.G., Chester, T.L., and Innis, D.P. (1995). Estimation of liquid-vapor critical loci for CO2-solvent mixtures using a peak-shape method. *Analytical Chemistry* 67 (2): 456–461.

59 Chester, T.L. and Haynes, B.S. (1997). Estimation of pressure-temperature critical loci of CO2 binary mixtures with methyl-tert-butyl-ether, ethyl acetate, methyl-ethyl ketone, dioxane and decane. *The Journal of Supercritical Fluids* 11: 15–20.

60 Yonker, C.R., Frye, S.L., Kalkwarf, D.R., and Smith, R.D. (1986). Characterization of supercritical fluid solvents using solvatochromic shifts. *The Journal of Physical Chemistry* 90 (13): 3022–3026.

61 Wen, D. and Olesik, S.V. (2000). Characterization of pH in liquid mixtures of methanol/H2O/CO2. *Analytical Chemistry* 72 (3): 475–480.

62 Qu, S., Du, Z., and Zhang, Y. (2015). Direct detection of free fatty acids in edible oils using supercritical fluid chromatography coupled with mass spectrometry. *Food Chemistry* 170 (0): 463–469.

63 Miller, L. (2014). Use of dichloromethane for preparative supercritical fluid chromatographic enantioseparations. *Journal of Chromatography A* 1363 (0): 323–330.

64 Prajapati, P. and Agrawal, Y.K. (2014). SFC-MS/MS for identification and simultaneous estimation of the isoniazid and pyrazinamide in its dosage form. *Journal of Supercritical Fluids* 95: 597–602.

65 West, C. and Lesellier, E. (2005). Effects of modifiers in subcritical fluid chromatography on retention with porous graphitic carbon. *Journal of Chromatography A* 1087 (1–2): 64–76.

66 Gong, X., Qi, N.L., Wang, X.X. et al. (2014). Ultra-performance convergence chromatography (UPC2) method for the analysis of biogenic amines in fermented foods. *Food Chemistry* 162: 172–175.

67 DaSilva, J.O., Coes, B., Frey, L. et al. (2014). Evaluation of non-conventional polar modifiers on immobilized chiral stationary phases for improved resolution of enantiomers by supercritical fluid chromatography. *Journal of Chromatography A* 1328 (0): 98–103.

68 Berger, T.A. and Deye, J.F. (1991). Role of additives in packed column supercritical fluid chromatography: suppression of solute ionization. *Journal of Chromatography A* 547 (0): 377–392.

69 Ashraf-Khorassani, M., Fessahaie, M.G., Taylor, L.T. et al. (1988). Rapid and efficient separation of PTH-amino acids employing supercritical CO2 and an ion pairing agent. *Journal of High Resolution Chromatography* 11 (4): 352–353.

70 Berger, T.A., Deye, J.F., Ashraf-Khorassani, M., and Taylor, L.T. (1989). Gradient separation of PTH-amino acids employing supercritical CO2 and modifiers. *Journal of Chromatographic Science* 27 (3): 105–110.

71 Steuer, W., Schindler, M., Schill, G., and Erni, F. (1988). Supercritical fluid chromatography with ion-pairing modifiers Separation of enantiomeric 1,2-aminoalcohols as diastereomeric ion pairs. *Journal of Chromatography A* 447 (0): 287–296.

72 Steuer, W., Baumann, J., and Erni, F. (1990). Separation of ionic drug substances by supercritical fluid chromatography. *Journal of Chromatography A* 500 (0): 469–479.

73 Janicot, J.L., Caude, M., and Rosset, R. (1988). Separation of opium alkaloids by carbon dioxide sub- and supercritical fluid chromatography with packed columns: application to the quantitative analysis of poppy straw extracts. *Journal of Chromatography A* 437 (0): 351–364.

74 Berger, T.A. and Deye, J.F. (1991). Effect of basic additives on peak shapes of strong bases separated by packed-column supercritical fluid chromatography. *Journal of Chromatographic Science* 29 (7): 310–317.

75 Zheng, J., Glass, T., Taylor, L.T., and Pinkston, J.D. (2005). Study of the elution mechanism of sodium aryl sulfonates on bare silica and a cyano bonded phase with methanol-modified carbon dioxide containing an ionic additive. *Journal of Chromatography A* 1090 (1-2): 155–164.

76 Zheng, J., Taylor, L.T., Pinkston, J.D., and Mangels, M.L. (2005). Effect of ionic additives on the elution of sodium aryl sulfonates in supercritical fluid chromatography. *Journal of Chromatography A* 1082 (2): 220–229.

77 Zheng, J., Taylor, L.T., and Pinkston, J.D. (2006). Elution of cationic species with/without ion pair reagents from polar stationary phases via SFC. *Chromatographia* 63 (5-6): 267–276.

78 Cazenave-Gassiot, A., Boughtflower, R., Caldwell, J. et al. (2009). Effect of increasing concentration of ammonium acetate as an additive in supercritical fluid chromatography using CO2-methanol mobile phase. *Journal of Chromatography A* 1216 (36): 6441–6450.

79 Geiser, F.O., Yocklovich, S.G., Lurcott, S.M. et al. (1988). Water as a stationary phase modifier in packed-column supercritical fluid chromatography: I. Separation of free fatty acids. *Journal of Chromatography A* 459 (0): 173–181.

80 Pyo, D. (2000). Separation of vitamins by supercritical fluid chromatography with water-modified carbon dioxide as the mobile phase. *Journal of Biochemical and Biophysical Methods* 43 (1–3): 113–123.

81 Camel, V., Thiébaut, D., Caude, M., and Dreux, M. (1992). Packed column subcritical fluid chromatography of underivatized amino acids. *Journal of Chromatography A* 605 (1): 95–101.

82 Salvador, A., Herbreteau, B., Lafosse, M., and Dreux, M. (1997). Subcritical fluid chromatography of monosaccharides and polyols using silica and trimethylsilyl columns. *Journal of Chromatography A* 785 (1): 195–204.

83 Salvador, A., Herbreteau, B., and Dreux, M. (1999). Electrospray mass spectrometry and supercritical fluid chromatography of methylated β-cyclodextrins. *Journal of Chromatography A* 855 (2): 645–656.

84 Ashraf-Khorassani, M., Taylor, L.T., and Seest, E. (2012). Screening strategies for achiral supercritical fluid chromatography employing hydrophilic interaction liquid chromatography-like parameters. *Journal of Chromatography A* 1229: 237–248.

85 dos Santos, P.A., Girón, A.J., Admasu, E., and Sandra, P. (2010). Green hydrophilic interaction chromatography using ethanol–water–carbon dioxide mixtures. *Journal of Separation Science* 33 (6-7): 834–837.

86 Scott, A.F. and Thurbide, K.B. (2017). Retention characteristics of a pH tunable water stationary phase in supercritical fluid chromatography. *Journal of Chromatographic Science* 55: 82–89.

87 Sen, A., Knappy, C., Lewis, M.R. et al. (2016). Analysis of polar urinary metabolites for metabolic phenotyping using supercritical fluid chromatography and mass spectrometry. *Journal of Chromatography A* 1449: 141–155.

88 Nováková, L., Grand-Guillaume Perrenoud, A., Nicoli, R. et al. (2015). Ultra high performance supercritical fluid chromatography coupled with tandem mass spectrometry for screening of doping agents. I: investigation of mobile phase and MS conditions. *Analytica Chimica Acta* 853 (0): 637–646.

89 Murakami, J.N. and Thurbide, K.B. (2015). Coating properties of a novel water stationary phase in capillary supercritical fluid chromatography. *Journal of Separation Science* 38 (9): 1618–1624.

90 Lisa, M. and Holcapek, M. (2015). High-throughput and comprehensive lipidomic analysis using ultrahigh-performance supercritical fluid chromatography – mass spectrometry. *Analytical Chemistry* 87 (14): 7187–7195.

91 Li, W., Wang, J., and Yan, Z.Y. (2015). Development of a sensitive and rapid method for rifampicin impurity analysis using supercritical fluid chromatography. *Journal of Pharmaceutical and Biomedical Analysis* 114: 341–347.

92 Levy, J.M. and Ritchey, W.M. (1987). Supercritical fluid chromatographic solvatochromic modifier polarity studies. *Journal of Separation Science* 10: 493–496.

93 Deye, J.F., Berger, T.A., and Anderson, A.G. (1990). Nile Red as a solvatochromic dye for measuring solvent strength in normal liquids and mixtures of normal liquids with supercritical and near critical fluids. *Analytical Chemistry* 62 (6): 615–622.

94 Olesik, S.V. (2015). Enhanced-fluidity liquid chromatography: connecting dots between supercritical fluid chromatography, conventional subcritical fluid chromatography, and HPLC. *LC/GC* 33: 24–30.

95 Åsberg, D., Enmark, M., Samuelsson, J., and Fornstedt, T. (2014). Evaluation of co-solvent fraction, pressure and temperature effects in analytical and preparative supercritical fluid chromatography. *Journal of Chromatography A* 1374 (0): 254–260.

96 Chester, T.L. (2012). Maximizing the speed of separations for industrial problems. *Journal of Chromatography A* 1261: 69–77.

7

Instrumentation for Preparative Scale Packed Column SFC

7.1 Introduction

This chapter covers instrumentation for preparative scale packed column supercritical fluid chromatography (SFC). Figure 7.1 shows a diagram of a modern preparative SFC instrument. Many of the components of a preparative SFC are the same as in an analytical SFC as described in Chapter 3. The major difference between analytical and preparative SFC is the need to collect product eluting from the column. Preparative SFC uses the same basic flow

Modern Supercritical Fluid Chromatography: Carbon Dioxide Containing Mobile Phases,
First Edition. Larry M. Miller, J. David Pinkston, and Larry T. Taylor.
© 2020 John Wiley & Sons, Inc. Published 2020 by John Wiley & Sons, Inc.

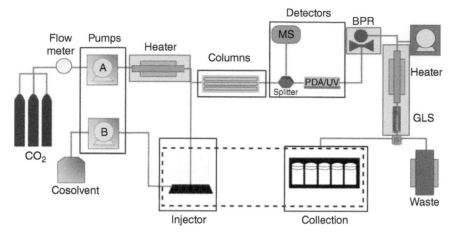

Figure 7.1 Preparative SFC schematic. *Source:* Used by permission of © Waters Corporation 2017.

path as preparative HPLC, including mobile phase pumping, sample injection, peak detection, and fraction collection. However, the use of carbon dioxide as a mobile-phase component leads to significant instrumentation changes to handle compression and expansion of CO_2. These differences are discussed throughout this chapter.

7.2 Safety Considerations

The major safety hazard relates to the use of CO_2 as a mobile phase. Preparative SFC equipment operates at high-flow rates, using large quantities of CO_2, and the risk of oxygen depletion must be addressed. The laboratories should have adequate ventilation to remove any CO_2 that may be released due to leaks or equipment malfunction. Depending on purification scale and quantities of CO_2 used, CO_2 sensors may be required to notify personnel of dangerous environments. Discussions with environmental, health, and safety personnel or other experts is recommended. Preparative SFC vendors can also provide guidance on best safety practices.

Prior to fraction collection, system pressure is reduced to atmospheric pressure, converting liquid CO_2 to a gas. All preparative systems have a gas–liquid separator to remove CO_2 from the organic solvent portion of the mobile phase. Even the best designed gas–liquid separators leave some CO_2 in the modifier solvent. Collection of this solvent into closed containers will result in container rupture. Collection into closed containers should therefore not occur. It is critical that collection containers have a vent of adequate size to allow release of any CO_2 that enters them. There have been reports of explosions of fraction containers; these were traced to insufficient CO_2 venting due to clogged or improperly sized vent tubing. Use of plastic coated glass or plastic containers for collection is also recommended. Finally fraction collection should occur in a ventilated cabinet or behind fume hood doors.

SFC purifications use an organic solvent as a modifier. Many of these modifiers are flammable and appropriate precautions should be in place to minimize risk. All operations using flammable organic solvent should occur in a properly designed laboratory that includes ventilation to avoid solvent exposure, sprinkler systems, secondary containment, and appropriate personal protective equipment.

SFC involves high pressure and preparative SFC systems should be designed taking into consideration the risk of high pressure operations [1]. This can include incorporation of rupture discs, as well as high pressure and high temperature alarms.

The safety hazards of preparative SFC are also discussed in Section 8.2 as they are worth reiterating.

7.3 Fluid Supply

7.3.1 Carbon Dioxide

While many compressible gases can be used as SFC mobile phases, CO_2 is used nearly exclusively for preparative SFC and is the only gas that is discussed in this chapter. Unlike other atmospheric gases, air separation is not the primary source of carbon dioxide. Though sometimes CO_2 is derived from directly combusting a fuel, the most economical way to produce carbon dioxide is to recover it as a byproduct from manufacturing processes or from natural wells. It is then purified and liquefied and sold by gas suppliers. The required purity of CO_2 varies depending on the application. Section 3.2 discusses CO_2 grades used for analytical SFC. The lowest purity typically used for analytical SFC is 99.99% (instrument grade). While these high-purity gases can be used for preparative SFC, they can be costly due to larger volumes of CO_2 used in preparative applications. Table 7.1 lists lower grades of CO_2 that are suitable for preparative SFC. The choice of CO_2 grade is made based on the application, purification scale, availability, and cost. An additional consideration is whether the system requires liquid or gaseous CO_2. CO_2 can be withdrawn from a cylinder or tank as either a liquid (using an educator tube) or a gas from the headspace above the liquid. Most preparative SFC systems utilize liquid CO_2 as it eliminates the additional energy to condense gas to a liquid. An additional disadvantage of gas withdrawal is the need to add a heating system to prevent freezing, which can occur with rapid gas withdrawal from a container.

Table 7.1 CO_2 grades commonly used for preparative SFC.

Product grade	Minimum purity (%)
Research	99.999
Instrument	99.99
Bone dry	99.9
Beverage grade	99.9
Food grade	99.5

Once a decision on CO_2 grade has been made, the next point to consider is how the material will be supplied. Most analytical scale SFC systems use cylinders. Cylinders are desirable as they are at a pressure (800–900 psi, 5.5–6.2 MPa) that allows direct connection to the system pump. Standard cylinders contain 60 lbs (27.2 kg) of CO_2. This quantity may be suitable for small scale (1 cm i.d. columns) preparative SFC but will be quickly depleted as purification scale increases. Many vendors offer "packs" that contain up to six cylinders. The outlets of these cylinders are connected to increase the amount of CO_2 that can be used before a new pack is needed.

For larger scale systems, or laboratories with multiple SFC purification systems, the use of cylinders is often not cost or time efficient. The use of liquid cryogenic tanks (60, 100, 250, and 500 L) is a good alternative, increasing supply as well as reducing CO_2 cost. Dewars typically deliver ~300 psi of gas, eliminating the ability to directly connect to system pumps. An intermediate pumping system is needed to increase CO_2 pressure to greater than 900 psi. The outlet of this pumping system is connected to the system pump. Pressurization skids are available from a number of vendors (Waters, Vatran, and Airgas).

A continuous CO_2 supply is an alternative used by a number of companies performing large numbers of SFC purifications. This configuration utilizes a large supply tank (1000 L to multi-ton) that is connected to a pressurization skid. The outlet of the pressurization skid is connected with high purity stainless steel tubing to the purification laboratories. This supply solution has the advantages of continuous supply (no need to change cylinders or dewars, eliminating chance of running out of supply), eliminating need to store and deliver cylinders/dewars, and the lowering the unit cost of CO_2. One disadvantage is the front end costs for procuring and installing a large tank. Also, the large supply tanks can only be filled with food or beverage grade CO_2 due to the existing commercial supply systems. Options for CO_2 supply for preparative SFC instrumentation are summarized in Figure 7.2.

In summary, there are a number of options for CO_2 supply for preparative SFC. We recommend discussing CO_2 supply with purification equipment

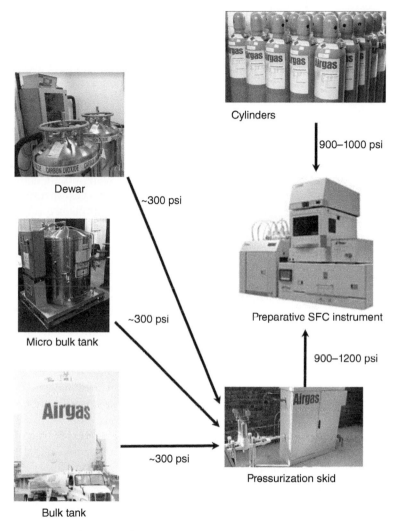

Cylinders

900–1000 psi

Dewar

~300 psi

Micro bulk tank

~300 psi

Preparative SFC instrument

900–1200 psi

Airgas

~300 psi

Pressurization skid

Bulk tank

Figure 7.2 CO_2 supply options for preparative SFC. *Source:* Photos courtesy of Airgas.

vendors as well as local gas suppliers to determine the best approach prior to purchase of preparative SFC equipment.

7.3.2 Mobile Phase Modifiers and Additives

Organic solvents used as modifiers in preparative SFC are usually "HPLC grade," although lower purity solvents can be utilized depending on the application. Many preparative SFC systems use UV detection/collection. In these cases, using solvents low in UV-absorbing impurities is important. The other consideration for solvent grade is the level of nonvolatile impurities. Products from SFC purifications are isolated in a solution of the mobile phase modifier, with the organic solvent ultimately removed via evaporation. Any nonvolatile impurities present in the modifier therefore concentrate in the isolated products. Most "HPLC" or "ACS grade" solvents have a specification for nonvolatiles of <10 mg/L. This level is sufficient for most preparative applications. For purifications requiring extremely high-product purity, purifications where the recovered product is at a low concentration, or isolation of low-level impurities, the use of high purity "HPLC-grade" solvents is always recommended. Container size for modifiers vary with purification scale. Small-scale prep SFC systems often utilize 4-L bottles, while larger systems can accommodate 5-gallon containers. For the largest systems modifier pumps are often connected directly to a stainless steel tank or solvent drum with a capacity of 30 gallons or larger.

Additives for preparative SFC are mostly small organic bases or acids. As a rule, these additives are almost always volatile. Nonvolatile additives can be used but would require a subsequent purification/extraction to remove from the isolated products. Mobile phase additives are generally "ACS Reagent" grade or better.

7.3.3 Carbon Dioxide Recycling

SFC purification has a reduced environmental impact relative to HPLC purification due to the use of carbon dioxide instead of hydrocarbon-based solvents. The "greenness" of preparative SFC can be further increased through CO_2 recycling. This approach decreases operating costs and reduces the size of CO_2 tanks required to supply the purification system(s). Except for analytical and small preparative instruments, CO_2 recycling should be used. Recycling units can be a component of the purification system or purchased as a separate unit. A well-designed unit will achieve a recycling efficiency of 90–95% and minimize carryover of modifier [2]. Because preparative SFC uses multiple injections and automated collection of purified products, if modifier is not satisfactorily removed in the CO_2 recycling process, the mobile-phase polarity will change over time, leading to reduced retention and resolution, impacting purity, and recovery. Even the best designed and optimized gas–liquid separator will contain trace levels of solute and modifier in the gas. Some vendors recommend the recovered CO_2 be further purified before being recycled [2].

CO_2 recycling occurs at a pressure of 40–50 bar, slightly below the typical pressure of CO_2 entering the preparative SFC instrument. The requirement of high pressures for recycling eliminates the possibility of recycling for instruments performing fraction collection at atmospheric pressure such as open bed collectors described later in Section 7.9.2.

7.4   Pumps and Pumping Considerations

7.4.1   CO$_2$ and Modifier Fluid Pumping

Accurate pumping of CO_2 is critical to successful operation of a preparative SFC system. Liquid carbon dioxide is very compressible compared to traditional liquids used in preparative HPLC. The compressibility of CO_2 decreases as it is cooled, so the pump heads of all preparative CO_2 pumps are cooled to 3–5 °C. In addition, heat exchangers are often installed in tubing feeding the pump. While Peltier coolers can be used for analytical SFC, they offer insufficient cooling capacity for preparative SFC systems. The increased cooling requirements are achieved with recirculating chillers.

Reproducible preparative chromatography requires consistent mobile-phase flow. Many preparative systems include a CO_2 mass flow meter with a feedback loop to pump control software. The use of a mass flow meter ensures a consistent amount of material is pumped into the purification column. Accurate flow also reduces mobile-phase density changes, which at low modifier percentages can have an impact on retention.

Modifier pumps in preparative SFC systems are typically the same design as those in preparative HPLC systems. Some vendors have added solvent selection valves to the modifier pumps, increasing flexibility for purification using multiple modifiers.

7.4.2   Pressures and Flow Ranges

Pressure limits for modern preparative SFC systems are typically 250–350 bar (25–35 MPa). The higher pressure limit allows the use of small particle size (5 µm) and increased flow rates that should lead to increased purities and purification throughput. The available pressure for chromatography is reduced by the back pressure regulator (BPR) setting. A BPR setting of 150 bar leaves only 200 bar pressure drop across the column and tubing for a system with a total pressure rating of 350 bar. As most preparative SFC columns have an operating limit of 200 bar or less, the current pump pressure limits rarely limit operating conditions.

Flow rates for preparative SFC vary depending on column size. Typical flow rate ranges for various preparative SFC columns are listed in Table 7.2. It is important to properly define the range of columns to be used with the system when choosing a preparative SFC system. Most pumping systems are accurate to 10–20% of maximum flow rate. In addition, larger systems often use larger i.d. tubing to minimize pressure drops. The increased tubing volume can lead to increased mixing, extra-column band broadening, and reduced separation when used with smaller i.d. columns at reduced flow rates.

Table 7.2 Column inner diameter (i.d.) and flow rates.

Column i.d. (mm)	Flow rate range (mL/min)
10	10–25
20	40–100
30	80–200
50	250–500
80	640–1280
100	1000–2000

7.5 Sample Injection

7.5.1 Injection of Solutions

Standard practice in preparative SFC is to dissolve material to be purified in an organic solvent. Preparative SFC injection has two configurations: (i) mixed stream injection and (ii) modifier stream injection. Schematics of these two options are shown in Figure 7.3. Mixed stream injection introduces the sample solution just prior to the column, after carbon dioxide and modifier solvent are mixed. This approach has the advantage of introducing sample into the chromatographic system just prior to the column but requires the injection loop to be decompressed prior to loading the sample solution. It also has a disadvantage of introducing air into the system if the injection loop is not totally filled prior to injection. Modifier stream injection introduces sample solution into the modifier flow path prior to mixing with carbon dioxide. Modifier stream injection has been shown to improve peak shape in preparative SFC under gradient conditions [3, 4] and isocratic conditions [5]. Modifier stream injection can be compared to "at-column dilution" (ACD) injection technique utilized for preparative reversed phase chromatography [6, 7]. The decreased chromatographic performance of mixed stream injection can be traced to the temporary localized increase in mobile phase strength due to the "slug" of solvent used for dissolution (which often is a stronger chromatographic solvent relative to CO_2), which can lead to peak distortion and reduced resolution. Additional discussion on preparative SFC injection options can be found in Section 8.3.6.

7.5.2 Extraction Type Injection

Ideally, the sample dissolution solvent should match polarity and composition of the mobile phase. The presence of CO_2 in the mobile phase eliminates this possibility in preparative SFC. An alternate injection technique that does not

Mixed stream injection schematic

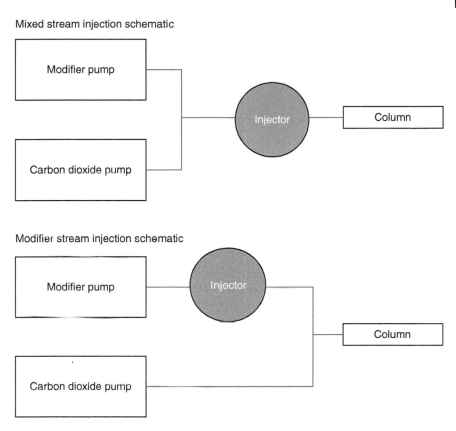

Figure 7.3 Instrument schematic for mixed stream and modifier stream injection.

rely on dissolution in organic solvent is using supercritical fluid extraction (SFE) for injection. While this process is more time consuming than injection of liquids, it can allow for injection of samples with poor solubility or injection of materials containing insoluble impurities. This approach has been used for chiral purifications [8] and achiral flash purifications [9]. Equipment sold by PIC Solution Inc. offers SFE as an injection option [10].

"Extraction" injection involves placing the sample to be purified in an extraction vessel and pressurizing the loaded vessel with the SFC mobile phase. This initiates sample solubilization. The sample is often mixed with an inorganic material such as silica to disperse the sample, increasing surface area, and improve the solubilization process. The extraction vessel is switched into the system flow path, allowing material dissolved in the SFC mobile phase to flow onto the preparative column, and ideally to focus on the head of the column. Additional information on extraction type injections can be found in Section 8.3.6.

7.6   Chromatographic Columns

As discussed previously, preparative SFC operates at high pressures. All columns must have both sufficient pressure rating and packing stability for use at high pressure. Stainless steel is the material of choice for preparative SFC columns. Many of the stationary phases used for preparative SFC are also sold for preparative HPLC. The pressure drop for HPLC can be lower than SFC; some preparative HPLC columns are packed in column hardware not rated for the higher pressures of preparative SFC. In the early days of preparative SFC, there were stories of column ends being pushed off column bodies under the higher pressures of SFC. Besides being a safety hazard, columns not designed for preparative SFC may also suffer poor column lifetime due to insufficient packing pressures during column preparation.

Smaller i.d. columns (1–5 cm i.d.) are most often purchased as prepacked columns. Larger i.d. columns (>5 cm i.d.) are not used as prepacked columns due to the difficulty in maintaining a well-packed column. It is advisable to use dynamic axial compression (DAC) technology as column diameter increases. DAC technology allows efficient packing and unpacking of columns with bulk stationary phases, and during operation keeps the column bed stable via a piston that moves inside the chromatographic column. The piston maintains a controlled pressure on the chromatographic bed during column operation, providing excellent bed stability. The piston is driven by a hydraulic jack, pushed by a liquid or a spring [11]. Most preparative column vendors are familiar with SFC and sell columns properly designed and packed for use under these conditions. It is recommended to inform vendors of plans to operate their columns under SFC conditions prior to purchase.

Temperature changes have been shown to have increased impact on selectivity and efficiency in SFC relative to HPLC [12]. Changes in temperature can affect mobile phase density which, especially at low-modifier percentages, can impact retention and selectivity. In addition, poor temperature control can result in temperature gradients within the column, resulting in poor peak shapes. Temperature control can be accomplished by preheating the mobile phase and/or heating the column in an oven. Some chiral separations can be improved by working at sub-ambient temperature. This can be achieved by cooling of the eluent via the heat exchanger after the pump.

7.7   Detection

While any detector used in analytical SFC can be used in preparative-scale SFC (see Chapter 4 for more information), the majority of SFC purifications use either UV/Vis or mass spectrometry detection. UV/Vis detectors, being non-destructive, can be plumbed into the main flow of the preparative system.

Unlike preparative HPLC systems, the UV/Vis flow cell in preparative SFC is under pressure and must be rated for higher pressure operation.

Preparative SFC using mass-directed collection is a technique that has gained popularity over the past 10 years [13–17]. Mass spectrometry is a destructive detection method. Because of this, preparative SFC systems with this technique incorporate a splitter to send a minor portion of the mobile phase to the MS with the remainder being sent to fraction collection. Often a make-up pump is used after the split to help move eluent to the mass spectrometer and to boost MS signal (see Section 4.3.1). For successful mass-directed collection, a delay volume must be built into the system after the splitter and before fraction collection. This delay volume, most often achieved through coiled tubing, allows the mass spectrometer to confirm the presence of the desired ion and then signal the fraction collector to collect the correct peak. The delay volume must be properly calibrated at various flow rates to ensure high-yield recovery [14].

7.8 Back Pressure Regulation

Proper backpressure regulation is a critical aspect of an SFC separation. Modern analytical and preparative SFC equipment uses a back pressure regulator (BPR) downstream of the UV detector to maintain the mobile phase as a single phase. Improper back pressure regulation can result in changes in mobile phase density, leading to irreproducible chromatography. BPRs for preparative SFC systems are designed to maintain consistent pressure within a run and between runs. While a large variety of BPRs are used in analytical SFC (see Section 3.8), most preparative SFC systems use a BPR with a needle and seat design as shown in Figure 7.4 [18]. The needle is moved, under the control of a pressure feedback loop, to maintain desired pressure. These BPRs are designed to accurately control pressure within a few psi over a range of flow rates, mobile phase densities and mobile phase compositions.

7.9 Fraction Collection

The use of carbon dioxide-containing mobile phases makes fraction collection particularly difficult. Standard collection processes used in preparative HPLC are not suitable for preparative SFC. As SFC mobile phase exits the BPR, it transitions from high pressure (100 + bar) to atmospheric pressure. This pressure change results in a volumetric expansion of 250–500-fold as carbon dioxide moves from a highly compressed liquid to a gas. The mobile phase rapidly transitions from a single phase to a biphasic mixture of liquid droplets and gas. The extent of expansion decreases as the percentage of modifier in the mobile

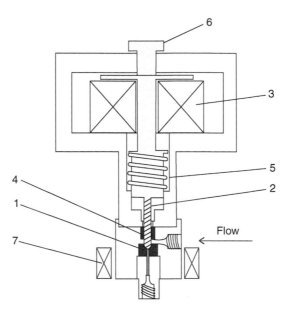

Figure 7.4 Cross-sectional view of needle and seat back pressure regulator. 1 = valve seat, 2 = valve needle, 3 = needle-drive solenoid, 4 = needle seal, 5 = return spring, 6 = gap adjustment screw, and 7 = heater. *Source:* Used by permission from Springer Nature: Springer Saito et al. [18].

phase increases. Another potential issue is Joule–Thomson cooling, which can cause tubing blockage due to dry ice formation. Expansion in an uncontrolled fashion can impact peak purity and isolated yields. A device to separate gas from liquid postBPR is required. A number of options have been developed for fraction collection and are discussed in the following sections.

7.9.1 Cyclone Collection

Cyclonic separation eliminates particulates from a stream of air, gas, or liquid. A drawing of a cyclone separator used in preparative SFC instrumentation is shown in Figure 7.5. SFC mobile phase that contains the analyte, liquid modifier, and gaseous CO_2 enters the side of the cyclone. The high pressure (typically 40–60 bar) pushes the heavier liquid portion (which contains the target compounds) to the sides of the cyclone, where it rotates around the inside of the vessel, being pushed down by gravity and gas flow until the liquid exits the bottom of the cyclone. The lighter CO_2 gas exits the top of the cyclone where it is vented or routed to a CO_2 recycler. The main advantage of a cyclone separator is recovery of CO_2 at high pressure which allows efficient, cost-effective recycling. If CO_2 recycling is not being performed, pressure within the cyclone can be maintained at 10–15 bar. Cyclone separators are more expensive relative to

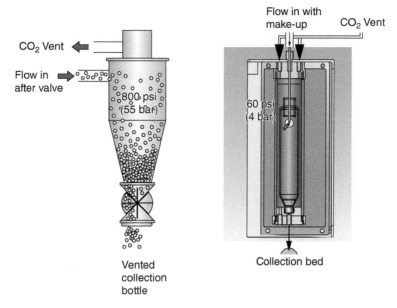

Figure 7.5 Collection options in preparative SFC. *Source:* Used by permission of © Waters Corporation 2017.

other collection options due to the use of high pressure stainless steel. There is a potential safety risk due to the liquid modifier exiting the separator at high pressures. Cyclone separators are ideally suited for separations with a minimum number of fractions are being collected (such as chiral purifications) into containers such as bottles or carboys. An alternate collection process should be explored when multiple fractions (>6) or collection into test tubes is required.

7.9.2 Open-Bed Collection

Initial preparative SFC design used cyclonic separation for the separation of CO_2 and modifier. This design was ideally suited for the first major use of preparative SFC, namely, batch chiral purifications. As the preparative field expanded into achiral purifications, there was a need for high throughput purification of multiple samples, requiring the collection of effluent fractions into many different test tubes or other small vessels. This form of fractionation is called open-bed collection. Cyclone separators are not compatible with open-bed collection.

The first description of tube collection, in 2001, used a modified Gilson 215 fraction collector, achieving recoveries of between 50 and 77% depending on the collection process [13]. Initial attempts for a commercial system allowing high throughput collection into tubes used a robot that moved collection tubes into and out of a chamber during the collection process [19]. While this process worked, the robot was expensive and incompatible with many of the high throughput LC purification processes in operation at that time. Subsequent research included installation of a larger i.d. tube at the end of the collection needle. This tube served to reduce mobile phase linear velocity and allowed high recoveries (87–92%) at flow rates up to 30 mL/min [14, 20].

The first commercial preparative SFC system allowing open-bed collection was introduced in 2009. Numerous vendors now offer open-bed collection with preparative SFC systems up to a flow rate of 100 mL/min including Waters, PIC Solutions, and JASCO. A gas–liquid separator that was reported to allow open bed collection at flow rates up to 350 mL/min was presented at a conference in 2012 but does not appear to have been commercialized [21]. An example of a gas–liquid separation (GLS) suitable for open-bed collection is shown in Figure 7.5. The GLS is a critical component of any open-bed collection system and must be designed to maintain peak integrity with minimal peak

Table 7.3 Preparative SFC instruments currently available in the market.

Manufacturer	System	Column sizes (mm)	CO_2 flow rate range (mL/min)	Modifier flow rate (mL/min)	Detection options
JASCO	SP-4000	4.6–10	3–50	3–50	UV, MS
	PR-4088	10–30	5–150	5–50	UV
Novasep	150	20–30	30–150	0–80	UV
	400	30–50	100–400	0–160	UV
	1000	50–80	250–1000	0–400	UV
	3000	80–100	600–3000	0–1000	UV
PIC solution	100	10–30	0–100	0–50	UV, MS
	200	20–30	0–200	0–100	UV
	400	20–50	0–400	0–250	UV
	600	30–80	0–600	0–250	UV
Sepiatec	Prep SFC Basic	4.6–10	0–20	0–20	UV
	Prep SFC 100	20–30	0–100	0–100	UV, MS
	Prep SFC 360	25–50	0–360	0–360	UV
Waters	Investigator	4.6–10	0.5–15	0.1–10	UV, MS
	SFC 80q	20	10–70	4–70	UV
	SFC 100	20–30	30–100	5–55	UV, MS
	SFC 200q	20–30	20–200	4–70	UV
	SFC 350	30–50	25–250	20–200	UV

broadening as products pass through the GLS as well as prevent cross-contamination within the unit. Additional information on various GLS designs can be found in patents [22] or vendor websites and literature.

7.10 Conclusion

The capability and sophistication of preparative SFC equipment has increased drastically since the first commercial instrument was introduced in the 1990s. The first systems were available from one vendor, had maximum flow rates of ~70 mL/min and offered only UV detection. As of 2017, numerous vendors, including major as well as smaller instrument manufacturers, offer equipment with a range of flow rates capable for purification of mg to kg quantities, offering both UV- and MS-based collection. A list of currently available preparative SFC systems can be found in Table 7.3. The equipment advances have allowed preparative SFC to move from a technique performing exclusively small-scale chiral purifications for the pharmaceutical industry to a technique suitable for both chiral and achiral purification at a range of purification scales and supports numerous industries.

References

1 Clavier, J. and Perrut, M. ed. (1996). Safety in supercritical operations. *Process Technology Proceedings,* Elsevier.

2 Colin, H., Ludemann-Hombourger, O., and Denet, F. (2005). Equipment for preparative and large size enantioselective chromatography. In: *Preparative Enantioselective Chromatography* (ed. G.B. Cox), 224–252. Oxford, UK: Blackwell Publishing Ltd.

3 Berger, T.A. and Fogelman, K.D. (2002). Method of sample introduction for supercritical fluid chromatography systems. US Patents 6428702B1.

4 Fogelman, K.D. (2004). HPLC 2004; June 2004; Philadelphia, PA.

5 Miller, L. and Sebastian, I. (2012). Evaluation of injection conditions for preparative supercritical fluid chromatography. *Journal of Chromatography. A* 1250: 256–263.

6 Neue, U.D., Mazza, C.B., Cavanaugh, J.Y. et al. (2003). At-Column Dilution for Improved Loading in Preparative Chromatography. *Chromatographia* 57: S-121–S-127.

7 Blom, K.F. (2002). Two-Pump at-Column-Dilution Configuration for Preparative Liquid Chromatography-Mass Spectrometry. *Journal of Combinatorial Chemistry* 4: 295–301.

8 Gahm, K.H., Tan, H., Liu, J. et al. (2008). Purification method development for chiral separation in supercritical fluid chromatography with the solubilities in supercritical fluid chromatographic mobile phases. *Journal of Pharmaceutical and Biomedical Analysis* 46 (5): 831–838.

9 Miller, L. and Mahoney, M. (2012). Evaluation of flash supercritical fluid chromatography and alternate sample loading techniques for pharmaceutical medicinal chemistry purifications. *Journal of Chromatography. A* 1250: 264–273.

10 Shaimi, M. and Cox, G.B. (2014). Injection by Extraction: A Novel Sample Introduction Technique for Preparative SFC. *Chromatography Today* November/December: 42–45.

11 Dingenen, J. (1998). Columns and packing methods. *Analusis* 26 (7): 18–32.

12 De Pauw, R., Choikhet, K., Desmet, G., and Broeckhoven, K. (2014). Temperature effects in supercritical fluid chromatography: A trade-off between viscous heating and decompression cooling. *Journal of Chromatography. A* 1365: 212–218.

13 Wang, T., Barber, M., Hardt, I., and Kassel, D.B. (2001). Mass-directed fractionation and isolation of pharmaceutical compounds by packed-column supercritical fluid chromatography/mas spectroscopy. *Rapid Communications in Mass Spectrometry* 15: 2067–2075.

14 Zhang, X., Towle, M.H., Felice, C.E. et al. (2006). Development of a Mass-Directed Preparative Supercritical Fluid Chromatography Purification System. *Journal of Combinatorial Chemistry* 8: 705–714.

15 McClain, R.T., Dudkina, A., Barrow, J. et al. (2009). Evaluation and Implementation of a Commercially Available Mass-Guided SFC Purification Platform in a High Throughput Purification Laboratory in Drug Discovery. *Journal of Liquid Chromatography and Related Technologies* 32 (4): 483–499.

16 Rosse, G. (2016). Modern SFC-MS Platform for Achiral and Chiral Separations. *SFC Day* San Diego (23 May 2016).

17 Francotte, E. (2015). SCF: a multipurpose approach to support drug discovery. *1st International Conference on Packed Column SFC China* Shanghai, China.

18 Saito, M., Yamauchi, Y., Kashiwazaki, H., and Sugawara, M. (1988). New pressure regulating system for constant mass flow supercritical-fluid chromatography and physico-chemical analysis of mass-flow reduction in pressure programming by analogous circuit model. *Chromatographia* 25 (9): 801–805.

19 Berger, T.A., Fogelman, K.D., Klein, K., et al. (2004). Automated sample collection in supercritical fluid chromatography. US Patents 20020144949A1.

20 Bozic, A. (2017). Inventor Gas-Liquid Separator. United States 10 August 2017.

21 Bozic, A. (2012). Conversions of a Standard Preparative HPLC into a Preparative SFC *SFC 2012* (October 2012) Brussels, Belgium.

22 Wang, Z., Zulli, S., Rolle, D., et al. (2014). Methods and devices for open-bed atmospheric collection for supercritical fluid chromatography. United States 7 August 2014.

8

Preparative Achiral and Chiral SFC – Method Development, Stationary Phases, and Mobile Phases

Modern Supercritical Fluid Chromatography: Carbon Dioxide Containing Mobile Phases,
First Edition. Larry M. Miller, J. David Pinkston, and Larry T. Taylor.
© 2020 John Wiley & Sons, Inc. Published 2020 by John Wiley & Sons, Inc.

8.1 Introduction

Chemical reactions rarely produce products of high purity; starting materials, reagents, and unwanted side products are often present and must be removed using a separation technique. Standard separation techniques include crystallization, distillation, precipitation, extraction, filtration, and preparative chromatography. Crystallization, distillation, precipitation, extraction, and filtration are beyond the scope of this chapter and will not be discussed. Purification using preparative chromatography can be traced to Twsett's initial research on chromatography of plant pigments in the early twentieth century [1, 2]. Preparative chromatography equipment has changed drastically over the past 100 years, moving from open glass tubes packed with large particle irregular stationary phases and using gravity to move mobile phase through the column, to today's technology that uses stainless steel columns packed with small particle stationary phases and high pressure pumps to control mobile phase flow. The first reported use of SFC was over 50 years ago [3]. In this article, the author proposed the use of SFC for purifications stating "the porphyrins could be recovered at the outlet valve." Reviews of early purification work (through mid-90s) can be found in [4, 5].

The main difference between analytical and preparative chromatography is the objective of the work. The objective of analytical chromatography is the generation of information about the sample (purity, enantiomeric excess, and impurity level) while the goal of preparative chromatography is the generation of pure substances. Preparative chromatography is used in a number of industries with the major users being pharmaceutical, food, and chemical industries [6–9]. Preparative chromatography in the food and chemical industries is performed only when alternate separation techniques are not suitable. Purification quantities in the food and chemical industries are very large; hundreds of thousands of metric tons of high fructose corn syrup and m-xylene are produced each year using continuous chromatography techniques.

Molecules synthesized in the pharmaceutical industry have higher molecular weights and are often more complex, or require higher purities than those in the food and chemical industries. Due to the increased molecular complexity and more rigorous demands, preparative chromatography is often a requirement for production of pure material. Preparative chromatography plays an important role in the pharmaceutical industry, from initial research where milligram quantities are purified through manufacturing where hundreds of metric tons per year are purified. The majority of larger scale (>1 kg) chromatographic purifications in the pharmaceutical industry use liquid chromatography. These larger scale purifications most often support pharmaceutical development and manufacturing. During pharmaceutical discovery, where thousands of compound per year with purification quantities ranging from mg to 1 kg are encountered, the technique of first choice is SFC (R. Schmidt, C. Ponder, and M Villeneuve, Supercritical fluid chromatography: how green is it? A life-cycle comparison of high performance liquid chromatography and supercritical fluid chromatography. *Oral presentation at SFC*, Boston, MA, USA. 2013) [10–19]. While the main focus of this chapter is SFC purifications in support of the pharmaceutical industry, the techniques and practices described are suitable for purification of any small molecule in a significant number of industries.

8.1.1 Advantages and Disadvantages of SFC vs. HPLC for Purification

Previous chapters have discussed the important advantages of SFC over HPLC. For preparative batch chromatography these advantages include:

- Increased flow rates
- Reduced pressure drop
- Reduced organic solvent usage
- Easier product isolation
- Lower separation costs

These advantages make SFC the technique of choice for purification in pharmaceutical discovery. One measurement of purification efficiency is throughput; how much pure material is produced per hour. SFC has advantages over HPLC that directly impact throughput. The increased flow rate reduces cycle time, allowing more material to be purified per hour. The reduced pressure drop of SFC allows use of smaller particles, increasing chromatographic efficiency, resulting in increased loadings, product purities, and yields. In SFC, the majority of the mobile phase is liquid CO_2, reducing the amount of organic solvent required for separation. Post separation, CO_2 is removed via a gas–liquid separator, generating products at higher concentrations relative to HPLC. This reduces the time and energy required for solvent removal post chromatography. The sum of these advantages results in a purification technique that has higher productivity, may produce material of higher purity, uses less solvent and requires less workup post purification. The end result is that SFC is faster and affords lower costs for small molecule purification. It is no wonder that SFC is now the first choice for small scale purification support within the pharmaceutical industry.

While SFC has many advantages over HPLC for small molecule purification, it does have some disadvantages. These disadvantages include (i) CO_2 supply, (ii) fraction collection, (iii) higher equipment costs, and (iv) potential for an oxygen deficient atmosphere. These issues will be addressed later in this chapter and in Chapter 7 on preparative SFC equipment. Another disadvantage is the inability to easily measure solubility in CO_2/modifier mixtures, which is also discussed later in this chapter.

8.1.2 Cost Comparison: Preparative HPLC vs. SFC

Preparative SFC relative to HPLC offers increased flow rates and reduced solvent volumes that translate to faster purifications and reduced costs for solvent purchase and disposal. In addition, there are significant cost savings resulting from the use of primarily alcohol-based solvents vs. hydrocarbon-based solvents. Reduced purification costs are one of the major drivers for numerous companies moving from HPLC to SFC for small molecule purifications. The costs savings have been documented for both chiral and achiral purifications. Miller et al. demonstrated cost savings for small scale chiral purifications (200–500 mg). At the 200-mg scale, purification costs were reduced from ~200 to 68 USD. At the 500-mg scale, the savings were even more impressive, decreasing from ~496 to ~166 USD [13]. Riley et al. discussed cost savings realized at Pfizer by the transition to preparative SFC. They observed an 83% reduction in solvent cost and a 65% time reduction compared to HPLC for a typical achiral purification. For chiral SFC purifications they observed a 96% reduction in solvent cost and an 85% time reduction vs. chiral HPLC [20].

Francotte discussed the transition from reversed phase to SFC-based purifications for achiral purifications at Novartis [14, 21]. Using mass-directed purification, achiral SFC has replaced reversed phase HPLC for at least 75% of the small molecule purifications in their laboratory. He has shown that preparative SFC is two to three times faster, has 50% lower solvent costs, reduced evaporation energy cost sevenfold, and reduced liquid waste 20% relative to preparative HPLC. None of the cost comparisons above included the additional costs for SFC vs. HPLC equipment or the costs associated with CO_2 distribution.

8.2 Safety Considerations

The major safety hazard for SFC purifications is related to the use of CO_2 as a mobile phase. Preparative SFC equipment uses CO_2 flow rates of 50–1000 mL/min. When using large volumes of CO_2, one must be careful to avoid oxygen depletion. The laboratory should be designed for the use of large volumes of CO_2 to ensure adequate safety. This includes adequate ventilation to remove CO_2 that may be released during a system or supply leak or other equipment malfunction. As purification scale or the number of instruments increases CO_2 sensors may be advisable. These sensors serve as a warning when a high CO_2 concentration is present and notify lab personnel to evacuate. Discussions with environmental health and safety or other experts is recommended. Preparative SFC vendors are another excellent source for information on best safety practices.

An additional safety hazard of preparative SFC is related to the collection of separation products post purification. Most preparative SFC systems use a gas–liquid separator that removes the majority of the CO_2 gas that is generated post column when the system pressure is reduced to atmospheric pressure. A well-designed preparative SFC system will allow venting of CO_2 gas from the gas–liquid separator. While a gas–liquid separator removes the majority of the CO_2, it does not remove it completely. The organic solvent exiting the gas–liquid separator still contains CO_2. It is important to avoid collection of product fractions in closed containers. A vent of adequate size to allow CO_2 to escape the collection vessels without a pressure increase is required. It is also imperative that the vent be constructed to eliminate possibility of kinking or blocking of tubing. Many vendors use a large i.d. (1/2 in. or larger) corrugated tubing to avoid potential for kinking of tubing, which could result in pressure increase in the collection vessel. It is also recommended to use plastic or plastic coated glass vessels for collection. There have been reported instances of explosion of collection vessels due to inadequate CO_2 venting. While these vessel explosions were traced to the use of vent tubing of inadequate size that allowed kinking or plugging, it is best to err on the side of caution. Fraction collection should occur in a ventilated cabinet or behind fume hood doors that will contain any vessel explosion that could occur.

The final safety consideration for preparative SFC relates to the use of flammable organic solvents. While this risk is lower for preparative SFC relative to preparative HPLC due to the reduced solvent volumes, depending on purification quantities it is still an important consideration. Any operation using flammable organic solvent should occur in a properly designed laboratory. This can include appropriate ventilation to avoid solvent exposure, sprinkler systems, secondary containment as well as appropriate personal protective equipment.

As with any laboratory procedure, discussion with safety experts internal or external to your organization is advised. For any laboratory new to preparative SFC a safety hazard analysis is recommended to minimize any safety risks.

8.3 Developing Preparative Separations

Development of a preparative separation involves the following steps:

1) Definition of purification objective
2) Development of analytical method
3) Preparative separation
4) Recovery of product from mobile phase

The following sections provide details on each of these steps as they relate to SFC purifications. The first step of a purification is to define the purification objective. The objective of the purification defines the work required for analytical method development and can help select operating conditions. Information required to define the purification objective include (i) chiral or achiral separation, (ii) impurity isolation, (iii) number of components to be isolated, and (iv) how much material will be purified. At this time purity and yield requirements should be determined. Other critical information includes (i) compound stability, (ii) solubility, and (iii) purity of crude material.

The next step is the development of an analytical separation that will be scaled to preparative loadings. During method development the stationary and mobile phases are chosen. For preparative separations, a stationary phase must be stable, available in larger column sizes, and have high loading capacity. Stationary phase stability is critical for preparative separations. Larger i.d. columns, especially packed with smaller particle phases, are expensive. Phases with excellent long-term stability provide long column lifetimes and reduce purification costs. Another consideration, although not as important for preparative SFC, is the ability to clean the columns to remove highly retained compounds. Work only with stationary phases that are available in larger quantities; either as larger i.d. prepacked columns or as bulk stationary phase. It is a waste of time to develop a separation on a stationary phase that is only available in analytical columns or in particle sizes not suitable for preparative work (<5 μm).

The most critical stationary phase requirement for preparative separations is high loading capacity. A stationary phase with a high loading capacity gives Gaussian peaks at higher amount of injected compound relative to a phase with low loadability. This allows increased amount of product to be produced per unit time. The importance of loading capacity on productivity is shown in Figure 8.1. While this example is for a preparative HPLC separation, the effects shown are applicable to preparative SFC. The racemate was resolved with methanol on two amylose based chiral stationary phases (CSP). For CSP1, the selectivity is 2.25, for CSP2, the selectivity is 2.68. Due to a larger selectivity value, one might predict that CSP2 would generate the highest productivity. When these methods were scaled up to preparative loadings it was shown that CSP1 had a higher loading capacity than CSP2. The increased loading produced a nearly twofold increase in productivity for CSP1 even though the analytical separation was inferior to CSP2. The importance of stationary phase loading capacity is discussed further in the chiral and achiral sections of this chapter.

The next consideration during the analytical method development is mobile phase. Mobile phase characteristics to consider include volatility, toxicity, cost, and viscosity. An ideal chromatographic solvent would be volatile, have low toxicity, be low cost, and have a low viscosity. The final step of any purification is recovery of product from the mobile phase. Most recoveries are performed

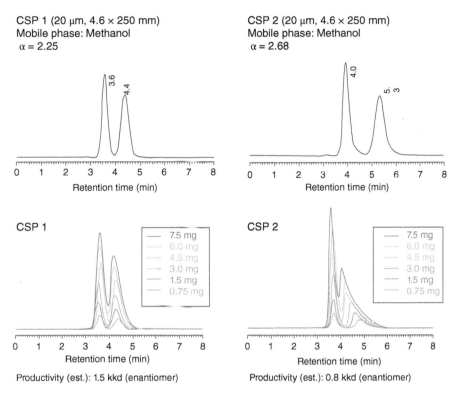

CSP 1 (20 μm, 4.6 × 250 mm)
Mobile phase: Methanol
α = 2.25

CSP 2 (20 μm, 4.6 × 250 mm)
Mobile phase: Methanol
α = 2.68

3.6
4.4

4.0
5.3

Retention time (min)

Retention time (min)

CSP 1

—— 7.5 mg
—— 6.0 mg
—— 4.5 mg
—— 3.0 mg
—— 1.5 mg
—— 0.75 mg

CSP 2

—— 7.5 mg
—— 6.0 mg
—— 4.5 mg
—— 3.0 mg
—— 1.5 mg
—— 0.75 mg

Retention time (min)

Retention time (min)

Productivity (est.): 1.5 kkd (enantiomer)

Productivity (est.): 0.8 kkd (enantiomer)

Figure 8.1 Impact of chiral stationary phase (CSP) loading capacity on purification productivities. kkd: kilograms product/kilogram CSP/day. *Source:* Reprinted with permission from —Caille et al. [22]. Copyright 2010 American Chemical Society.

using some form of solvent distillation at higher temperatures and reduced pressures (i.e. rotary evaporation, falling film evaporators). It is imperative that the mobile phase used for purification be volatile. Higher volatility solvents reduce time and energy requirements for evaporation, leading to quicker and lower cost purification processes. Volatility of any mobile-phase additives should also be considered. Solvent toxicity is not critical in analytical chromatography where smaller volumes of solvent are used. When scaled to preparative loadings, these separations can require liters to hundreds of gallons of solvent, depending on the amount of material to be purified. For this reason, solvent toxicity is an important criteria for any analytical method being developed for preparative scale-up. The third consideration is cost. Avoiding high cost solvents reduces laboratory operating costs. The final consideration is viscosity. Higher viscosity solvent limit the purification linear velocity, resulting in increased purification times and reduced productivities. It is for these reasons that methanol is often the mobile-phase modifier of choice for SFC purifications. It is relatively volatile, has moderate toxicity, is inexpensive relative to other solvents, and has lower viscosity relative to other SFC modifiers such as ethanol and isopropanol.

Another factor to consider during method development is solubility. As a general rule, a mobile phase offering higher compound solubility has a higher preparative productivity than a mobile phase with a lower solute solubility. In addition, higher solubility often translates to reduced solvent requirements for the purification. The impact of solubility on purifications is discussed later in this chapter.

The final mobile phase consideration during method development is compound stability. During preparative purifications purified compounds can remain in the mobile phase for hours or days until solvent removal is complete. Compounds must be stable in these solvents both at room temperature as they wait for evaporation, as well as at the elevated temperatures often used for the evaporation process. It is disheartening to spend one's valuable time to produce pure material only to have it degrade or racemize while waiting for, or during the evaporation step.

The above recommendations are only guidelines. Depending on the separation it may be necessary to use a modifier that has higher cost and viscosity. For example, a separation using isopropanol may be much improved relative to methanol. Or solubility in isopropanol may be significantly higher than in methanol. In this case, use of isopropanol can result in higher productivity or reduced purification times even with lower flow rates and increased evaporation times.

What are the most important chromatographic characteristics of an analytical method for scale-up to a highly productive preparative separation? Forssen et al. [23] performed a theoretical study to answer this question. They calculated the following:

- Column length: the shorter the better
- Selectivity: the higher the better
- Retention factor for the first eluting component: the lower the better
- Stationary phase capacity: the higher the better

It should be noted that these are only guidelines. Optimum conditions vary depending on whether the product of interest elutes first or second as well as other related purification objectives.

One final reminder for the method development process: For an analytical method, separation of all compounds in the mixture is a requirement. This is not a requirement for an analytical method being developed for scale-up to preparative loadings. The only requirement for this analytical method is separation of the desired compound from other compounds in the mixture. It is not a requirement that the other compounds be separated from each other. In fact from a purification viewpoint, it is desirable to not have these compounds resolved from each other. This can lead to a shorter purification cycle time and increased purification throughputs.

Once an analytical method is developed, it is time to scale to preparative loadings. As discussed earlier, the amount of material purified does not qualify a separation as a purification process. Instead it is the fact that material is collected from the column for some subsequent use. The quantity can be as small as 100 µg for structure elucidation, or larger quantities for subsequent synthetic steps or biological studies. The size of the preparative column chosen depends on the amount of material to be purified. Although preparative separations can be performed on analytical scale columns (4.6 mm i.d.), this scale is rarely used due to lack of fraction collection capabilities for analytical SFC systems. This is one major difference from HPLC where collection is possible by simply diverting flow directly from detector output to a collection vessel. The presence of CO_2 in the mobile phase is a complicating factor for fraction collection in SFC. This topic is discussed in detail in Chapter 7 on preparative SFC equipment.

The majority of preparative SFC purifications are performed using 2 or 3 cm i.d. columns [12, 15, 17, 19, 24–26]. The largest prepacked columns for preparative SFC have internal diameters of 5 cm. Larger diameter preparative SFC columns can be packed using bulk packings and dynamic axial compression (DAC) technology [27–29]. The amount of material injected onto a preparative column should be in balance with desired purity and yield of the purification product(s). With smaller amounts of material injected, it is possible to recover product at high purity and yield. Increasing injection quantities results in a balancing game between purity and yield. This is the classical purification triangle with time, purity, and yield at the three apexes. It is only possible to maximize two of the parameters at the expense of the third. The needs of each purification project dictate the optimization process.

It is also possible to upgrade purity through crystallization of chromatographic products. Chromatographic products with low purity can be crystallized to achieve desired purities, leading to increased overall productivity [30–33]. This approach is contingent on the development of a crystallization process that is straightforward and results in minimal loss of product. Coupling of chromatographic purification with crystallization is used mainly for chiral separations where larger quantities (kg to metric ton) of material are to be purified. For smaller quantities, it is often more time efficient to achieve desired purity using only chromatographic purification.

How much material can be injected onto a preparative column and still achieve desired resolution? The answer to this question depends on a number of factors and cannot be determined without experimental data. Maximum possible loading depends on a number of parameters including (i) chiral or achiral separation, (ii) selectivity and resolution of analytical method, (iii) compound solubility, (iv) adsorption isotherms, and (v) particle size. Assuming the same column size, one can typically load more material on an achiral column than on a chiral column. In general, a higher selectivity/resolution allows more material to be injected before peak overlap is observed. Achieving maximum selectivity through extensive stationary phase screening is critical to developing high productivity purifications. All things being equal, a method offering better solubility and larger saturation capacities allows increased amounts of material per injection. Finally, smaller particle size results in sharper peaks and larger injection quantities relative to larger particle size. A typical approach is to use a small injection volume (50–100 µL) to confirm proper system operability and then increase the injection size until touching bands are obtained as shown in Figure 8.2.

While the majority of SFC purifications use ultraviolet/visible absorbance (UV) detection, other detectors such as evaporative light scattering detection (ELSD), refraction index detection (RI), and corona charged aerosol detection (Corona CAD) have been used. Destructive detectors such as mass

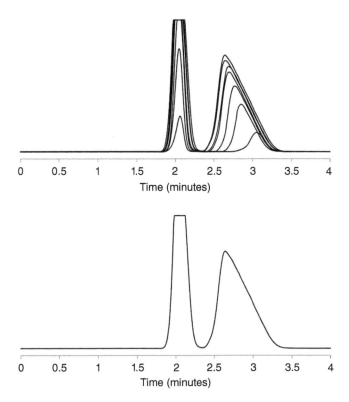

Figure 8.2 Touching band example. Top chromatogram: UV trace for 1, 3, 5, 7, 8, 9, and 10 mL of 35 mg/mL solution. Bottom chromatogram: UV trace for 10 mL of 35 mg/mL solution.

spectrometry (MS) can be used if splitters are used to send only a portion of the flow to the detector. In analytical chromatography using UV/visible detection, the wavelength is chosen to maximize detector signal. This is done to minimize detection limits and ensure all components in a mixture are detected. In preparative chromatography if a wavelength offering maximum response is used, then the detector can quickly become saturated. This impact can be offset somewhat by using smaller flow cells designed for preparative separations (see Chapter 7 for more details). Even with a preparative flow cell, detector overload can easily occur and peak separation cannot easily be seen. It is standard practice in preparative chromatography to detune the wavelength away from maximum absorbance to reduce UV signal intensity. Use of a photodiode array detector or a multi-channel UV detector in analytical method development can allow the user to determine an appropriate wavelength for preparative work.

Even when using a wavelength offering lower sensitivity or a preparative flow cell to reduce detector response, peaks in preparative chromatography can differ greatly from those seen in analytical chromatography. Peak shape in preparative chromatography is determined by stationary phase particle size, injection amount, injection solvent, adsorption isotherms, and the level and number of additional components in the sample. Poor peak shape in preparative chromatography is not an indication of poor chromatography. This is illustrated in the preparative traces for CSP1 in Figure 8.1. Evaluation of the UV trace would seem to indicate large overlap between the two enantiomers at higher load. In reality, fractionation across the preparative peaks would show that peak 1 can be isolated at high purity if it is collected to the point where peak 2 begins to elute (valley between the two peaks). Also, peak 1 does not tail as much into peak 2 as expected due to a phenomenon called sample self-displacement [34]. In sample self-displacement the second eluting peak compresses the elution band of the first eluting peak, reducing tailing and allowing peak 2 to be isolated at high purity with minimal yield loss.

8.3.1 Linear Scale-Up Calculations

For a consistent scale-up it is important to maintain identical mobile phase linear velocities between the analytical and preparative columns. Flow rate is set using the following equation:

$$Flowrate_\text{p} = flowrate_\text{a} \times \frac{\left(diameter_\text{p}\right)^2}{\left(diameter_\text{a}\right)^2}$$

Assuming identical column lengths with identical stationary phases, using the above equation to set a preparative flow rate results in identical retention times between the two column diameters. If column lengths are not identical, retention will vary depending on the ratio of the lengths. For example, if the analytical column is 10 cm in length and the preparative column is 25 cm in length, with constant linear velocity and assuming an isocratic mobile phase, retention will be $2.5 \times (25 \, \text{cm}/10 \, \text{cm})$ longer on the preparative column relative to the analytical column. The above equation is used for scaling between any columns of different sizes, not just analytical and preparative columns.

Depending on the amount of material to be purified it may be necessary to perform initial loading studies on a small i.d. column and then scale-up the optimized separation to a larger column. This technique is routinely used in preparative HPLC and has been shown to be accurate up to a one million fold scale-up [35]. The same approach is useful for preparative SFC scale-up. If particle size and column length are the same in both columns, and the same linear velocity is used for the separation, identical results should be obtained between the two columns. Load is scaled as a function of column size using the following equation:

$$Load_{col2} = Load_{col1} \times \frac{D_{col2}^2 \times L_{col2}}{D_{col1}^2 \times L_{col1}}$$

where D is column diameter and L is column length.

8.3.2 Scaling Rule in Supercritical Fluid Chromatography

If column characteristics (particle size, dimension) are identical and mobile phase linear velocities are kept constant between an analytical and preparative column, it is possible to directly scale from analytical to preparative. Scale-up becomes more complicated when these parameters are modified. In preparative LC, scaling rules have been developed that allow transfer of methods from small particle analytical columns to larger particle preparative columns [36, 37]. The main difference between LC and SFC is the much higher mobile phase compressibility of SFC. For a compressible fluid, changes in pressure lead to changes in density. In SFC, mobile-phase density impacts retention and sometimes selectivity. Changing particle size as well as column length changes average column pressure, which in turn changes mobile-phase density and potentially compound retention and selectivity. For this reason, the standard scaling rules developed for preparative LC may not be suitable for preparative SFC.

Tarafder et al. have studied this phenomenon and developed scale-up rules relevant to isocratic SFC purifications [38, 39]. They have shown for accurate reproduction of performance during the scaling process, it is necessary to maintain the same density variation profile in both the analytical and preparative system. The effect of not matching average densities across two columns is shown in Figure 8.3. The only change between the two chromatograms is particle size (1.7 µm for top chromatogram and 5 µm for bottom chromatogram).

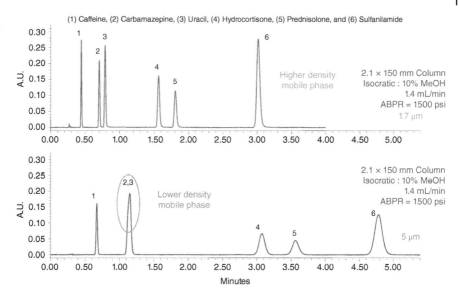

(1) Caffeine, (2) Carbamazepine, (3) Uracil, (4) Hydrocortisone, (5) Prednisolone, and (6) Sulfanilamide

Figure 8.3 Impact of not matching average densities. Increasing the particle size while keeping all other experimental conditions the same, significantly affecting retention factors, selectivity, and column efficiency in SFC. *Source:* Reprinted with permission from Tarafder et al. [38]. Copyright 2014 Elsevier.

The change in particle size results in the columns experiencing different pressure drops. For the 1.7-µm column the pressure drop is 5888 psi, for the 5-µm column the pressure drop is 2004 psi. As SFC mobile phases are compressible, and mobile-phase density varies with pressure, the 1.7-µm column has a mobile phase of higher density relative to the 5-µm column. The lower mobile phase density for the 5-µm column results in increased retention as well as changes in selectivity (compounds 2 and 3). Under LC conditions retention and selectivity would not have changed. The only change would have been reduced efficiency due to larger particle size.

To achieve comparable performance between SFC columns with different operating pressure, and thus different mobile phase densities, the average density between the two columns must be matched. For this calculation, densities are approximated using system pressure readouts and assuming a linear pressure drop across the column. Densities of the CO_2/methanol mixture are calculated using *REFPROP* software from NIST. As mentioned earlier, different system pressures result in different mobile phase densities. Maintaining average densities across columns with different pressure drops is achieved by increasing the automatic back pressure regulator (ABPR) setting for the preparative SFC system. This is illustrated in Figure 8.4 for the separation on columns with varying particle sizes and column dimensions. Matching mobile

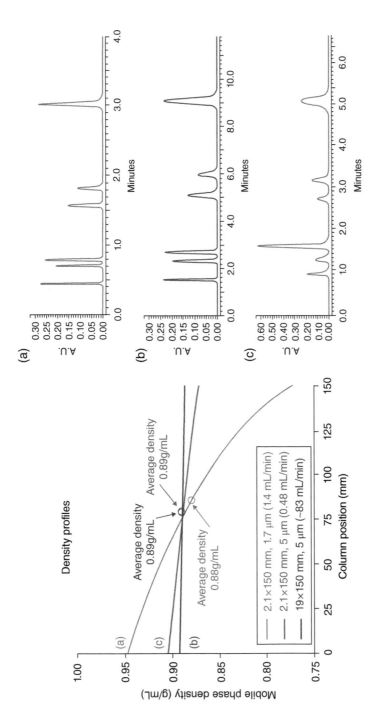

Figure 8.4 Use of matching average densities. The left side of the figure shows estimated density profiles across three different columns with three different flow rates. The ABPR pressures were adjusted such that the average densities of all operations were similar. The right side of the figure shows the chromatograms corresponding to these experiments. (ABPR settings: (a) 1500 psi, (b) 3600 psi, and (c) 3191 psi.) *Source: Reprinted with permission from Tarafder et al. [38]. Copyright 2014 Elsevier.*

phase densities and comparable chromatographic results are obtained by increasing the ABPR setting to 3600 psi for the 5-µm analytical column and to 3191 psi for the preparative column.

8.3.3 Metrics for Preparative Separations

A number of metrics have been proposed to quantify a preparative separation [40]. These include (i) throughput, (ii) production rate, (iii) productivity, (iv) specific productivity, and (v) cost. The most useful metric is productivity, also known as production rate. Production rate is a measure of the amount of material that can be purified on a given quantity of packing material in a given time. It is often measured as kkd; kg product/kg packing material/day. At times productivity will be calculated as kkd for feedstock, not product. Using kg product for productivity calculation is an improved metric as it takes into account the purification yield. An example of production rate calculation is shown in Figure 8.5. In this example, 1.1 g of racemate can be injected with a cycle time of 4.5 minutes on a column packed with 295 g CSP (assuming packing density of 0.6 g/mL). This yields a production rate of 1.19 kkd for the racemate.

Another useful metric for preparative separations is solvent usage. Solvent usage is easily calculated by dividing solvent used per hour by gram of feed processed or product produced per hour. For the separation shown in Figure 8.5, solvent usage is 0.43 L/g racemate. The overall most important metric for purification is the cost per kilogram of final product. This cost includes all components of the separation including equipment, operators, facility, and

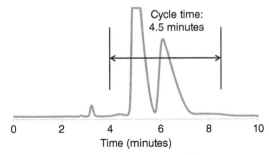

Column: 5 cm i.d. × 25 cm, containing ~295 g CSP
350 ml/min, 30% 1:1 methanol:dichloromethane
1100 mg/injection

Injection Details: 1.1 g/4.5 minutes = 14.66 g/hour = 352 g/day
Solvent Details: 350 ml/min = 21 L/hour; 30% modifier = 6.3 L solvent/hour

Productivity (kkd racemate) = 0.352 kg product /0.295 kg CSP = 1.19 kkd
Solvent Usage (L/g racemate) = 6.3 L/14.66 g = 0.43 L/g racemate

Figure 8.5 Productivity calculation example.

operating costs. This metric is difficult to calculate for smaller scale separations and is only used for large scale manufacturing purification processes.

8.3.4 Options for Increasing Purification Productivity

8.3.4.1 Closed-Loop Recycling

There are currently dozens of stationary phases that can be utilized for SFC purifications. Even with all these possibilities there are still times when the best analytical method does not provide resolution or selectivity sufficient to meet purification demands. Closed-loop recycling and peak shaving are techniques shown to improve throughput for preparative liquid chromatography [41]. These techniques are almost exclusively used for chiral purifications, but in theory can also be used for achiral purifications. Closed-loop recycling is not routinely used in preparative SFC due to difficulty in maintaining pressures, and thus densities and volumes, within all components of a preparative SFC instrument. Closed-loop recycling is a technique where the eluent exiting the chromatographic column is pumped back onto the column to increase separation. This recycling is repeated until sufficient resolution is obtained. Peak shaving is a technique where only the leading and tailing edges of an elution band (where pure first eluting enantiomer and pure second eluting enantiomer elute) are collected and the impure portion of the elution band is recycled. With the difficulty of recycling within a preparative SFC system, a related approach is to collect the impure band of the elution profile and then reprocess to improve isolated yields.

8.3.4.2 Stacked Injections

An additional technique for increasing purification throughput is stacked injections. This technique is known by many names, including overlap injections and boxcar injections. With this technique, a second injection is made prior to the elution of all peaks from the first injection. An example of this technique is shown in Figure 8.6. Chromatogram A shows a single preparative injection of racemate with a total run time of approximately 110 seconds. Examination of the chromatogram shows no product eluting during the first 50 seconds of the separation with the elution time for the enantiomers being only 60 seconds. To fully optimize this separation, the stacked injection technique was used and an injection performed every 60 seconds. The separation of five stacked injections is shown in chromatogram B. Using this technique the time required for separation was reduced by 45% with a corresponding reduction in solvent requirement of 45%.

8.3.5 Importance of Solubility on Preparative Separations

As mentioned earlier, to achieve maximum purification productivity the sample should have high solubility (>25 mg/mL) in the chromatographic mobile

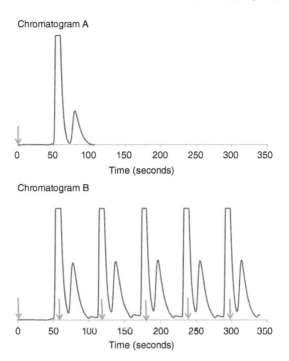

Figure 8.6 Stacked injection example. Chromatogram A: Single injection at time zero. Chromatogram B: Stacked injection, injection made every 60 seconds. Injection indicated by arrow.

phase. Measurement of solubility in CO_2 based solvents is complex and rarely performed [42, 43]. In a standard preparative SFC experiment, the compound being purified does not contact CO_2 until the injection step. Without CO_2 solubility measurements it is difficult to predict what will happen when the compound first interacts with CO_2 during the injection step. In a worst case scenario, the compound will have such low CO_2 solubility that it will precipitate out of solution, depositing on the inlet frit or in the column. This can lead to a system pressure spike or increase, which will eventually lead to system shutdown. While this does occur, it is rare.

Empirically compounds tend to exhibit lower solubilities in supercritical fluid media than in neat organic solvents [44]. In a best case scenario, solubility in the SFC mobile phase modifier would drive compound solubility, with CO_2 having little impact. This is not always the case. Figure 8.7 shows the solubility of caffeine and theophylline in various CO_2 and organic solvent combinations. Graph A for caffeine shows solubility in acetonitrile decreased as CO_2 was added. For methanol, ethanol, and isopropanol, solubility increases with addition of CO_2 (to 50%) and then decreases as the CO_2 fraction increases. The impact of CO_2

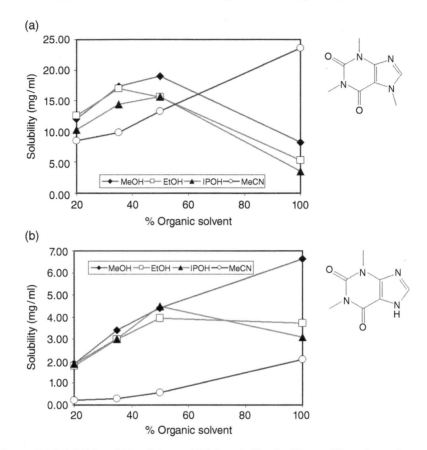

Figure 8.7 Solubilities of (a) caffeine and (b) theophylline in different CO_2 and organic solvent combinations. *Source:* Reprinted with permission from Gahm et al. [44]. Copyright 2011 Wiley.

on the solubility of theophylline (graph B) is different from that observed with caffeine with acetonitrile and isopropanol showing a decrease in solubility with CO_2 addition. This data shows the difficulty in predicting solubility in CO_2/organic solvents based on organic solvent solubility alone.

Figure 8.8 is an example in which poor CO_2 solubility had a detrimental impact on a preparative resolution. Analytical separation of this racemate using 15% ethanol as modifier afforded baseline separation of the enantiomers. When this method was scaled directly to preparative SFC, poor peak shape was observed even with only 1 mg injected onto the preparative column. In this separation, retention times decreased as the percentage of ethanol in the mobile phase increased, which is expected. However, this separation is unusual because even through the retention times decreased, resolution increased.

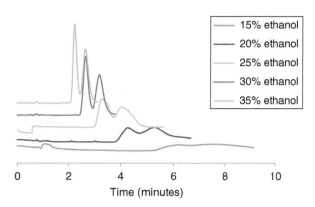

Figure 8.8 Separation of 1 mg racemate (proprietary structure). SFC conditions: Chiralpak AY, 5 μm, 2×25 cm, 80 mL/min, 100 bar, 290 nm. See the figure for mobile-phase conditions.

This is not normally seen in preparative chromatography. With 15% ethanol the majority of the mobile phase is CO_2. With the racemate having poor CO_2 solubility it cannot remain in solution and broad peaks are observed. As the ethanol percentage in the mobile phase increases, solubility of the racemate increases and peak shapes improves, while at 25% ethanol near baseline separation of the two enantiomers is observed.

While solubility can impact purification productivity, it is often not an issue, especially for small scale purifications. Some suggestions to minimize solubility impact include the following:

- If poor organic solvent solubility is observed, it will probably not improve in the presence of CO_2.
- If multiple mobile phases show approximately the same resolution, choose the one with highest modifier percentage.
- If poor solubility is observed in standard SFC modifiers, explore alternate modifiers such as dichloromethane, tetrahydrofuran, methyl tert-butyl ether, or ethyl acetate which may offer improved solubility.
- For large scale or time consuming separations it may be worthwhile to measure CO_2 solubility.

8.3.6 Preparative SFC Injection Options

Standard LC equipment introduces sample into the chromatographic system after mixing of mobile phase and just before the column. SFC equipment has two configurations: (i) mixed stream injection or (ii) modifier stream injection. Schematics of the two configurations as well as advantages and disadvantages of the two injection options are discussed further in Chapter 7. Miller et al. published a comprehensive evaluation of mixed stream and modifier stream injection [45]. They showed that for the majority of the compounds evaluated, modifier stream injection gave better resolution relative to mixed stream injection. The improvement in resolution for modifier stream injection increased as injection volume increased. The benefit of modifier stream injection over mixed stream injection is shown in Figure 8.9. Both chromatograms show the chiral resolution of 100 mg propranolol dissolved in 5 mL methanol. The resolution using mixed stream injection (chromatogram A) is drastically lower than modifier stream injection (chromatogram B). In mixed stream injection, there is a temporary localized increase in mobile phase strength due to the methanol used for dissolution, which can lead to peak distortion and reduced resolution. This difference is more pronounced as injection volume increases and as retention time decreases.

Ideally for chromatographic analysis and purification, the sample should be dissolved in the identical solvent mixture as the mobile phase. Due to the presence of CO_2 in the mobile phase, this option is not possible in preparative SFC. In preparative HPLC, dissolution in the mobile phase is not always possible due to limited solubility of the sample. In this case, it is standard practice in both preparative SFC and HPLC to use dissolution solvents different from the chromatographic mobile phase. This practice allows purification of compounds with poor solubility in the mobile phase, or reduces injection volume

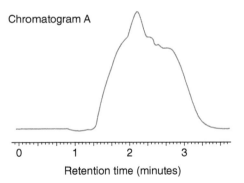

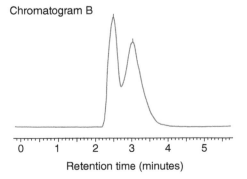

Figure 8.9 Preparative separation of 100 mg of propranolol dissolved in 5 mL methanol. Chromatogram A: Mixed stream injection. Chromatogram B: Modifier stream injection. *Source:* Reprinted with permission from reference [44]. Copyright 2012 Elsevier.

of compounds with minimal solubility. In HPLC, injection from stronger solvents than the mobile phase can lead to poor peak shape, sample breakthrough, and reduced loadings [46]. Miller et al. evaluated injection solvents to see if the same effect was observed in preparative SFC [45]. Their studies (summarized in Table 8.1) showed that dissolution solvent had minimal impact on plate count, retention time, or resolution regardless of the dissolution solvent strength for SFC purification. This is relative to HPLC purifications where the

Table 8.1 Effect of dissolution solvent on preparative SFC separation of trans stilbene oxide[a]

Dissolution solvent	Rt_1	Rt_2	N_1	N_2	Rs
Methanol	2.55	3.96	620	358	1.88
Ethanol	2.56	3.99	522	334	1.87
Acetonitrile	2.55	3.94	750	418	1.97
Isopropanol	2.59	4.08	507	300	1.86
75/25 isopropanol/heptane	2.60	4.06	548	329	1.88
50/50 isopropanol/heptane	2.60	4.07	642	355	2.02
25/75 isopropanol/heptane	2.59	4.05	691	388	1.96
10/90 isopropanol/heptane	2.59	4.02	722	399	2.05
heptane	2.57	4.00	762	399	2.03
50/50 methanol/dimethylether	2.59	4.01	421	436	1.85
75/25 methanol/dichloromethane	2.62	4.06	690	420	1.96

[a] Chiralpak AD-H, 3 × 25 cm, 126 mL/min, 25% methanol w/0.2% diethylamine, 80 mg injection (4 mL @ 20 mg/mL), mixed stream injection.
Used with permission of [45].

Effect of dissolution solvent on preparative HPLC separation of trans stilbene oxide[b]

Dissolution solvent	Rt_1	Rt_2	N_1	N_2	Rs
Heptane	3.22	8.27	3229	718	6.83
10/90 isopropanol/heptane	3.22	8.22	2740	826	6.97
25/75 isopropanol/heptane	3.19	8.13	2691	885	7.37
50/50 isopropanol/heptane	3.17	8.18	2192	834	7.32
75/25 isopropanol/heptane	3.10	8.11	1980	840	7.16
Isopropanol	3.06	8.16	391	718	6.19
Ethanol	1.19	8.37	1600	613	6.19

[b] Chiralpak AD-H, 3 × 15 cm, 42 mL/min, 10/90 (v/v) isopropanol/heptane, 50 mg injection (2.5 mL @ 20 mg/mL).

impact of stronger dissolution solvents resulted in broader peaks and reduced plate counts, especially for the first eluting peak.

An alternate sample injection approach that does not rely on compound dissolution prior to introduction to the chromatographic system is coupling supercritical fluid extraction (SFE) to a preparative SFC. This approach allows injection of material with poor organic solvent solubility or injection of materials containing insoluble impurities. This approach has been used for chiral purifications [44] as well as achiral flash purifications [47]. The technique is a feature on preparative SFC equipment sold by PIC Solution Inc. [48]. "Extraction" injection involves placing the sample to be purified (often as a mixture with silica or other inorganic material) in an extractor that is pressurized with the SFC mobile phase. Periodically the extractor is opened to the column, allowing sample to elute from the extractor onto the preparative column for purification. This approach avoids the potential of sample coming out of solution upon contact with CO_2. The only material injected into the preparative system is that which dissolves in the mobile phase. One disadvantage to this technique is the inability to easily measure how much material is being injected onto the column, and thus determine when the purification will be complete. It may be necessary to continue to inject material from the extractor until no additional material is eluting from the column.

8.4 Preparative Chiral SFC Purifications

8.4.1 Chiral Stationary Phases (CSPs) for Preparative SFC

Chromatographic resolution of enantiomers is critical to many areas of pharmaceutical discovery and development. Analytical-scale separations of chiral compounds via HPLC have been used in the pharmaceutical industry since the first separations of enantiomers were reported in 1985 [49]. SFC has become the predominant technique for preparative chromatographic purification of enantiomers over the past 25 years [10, 12, 13, 50, 51]. It has been reported that chiral separations are the niche application for packed column SFC that is responsible for keeping the technique visible from the mid-1990s to the late 2000s. Within many pharmaceutical companies SFC is used for greater than 90% of all preparative chromatographic separations, especially during discovery when the quantities to be resolved are small (<1 g) and the number of racemates to be resolved are large.

Chromatographic purification of racemates involves four steps. They are (i) selection of a chiral stationary phase (CSP) and mobile phase, (ii) choice of separation technique, (iii) preparative separation, and (iv) recovery of product from the mobile phase. By far the most important step is selection of the chiral phase and mobile phase. The selection of the separation conditions has a direct impact on productivity, the amount of solvent required, and ultimately the cost for the separation. For a highly productive separation one desires a retention factor of less than five for the second eluting enantiomer as well as maximum selectivity. The low retention factor reduces cycle time for the separation, allowing maximum productivity. The maximum selectivity allows larger quantities of racemate to be loaded onto the column.

There are currently hundreds of CSP available on the market. The main types of CSPs are polysaccharides [52–54], Pirkle type [55], protein based [56], cyclodextrin [57, 58], and macrocyclic glycopeptides [59]. Each of these phases has different characteristics. Additional details on each phase can be found in Chapter 5. While all of these phases are useful for analytical separations, not all are suitable for preparative separations. For a preparative separation a CSP must be stable, available in larger column sizes or as a bulk packing, and have high loading capacity. The characteristics of the types of CSPs are summarized in Table 8.2. Loading capacity is the most critical requirement for a highly productive purification. A CSP that has a high loading capacity generates Gaussian peaks at higher amount of injected racemate relative to a CSP with low loading capacity. This allows increased amounts of pure racemate to be produced per unit time. While productivity is dependent on the racemate to be resolved, as well as loading capacity, the general trend is polysaccharide > Pirkle type > macrocyclic glycopeptides > cyclodextrin > protein based. The majority of SFC preparative separations are performed using polysaccharide based or Pirkle type phases. Structures of these CSPs can be found in Chapter 5. The main limitation of polysaccharide based CSPs is their limited solvent compatibility. Most polysaccharide based chiral selectors are adsorbed on silica and can be dissolved and washed off the silica if the wrong solvent is used. This limitation has been reduced with the introduction of immobilized polysaccharide based CSPs, which will be discussed later in this chapter.

Table 8.2 Characteristics of chiral stationary phases (CSP).

	Polysaccharides	Immobilized Polysaccharides	Pirkle type	Protein based	Cyclodextrin	Macrocyclic Glycopeptides
Solvent limitations	Severe	None	None	Reverse phase only	None	None
Loadability	High	Medium	Medium	Very Low	Low	Medium
Range of resolution	High	High	Low	Medium	Low	Medium
Large column sizes/bulk availability	Yes	Yes	Yes	No	2 cm i.d. and less	2 cm i.d. and less

8.4.2   Method Development for Chiral Purifications

During pharmaceutical discovery, thousands of different compounds are synthesized and tested in order to progress the best molecule into clinical studies. Depending on the chemical space being evaluated, this can result in the need to resolve several hundred to several thousand different racemates for a typical discovery project. The nature of drug discovery is that most molecules are synthesized only once, usually at 25–100 mg scale. Optimization of the chiral purification process for drug discovery requires rapid method development and purification; SFC excels in both areas.

The high flow rates possible in SFC result in reduced time for screening of chiral phases and modifiers. Flow rates of 4–5 mL/min are standard for 4.6 mm i.d. columns. This compares to 1–2 mL/min for HPLC screening with columns of the same dimensions and particle size. Using SFC, a typical screening method is often less than five minutes [12, 13, 60–77]. Most analytical SFC equipment offers automated column and modifier switching valves to automate the method development process. Further details on analytical SFC equipment can be found in Chapter 3. An example of an SFC method development approach is summarized in Table 8.3. Studies in Amgen laboratories have shown little difference between ethanol and methanol as a modifier for SFC enantioseparations. Ethanol is only evaluated if acceptable separation is not obtained with methanol or isopropanol. Due to the wide polarity range of

Table 8.3 SFC method development conditions.

	SFC conditions
Column dimensions (mm)	4.6 × 100
Chiral stationary phases	Chiralpak AD-H
	Chiralpak AS-H
	Chiralcel OD-H
	Chiralcel OJ-H
	Chiralpak IC
Flow rate (mL/min)	5.0
Modifiers	Methanol w/ 0.2% diethylamine
	Isopropanol w/ 0.2% diethylamine
Gradient conditions	5–55% over 3.5 minutes
	55% for 1 minute
Temperature	40 C
Run Time	4.5 minutes

Source: Used with permission of [12].

molecules being analyzed, initial evaluation is performed under gradient conditions. Once the best CSP/modifier combination has been identified an isocratic method is quickly developed prior to preparative separation. A guide (Figure 8.10) has been developed to allow selection of appropriate isocratic conditions based on gradient retention time. The chart is used in the following manner; the average retention time for two enantiomers separated under gradient conditions is calculated. Assume for this example the average retention is 2.5 minutes. Find 2.5 minutes on the x-axis (R_t gradient) and move up the y-axis to determine approximate retention time under various isocratic conditions. A gradient retention time of 2.5 minutes would translate to approximately 0.5, 1.1, 1.5, or 2.3 minutes for 30, 25, 20, and 15% modifier under isocratic conditions. We have found an analytical retention time of approximately 1.5 minutes, using 4.6 mm × 10 cm analytical columns, gives adequate separation without excessive retention when scaled to preparative columns of 15 or 25 cm in length. For this example, isocratic conditions of 20% modifier would be evaluated at analytical scale prior to the preparative separation.

8.4.3 Preparative SFC Examples

8.4.3.1 Milligram Scale Chiral Purification

As discussed previously, SFC is an excellent technique due to the high speed of analysis and purification. This was demonstrated well for the separation of the enantiomers of compound 1 (Figure 8.11). One hundred and twenty milligrams of this racemate was synthesized and required preparative enantioseparation prior to biological testing. Using a method development process described in Section 8.4.2, an analytical separation was quickly developed (Figure 8.11, chromatogram A). This method was scaled to preparative loading using a 2-cm-i.d. column and a loading of 20 mg (Figure 8.11, chromatogram B). It is evident from chromatogram B that loading was not optimized for this

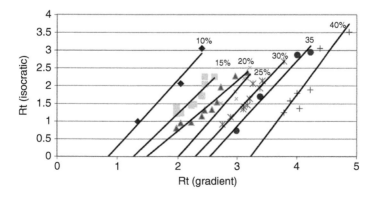

Figure 8.10 Guide for choosing isocratic conditions based on SFC gradient retention.

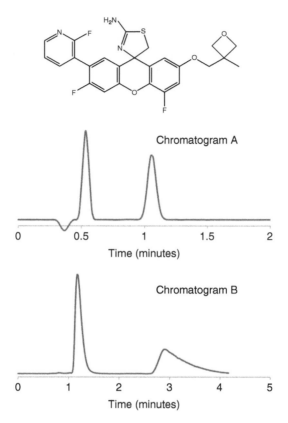

Figure 8.11 SFC separation of compound 1. Chromatogram A: Analytical separation: Chiralpak AD, 5 µm, 4.6 × 100 mm, 40% isopropanol with 0.2% diethylamine, 5 mL/min, 100 bar back pressure. 10 µL of 1 mg/mL methanol solution injected. Chromatogram B: Preparative separation: Chiralpak AD, 5 µm, 2 × 15 cm, 40% isopropanol with 0.2% diethylamine, 80 mL/min, 100 bar back pressure and detection at 270 nm. One-hundred and twenty milligram racemate dissolved in 18 mL of 50/50 (v/v) methanol/dichloromethane, 3 mL was injected.

preparative separation as more sample could be injected without compromising the separation. This is often the case during medicinal chemistry purification support where sample size limits the loading experiments that can be performed. Even with nonoptimized loading, the entire sample was purified in under 30 minutes.

8.4.3.2 Gram Scale Chiral Purification
The example above illustrated the advantages of SFC for preparative resolution of racemates during early stages of drug discovery when quantities of less than 200 mg are often synthesized. SFC is also often the technique of choice when

larger quantities of racemate need resolution. This is illustrated for the resolution of compound 2 in Figure 8.12. Larger quantities of this racemate (24.5 g) were resolved as this compound was an early intermediate to a large series of final compounds being prepared for testing. Resolution of this amount of racemate avoided the need to develop analytical and purification methods for approximately 50 final compounds. This approach greatly reduces the amount of analysis and purification required, but it is imperative that no racemization occurs during subsequent synthetic steps.

The best analytical SFC method is shown in Figure 8.12, chromatogram A. Due to the larger sample size the preparative method was scaled to a 5-cm-i.d column. This reduced the purification time approximately sixfold vs. scaling to a 2-cm-i.d. column. The preparative separation of ~260 mg racemate is shown in Figure 8.12, chromatogram B. The sharp peaks allowed a cycle time of only 1.5 minutes to be used for processing the racemate. Using this approach the entire 24.5 g of racemate was resolved in two hours.

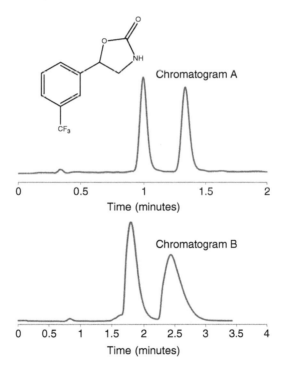

Figure 8.12 SFC separation of compound 2. Chromatogram A: Analytical separation: (S,S) Whelk-O, 5 μm, 4.6 × 100 mm, 15% isopropanol, 5 mL/min, 100 bar back pressure. Ten microliter of 1 mg/mL methanol solution injected. Chromatogram B: Preparative separation: (S,S) Whelk-O, 5 μm, 5 × 15 cm, 15% isopropanol, 350 mL/min, 100 bar back pressure and detection at 215 nm. 24.5 g racemate dissolved in 120 mL of 50/50 (v/v) isopropanol/dichloromethane, 1.5 mL was injected.

8.4.4 Impact of Solubility on Productivity

Preparative separations require the introduction of larger amounts of chemical onto the separation column. To avoid purification issues, it is important to not only inject the sample onto the purification column, but also have the material remain in solution as it travels through the column. If material comes out of solution within the chromatography system, tailing peaks, and/or increased system pressure can occur. The presence of CO_2 in the mobile phase negates the ability to dissolve sample in the mobile phase. The common approach for preparative SFC is to dissolve the racemate only in the modifier. Unfortunately, many pharmaceutical compounds do not exhibit high solubility in methanol. An approach used by some pharmaceutical purification laboratories is to dissolve racemates in a 1:1 mixture of methanol and dichloromethane (DCM). This approach often results in acceptable peak shapes and allows rapid resolution. While the use of DCM/methanol mixtures allows introduction of racemate onto the column (provided solubility in methanol or methanol/CO_2 mixtures is low), poor peaks shape including peak splitting and broadening can still result. On rare occasion, sample precipitation can occur in the column once dichloromethane is pumped away from the racemate, leading to high pressure and system shutdowns. An example of peak splitting that can occur due to poor solubility under preparative conditions is shown in Figure 8.13 for Compound 3. For this racemate, a maximum of 15 mg could be injected; higher amounts resulted in peak overlap and reduced purities and/or yield. Even at 15 mg injected, peak widths of 1–1.5 minutes are observed, greatly limiting the productivity of the separation.

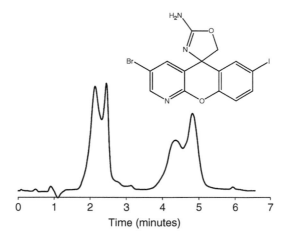

Figure 8.13 SFC separation of compound 3. Preparative separation: Chiralpak AD-H, 2×25 cm, 35% methanol w/0.2% diethylamine, 80 mL/min, 15 mg racemate injected, dissolved in 1 mL of 10/10/1 (v/v/v) methanol/dichloromethane/dimethylsulfoxide. *Source:* Reprinted with permission from Miller [78]. Copyright 2012 Elsevier.

8.4.5 Use of Immobilized Chiral Stationary Phase (CCP) for Solubility-Challenged Samples

Polysaccharide-based CSPs have been used for almost 30 years for the preparative resolution of racemates [12, 79–90]. While these phases have proven extremely useful, they have severe solvent limitations due to the derivatized cellulose or amylose being coated on rather than bonded to the silica gel. The derivatized polysaccharide is soluble in many organic solvents. Contact with these solvents can result in dissolution of the polysaccharide and loss of resolution and/or column destruction. Numerous approaches have been developed to make the polysaccharide insoluble, thus removing the solvent restrictions [91–97]. During the past eight years a series of immobilized cellulose/amylose CSP (Chiralpak IA, IB, IC, ID, IE, IF, and IG, manufactured by Daicel) have been introduced. These CSP have no solvent restrictions, and have proven useful in offering unique selectivity when solvents other than traditional solvents used for chiral separations (i.e. other than alkanes, alcohols, and acetonitrile) are required for resolution [98–100]. These CSPs have also proven useful for preparative resolutions where the racemate to be separated is poorly soluble in traditional SFC modifiers such as alcohols or acetonitrile [78, 101].

8.4.5.1 Immobilized CSP Example #1

The impact of poor racemate solubility on a preparative separation is illustrated in Figure 8.13. While poor peak shape was observed for this racemate, due to the small amount of racemate to be resolved (~6 g), it was possible to process using these conditions. However, a large number of injections (400) and ~30 hours of separation time were required. What is the solution to this type of separation problem when the quantity of racemate to be resolved reaches a level where a brute force approach is not acceptable? In these cases, the use of nontraditional modifiers and immobilized CSPs may be the answer.

This approach was used when larger quantities of Compound 3 required resolution. Solubility studies showed improved solubility in dichloromethane. Thus, a modifier mixture of methanol/dichloromethane was explored. The CSP evaluated was Chiralpak IA, which is the immobilized equivalent of Chiralpak AD. The analytical separation is shown in Figure 8.14, chromatogram A. Separation with the Chiralpak IA CSP afforded lower selectivity compared to the Chiralpak AD CSP (1.59 vs. 2.33), but it was felt the improved solubility would allow higher preparative loadings even with reduced selectivity. The separation of 160 mg racemate on a 5-cm-i.d. × 25 cm Chiralpak IA column is shown in Figure 8.14, chromatogram B. A late eluting achiral impurity (~9 minutes) required a longer cycle time for removal. Using stacked injections 16 g of racemate were processed in under six hours and generated both enantiomers at >99% ee and >90% yield.

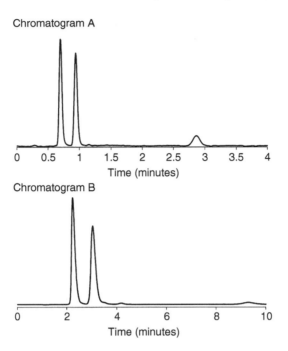

Figure 8.14 SFC separation of Compound 3. Chromatogram A: Analytical separation: Chiralpak IA, 4.6×100 mm, 30% 1:1 methanol:dichloromethane w/ 0.2% diethylamine, 5 mL/ min, 100 bar back pressure. Chromatogram B: Preparative separation: Chiralpak IA, 5×15 cm, 30% 1:1 methanol:dichloromethane w/0.2% diethylamine, 350 mL/min, 160 mg racemate injected, dissolved in 4 mL of 10/1 (v/v) dichloromethane/dimethylsulfoxide. *Source:* Reprinted with permission from Miller [78]. Copyright 2012 Elsevier.

The two purification methods are compared in Table 8.4. Switching from methanol to a methanol : dichloromethane modifier resulted in a greater than threefold increase in productivity and a nearly eightfold decrease in solvent usage. If the achiral impurity in the 16-g lot is ignored, or removed prior to the chiral separation, a more accurate comparison between the methods can be made. This allows the cycle time to be reduced from 3 to 1.75 minutes. A greater than fivefold increase in productivity (0.078 to 0.46 kkd) and a nearly 14-fold decrease in solvent usage (7.4 to 0.54 L/g racemate) as obtained under these conditions. The use of a nontraditional methanol: dichloromethane modifier allows 16 g of racemate to be resolved in 3 hours compared to greater than 17 hours with a methanol modifier. Solvent consumption was reduced from 118 to 15 L.

8.4.5.2 Immobilized CSP Example #2

In another example, 762 g of racemate (Compound 4, proprietary structure) enriched in the active enantiomer (ee ~20%) required resolution to meet purity specifications of >98% ee. Following analytical method development, conditions which afforded high selectivity ($\alpha = 1.96$) were achieved (Figure 8.15,

Table 8.4 Comparison of purification methods for compound 2.

CSP	Mobile phase	Productivity (kkd)	Solvent usage (L/g racemate)
Chiralpak AD	35% MeOH w/ 0.2% DEA in CO2	0.078	7.4
Chiralpak IA	30% 50:50 (v/v) DCM/ MeOH in CO_2	0.268	0.94
[a]Chiralpak IA	30% 50:50 (v/v) DCM/ MeOH in CO_2	0.46	0.54

MeOH = methanol
DCM = dichloromethane
kkd = kilograms racemate/kg CSP/day
[a] Productivities and solvent usage obtained if achiral impurity not present in racemate and cycle time was shortened from 3 to 1.75 minutes.

chromatogram A). Using a 5-cm-i.d. preparative column good separation was achieved for a 100 mg injection (Figure 8.15, chromatogram B) but multiple broad tailing peaks were observed when 600 mg racemate was injected (Figure 8.15, chromatogram C). The poor peak shape was attributed to low mobile phase solubility.

During sample dissolution it was determined that high solubility was achieved using a methanol : dichloromethane mixture. Analytical method development using immobilized CSPs (Chiralpak IA, IB, IC, ID, IE, and IF) as well as a Pirkle type CSP ((S,S) Whelk-O) which is covalently bound, and a modifier of methanol : dichloromethane was performed. The best selectivity was observed using the (S,S) Whelk-O CSP and a modifier of 30% 1:1 methanol:dichloromethane (Figure 8.16, chromatogram A). While the Whelk-O method afforded lower selectivity relative to the Chiralpak AD method (1.27 vs. 1.96), it was thought the improved racemate solubility would afford improved peak shape and increased productivity. The preparative resolution of 1100 mg of racemate using these conditions is shown in Figure 8.16, chromatogram B. The nontraditional modifier afforded improved chromatographic solubility and allowed 762 g of racemate to be processed in approximately 52 hours. Under these conditions, a productivity of 1.19 kkd (kilograms racemate/kg CSP/day) and a solvent usage of 0.43 L/g racemate was obtained.

8.4.6 Coupling of Chiral and Achiral Columns for SFC Purifications

Achiral impurities present in racemates can complicate the analysis or purification process. A number of scientists have investigated the coupling of achiral and chiral columns for analysis [103–105]. For racemates containing achiral impurities, standard approaches are to (i) purify prior to chiral purification or (ii) attempt to develop a chiral SFC method that resolves the achiral impurity as well as the

Chromatogram A

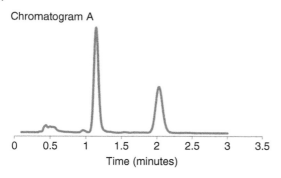

Chromatogram B

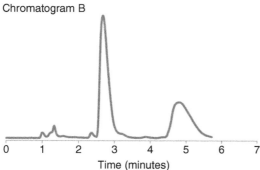

Chromatogram C

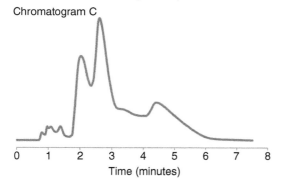

Figure 8.15 SFC separation of compound 4. Chromatogram A: Analytical separation: Chiralpak AD-H, 5 μm, 4.6 × 100 mm, 60% methanol, 5 mL/min, 100 bar back pressure. Ten microliter of 1 mg/mL methanol solution injected. Chromatogram B and C: Preparative separation: Chiralpak AD-H, 5 μm, 5 × 15 cm, 60% methanol, 325 mL/min, 100 bar back pressure and detection at 295 nm. Chromatogram B: 100 mg racemate dissolved in 1 mL of 50/50 (v/v) methanol/dichloromethane was injected. Chromatogram C: 600 mg racemate dissolved in 6 mL of 50/50 (v/v) methanol/dichloromethane was injected. *Source:* Reprinted with permission from Miller [102]. Copyright 2014 Elsevier.

enantiomers. Recent research has investigated coupling achiral and chiral columns in attempts to eliminate preSFC purifications or complicated method development to resolve both achiral impurities and enantiomers. Zeng et al. described a 2D SFC system that isolated pure racemate in the first dimension using a 2-ethylpyridine column; after trapping the racemate it was then transferred to a chiral column for preparative resolution [69]. Ventura documented a process for identification of complimentary achiral columns that could be directly coupled to chiral columns [106]. Either approach may not suitable as a higher throughput purification approach due to the increased method development time. These approaches may be more suitable for larger scale purifications.

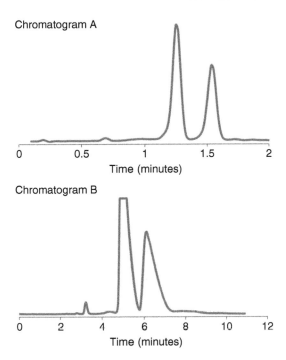

Figure 8.16 SFC separation of compound 4. Chromatogram A: Analytical separation: (S,S) Whelk-O, 5 μm, 4.6×100 mm, 30% 1:1 methanol:dichloromethane, 5 mL/min, 100 bar back pressure. Ten microliter of 1 mg/mL methanol solution injected. Chromatogram B: Preparative separation: (S,S) Whelk-O, 5 μm, 5×25 cm, 30% 1:1 methanol:dichloromethane, 350 mL/min, 100 bar back pressure and detection at 295 nm. 1100 mg racemate dissolved in 10 mL of 50/50 (v/v) methanol/dichloromethane was injected. *Source:* Reprinted with permission from Miller [102]. Copyright 2014 Elsevier.

8.5 Preparative Achiral SFC Purifications

8.5.1 Introduction to Achiral SFC Purifications

SFC has been routinely used for preparative enantioseparations for the past 20 years. It is only during the past 5–10 years that SFC has started to be used for achiral SFC purifications. In the past, the majority of achiral purifications in the pharmaceutical industry were performed using flash or reversed phased chromatography. The major limitation of flash chromatography is the lower efficiency obtained through the use of larger particle size stationary phases. While reversed phased chromatography has advantages of efficiency and relatively easy scale-up from analytical separations, it does have some disadvantages. The main disadvantage is the use of aqueous-based mobile phases. Most pharmaceutical compounds have poor solubility in aqueous based mobile phases, leading to poor peak shape and/or poor loading under preparative purification conditions. This can impact purity and/or yield of the product as well as purification productivities. A major green disadvantage of aqueous mobile phases is the increased time and energy required for distillation of product post purification. The use of SFC for achiral purifications greatly reduces the amount of solvent required for purification, as well as the energy and time required for distillation to recover products. The mobile phase : stationary phase interactions are very different in SFC relative to reversed phase chromatography. As SFC uses polar stationary phases and nonpolar mobile phases, it is considered a normal phase separation. For this reason, SFC often offers orthogonal separations and selectivity relative to reversed phased HPLC [107, 108]. This is well illustrated in Figure 8.17. It is important to remember that SFC is not suitable for all small molecule purifications. For highly basic molecules (pKa > 8), it has been proposed that SFC is not the ideal technique for purification (due to poor peak shape, long retention) and reversed phase purification should instead be used [21]. The recent explosion in the use of SFC for achiral purifications is due to the advantages listed above.

8.5.2 Stationary Phases for Achiral Preparative SFC

Numerous major chromatography vendors have marketed large number of achiral stationary phases for use in SFC over the past decade. The number of phases used for analytical achiral SFC is greater than one hundred. The number of phases available in preparative columns is much lower, around 15–20. Extensive work has been performed to characterize achiral stationary phases for analytical SFC [109–116] but only anecdotal information on characterization of achiral phases for purification has been published or presented at scientific meetings [21, 117, 118]. The majority of available preparative SFC achiral stationary phases are silica-based and any chemical modifications involve the addition of small organic molecules (ethyl pyridine, amino phenyl, etc.); unlike chiral stationary phases, loading capacities are approximately the same for the majority of achiral stationary phase. The only consideration when choosing a stationary phase for potential purification work is that the phase is available in larger i.d. prepacked columns or is available as bulk packing. A good review of the history of achiral stationary phases can be found in [114, 119].

8.5.3 Method Development for Achiral Purifications

Method development approaches for achiral purifications resemble those used for chiral purifications; a series of stationary phases and/or modifiers are evaluated and the best method selected for scale-up to preparative loadings. One major difference relative to chiral method development is that most users evaluate only methanol as a modifier [120, 121], especially during initial method development. In the pharmaceutical industry the number of

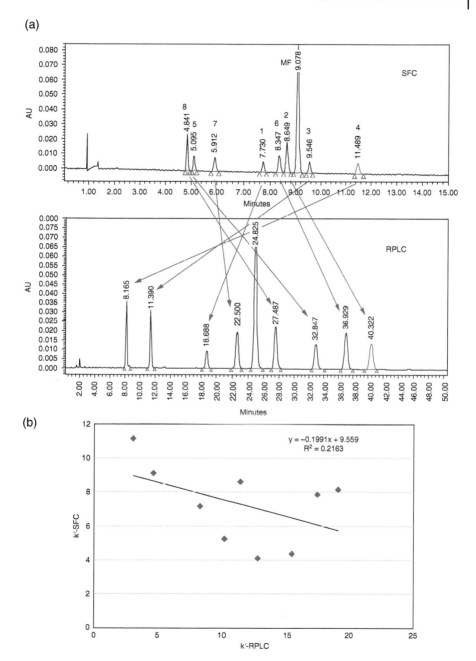

Figure 8.17 The orthogonal selectivity of SFC method vs. RP-HPLC method. SFC conditions: 100 bar, 30 C, 3.0 mL/min, modifier: methanol 5–15% in 15 minutes, silica column. RP-HPLC conditions: 5 C, 1.5 mL/min, water/acetonitrile (58:42 (v/v) to 48:52) in 60 minutes. Ultrasphere ODS column. *Source:* Reprinted with permission from Wang et al. [108]. Copyright 2011 Elsevier.

compounds to be purified can be large (up to hundreds/day). To meet these demands it is necessary to streamline the method development and purification workflows. Streamlining method development is often accomplished by identifying four to six achiral stationary phases and one modifier for evaluation. Additional column chemistries and/or modifiers can be evaluated if adequate separation is not obtained. Pfizer scientists reported on the use of method selection software which ranked the separations observed during method development process, greatly reducing the time required for method evaluation [120]. One disadvantage of achiral SFC is the lack of a generic stationary phase that is suitable for a wide variety of compounds, analogous to a C8 or C18 column for reversed phase chromatography.

8.5.4 Achiral SFC Purification Examples

8.5.4.1 Achiral Purification Example #1

The advantage of SFC for achiral purifications is well illustrated in the following example. The target molecule (Compound 5, confidential structure) was synthesized and then purified using reversed phase mass-directed preparative HPLC. Synthesis generated two positional isomers that closely eluted by HPLC, limiting the post purification purity to 70% (Figure 8.18, chromatogram A). Normal phase generally resolves structural isomers better than reversed phase chromatography. As SFC is a normal phase separation process, it was explored for this separation. Using a previously described method development protocol [122], the analytical separation shown in Figure 8.18, chromatogram B was developed. The preparative separation of 10 mg is shown in Figure 8.18, chromatogram C. Due to the small amount of sample to be purified (<100 mg), loading was not optimized for this separation. Even without optimization the separation was completed in approximately 30 minutes, generating both isomers at greater than 99% purity. In addition, product fractions were isolated in less than 250 mL of methanol, allowing distillation to be complete in approximately 15 minutes. Total time for the purification, from sample dissolution to generation of dry products was less than two hours.

8.5.4.2 Achiral Purification Example #2

As previously discussed, SFC is a normal phase process and can offer orthogonal separations to those developed using reversed phase conditions. This is well illustrated in the purification of compound 6 (confidential structure). Standard procedure in many pharmaceutical medicinal chemistry laboratories is to use reversed phase liquid chromatography to analyze and ultimately purify synthetic molecules [123, 124]. This approach was not feasible for compound 6 due to poor resolution of product and a major impurity by LC (Figure 8.19, chromatogram A). However, the product and impurity were easily resolved when the synthetic mixture was subjected to SFC using a diol stationary phase

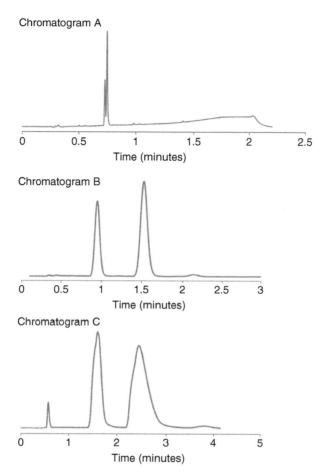

Figure 8.18 Analytical and preparative separation of compound 5. Chromatogram A: Analytical HPLC separation of compound 4. Analysis conducted on HALO C18, 2.7 um, 3 mm i.d. × 50 mm, flow rate of 2 mL/min. Solvent A: water with 0.1% (v/v) trifluoroacetic acid. Solvent B: acetonitrile with 0.1% (v/v) trifluoroacetic acid. Gradient from 5% to 95% B over 1.5 minutes, hold at 95% B for 0.3 minutes. Chromatogram B: Analysis conducted on Chiralcel OJ-H (100 mm × 4.6 mm i.d.) with a mobile phase of 55/45 methanol with 0.2% diethylamine/CO_2. A flow rate of 5 mL/min was used. Chromatogram C: Purification conducted on Chiralcel OJ-H (150 mm × 20 mm i.d.) with a mobile phase of 55/45 methanol with 0.2% diethylamine/CO_2. A flow rate of 80 mL/min, detection at 256 nm and a loading of 10 mg were used. *Source:* Miller and Peterson [122] Reproduced by permission of The Royal Society of Chemistry.

(Figure 8.19, chromatogram B). The SFC method was quickly and easily scaled to a 2-cm-i.d. preparative column and the 900-mg mixture purified using a series of 10 mg injections (Figure 8.19, chromatogram C). Product with purity of >99% was obtained.

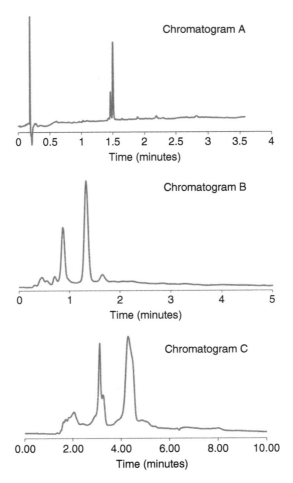

Figure 8.19 Separation of compound 6. Chromatogram A: LC Separation conducted on Gemini (get remainder of conditions). Chromatogram B: Analysis conducted on diol column (100 mm × 4.6 mm i.d.) with a mobile phase of 20/80 methanol with 0.2% diethylamine/CO2. A flow rate of 5 mL/min was used. Chromatogram C: Purification conducted on diol column (250 mm × 20 mm i.d.) with a mobile phase of 20/80 methanol with 0.2% diethylamine/CO$_2$. A flow rate of 80 mL/min, detection at 254 nm and loading of 10 mg were used.

8.5.5 Purifications Using Mass-Directed SFC

The vast majority of SFC purifications use UV to detect peaks, and to direct fraction collection. UV based fraction collection works but can be difficult for some purifications. This includes target compounds with no UV chromophore, or when using mobile phases that have high UV background (such as toluene, ethyl acetate, or dichloromethane) that prevent monitoring at lower wavelengths. As collection based on UV detection requires setting of a collection threshold, for complex samples this can result in collection of numerous peaks. These fractions must then be analyzed to determine which contain the desired product. Mass-directed purification has been developed as an alternative to UV-based collection [123, 125–131]. Mass-directed purification uses mass spectrometry (MS) as a detector. The MS signal directs fraction collection so that only the desired product is collected. The main advantage of this approach is that the only material collected is the product of interest. LC based mass-directed purification has been used for almost 20 years and has led to the purification of hundreds of thousands of compounds worldwide.

The first reported use of mass-directed SFC was published in 2001 [132]. Using a custom modified system they reported on the purification of a dozen drug-like compounds. The use of custom modified SFC for mass-directed purification of acid and base liable compounds not suitable for reversed phase HPLC was reported by Arqule in 2006 [133]. The year 2008 saw the introduction of the first commercially available mass-directed SFC purification platforms. The first models had maximum flow rates of 30 mL/min. while recent models allow flow rates up to 100 mL/min [134].

The advantages of a mass-directed purification approach are illustrated in Figure 8.20. The sample contained two products of interest (retention times of ~1.9 and 2.1 minutes in Figure 8.20, chromatogram A). If UV-based collection were used the products at ~4 minutes would be isolated in addition to the two desired products, with each fraction requiring subsequent analysis to determine which is the desired product. The use of mass-directed purification simplifies the collection process. The mass chromatograms at m/z 223 and m/z 237 are shown in Figure 8.20, chromatograms C and D. Use of MS to direct collection resulted in only two fractions being collected. Mass-directed SFC is routinely used for achiral purifications in the pharmaceutical industry and has proven invaluable when large numbers of samples must be purified as part of the drug discovery process [21, 121].

8.5.6 Impurity Isolation Using Preparative SFC

Isolation of impurities from reaction mixtures is a critical step in determining reaction mechanisms and can help to improve synthetic yields. Regulatory agencies require characterization of low level impurities as a requirement of the drug development process. Current ICH (International Council for Harmonization of Technical Requirements for Pharmaceutical for Human Use) guidelines set thresholds, based on daily dose, above which impurities must be identified. Impurity isolation is also a critical step of degradation studies, which are performed on active pharmaceutical ingredients (API) as well as formulations.

There are numerous approaches used to identify impurities. One approach is to use MS to obtain molecular mass as well as fragmentation data. Coupled

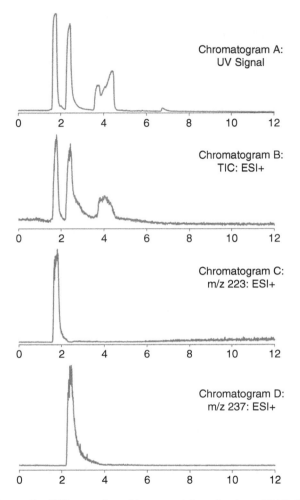

Figure 8.20 Preparative SFC separation of two proprietary structures. Viridis 2-EP, 3 × 10 cm, 100 mL/min, 100 bar BPR. Modifier: methanol. Gradient Conditions: 5–55% over 10.5 minutes, held at 55% for 1 minute. Chromatogram A: UV trace (total response from 210 to 400 nm). Chromatogram B: ESI+ total ion current trace. Chromatograms C and D: mass chromatograms of selected m/z ranges (223 and 237).

with knowledge of the synthetic process, it may be possible to propose a structure for the impurity. The proposed molecule can be synthesized and then confirmed as the impurity via chromatographic analysis. For complex molecules, synthesis can be a time consuming process. Often it can be easier to isolate the impurity and confirm structure using characterization tools such as NMR and high resolution MS. Impurity isolation is often accomplished using preparative chromatography. There are two options regarding sample origin for impurity isolation. The first is to work with the compound that contains the impurity at low levels. Due to the lower levels (often less than 0.5%) this requires larger quantities of feedstock to isolate the material in quantities required for identification (1–5 mg). An advantage of this approach is the sample is mainly one component, which if easy to remove can result in an easier isolation. A second option is to work with a feedstock that is enriched in the impurity. An excellent source of enriched material is mother liquors generated from crystallizations. As crystallizations are designed to reject impurities, the mother liquors will often contain elevated levels of impurities of interest. This will result in lower quantities of material to be processed to produce sufficient quantities for identification. One downside of this approach is the sample mixture may be more complicated, containing numerous compounds that need to be resolved from the product of interest.

If the analytical HPLC method used for compound analysis is suitable for preparative chromatography, isolation can be straightforward. The HPLC method can be scaled to preparative loadings and the impurity of interest isolated in one step. Often direct scale-up is not possible. For example, many HPLC methods use nonvolatile buffers that cannot be easily removed from the isolated impurities, or use a stationary phase that is not available in preparative columns. In addition, the resolution between the impurity of interest and other sample components may be insufficient for isolation. In this case it is necessary to use other purification methods, while using the original HPLC method to track the impurities through these alternate methods.

Impurity isolation, especially if more than one product is desired, is often a multi-step process [135]. Initial steps will often be performed using a lower efficiency separation process such as flash chromatography. The fractions that result from the first purification step are then processed using a higher efficiency process. Preparative reversed phase HPLC is frequently used for impurity isolation. This approach works well, but does have some potential limitations. Post purification, it is necessary to remove the mobile phase to produce material for identification. Removal of aqueous-based mobile phases can require higher temperatures or increased evaporation time. This can lead to degradation and eventual rework to produce material of sufficient purity depending on the stability of the impurity. RP-HPLC also uses additives such as trifluoroacetic acid that may lead to compound stability issues or cause difficulties with the identification process. SFC has proven to be a valuable tool for impurity isolation. Due to the unknown stability of impurities, rapid purification and evaporation is required to ensure compound integrity. The rapid flow rates, use of highly volatile solvents like methanol, and the higher concentrations of impurities post chromatography (relative to HPLC) can help to minimize any potential degradation. In addition, the higher flow rates of SFC relative to HPLC can reduce time requirements for impurity isolations. Riley et al. discussed the use of SFC for impurity isolation at SFC 2008 [20]. For the isolation of an impurity at 0.3%, the use of SFC allowed the impurity to be isolated in sufficient quantities for NMR analysis in six hours (method development and purification time) and a solvent cost of less than 5 USD. This was a 90% solvent cost reduction relative to HPLC. They also discussed an impurity isolation process they titled "API Extraction." This technique is useful for enrichment of low level impurities to a level sufficient for LC-NMR identification. The goal of API Extraction is to remove the API, the main component of the sample at 97–98%. This technique was shown to increase impurity levels approximately 40-fold to a level sufficient for acquisition of high quality NMR data using LC/NMR.

8.5.6.1 Impurity Isolation Example

To fully characterize a synthetic step, the isolation of six to seven impurities was required. A decision was made to work with a mother liquor sample generated during crystallization to isolate the penultimate product. Figure 8.21, chromatogram A shows the HPLC separation of the mother liquor sample. Low water solubility and the presence of perchloric acid eliminated scale-up of the HPLC

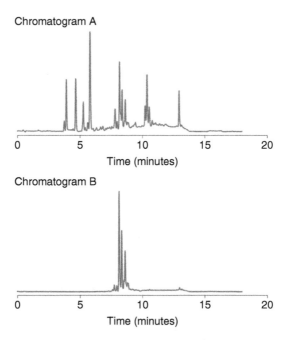

Figure 8.21 HPLC analysis of mother liquors. Analysis conducted on XBridge C18, 3.5 μm, 3 × 100 mm, flowrate of 1 mL/min. Solvent A: water with 0.l% perchloric acid. Solvent B: acetonitrile with 0.1% perchloric acid. Gradient: 50–100% B over 9 minutes, hold at 100% B for 3 minutes, 100–50% B over 0.3 minutes, hold at 50% B for 2.5 minutes. Chromatogram A: original mother liquor sample. Chromatogram B: fraction enriched in 8 minute impurities.

method for any isolation work. Due to the wide polarity range for the molecules of interest an initial flash purification step using silica gel and methanol/dichloromethane gradient was used to separate the mother liquor sample into numerous fractions enriched in various impurities. From this initial purification and additional flash purifications a number of the desired impurities were isolated. The initial flash purifications were unable to isolate the peaks with retention times of approximately 8 minutes by HPLC. HPLC analysis of the fraction enriched with the eight minutes peaks is shown in Figure 8.21, chromatogram B. The fraction contained three main peaks at approximately 46, 25, and 16 area percent. For isolation of these peaks, preparative SFC was investigated.

Initial work involved achiral SFC method development to determine SFC conditions that resolve the three peaks. The analytical SFC separation is shown in Figure 8.22, chromatogram A. As expected the elution pattern for the SFC separation was different from HPLC and offered increased resolution. This method was scaled to a 3 cm i.d. preparative column. The preparative SFC separation of 100 mg of this mixture is shown in Figure 8.22, chromatogram B. Using UV-based collection, all three products could be isolated in sufficient purity and quantity for identification. The mass chromatograms for the three products are shown in Figure 8.22, chromatogram C. The first two products have the same mass, eliminating the possibility of using mass-directed purification for this separation. Due to the close eluting peaks, and lack of baseline resolution between these peaks in the mass ion chromatogram, use of mass-directed purification was not used as it would result in both peaks being collected into the same fraction.

8.5.7 SFC as Alternative to Flash Purification

Flash chromatography [136–138] is the purification method of choice, and is the major source of solvent waste in a pharmaceutical medicinal chemistry laboratory. While flash chromatography can be made greener by solvent replacement strategies [139], the process still uses substantial amounts of solvent. While preparative SFC is the technique of choice for chiral separations and is beginning to see great use for achiral purifications, to date it has not been used as an alternative for flash LC purifications.

During the past five years initial exploratory work on what has been termed "flash SFC" has been performed [47, 140]. The result of this work shows that flash SFC is a potential alternative for flash LC purifications. The recent introduction of flash SFC equipment is another step toward the realization of the promise of flash SFC [141]. Flash SFC has the advantages of greatly reduced solvent usage (up to fourfold less) and increased purification productivity due to higher chromatographic efficiency seen at elevated flow rates. Another advantage of flash SFC is that products are up to four times more concentrated relative to flash LC, reducing time and energy requirements for product isolation [47].

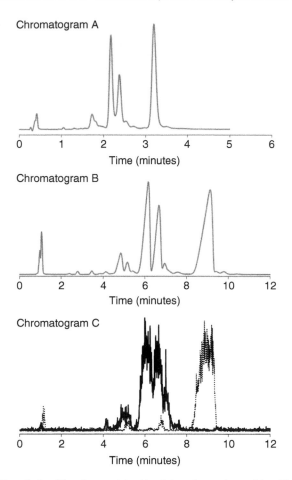

Figure 8.22 SFC analysis of fraction enriched in eight minutes impurities. Chromatogram A: analytical separation, separation conducted on DEAP column, 3 µm, 4.6 × 100 mm with a mobile phase of 40/60 methanol/CO_2. Chromatogram B: Preparative SFC separation of 100 mg (2 mL in methanol) using DEAP column, 3 × 150 mm, 5 µm, 100 mL/min., 100 bar BPR, modifier: 30% methanol. Chromatogram C: mass chromatograms of selected m/z ranges; 628 (solid line), 614 (dashed line).

The advantages of flash SFC are demonstrated through the separation of a mixture of carbamazepine, flavone, and nortriptyline HCl. The flash LC, and flash SFC separation of 1200 mg of this mixture is shown in Figure 8.23. Comparison of the flash LC and flash SFC separations is shown in Table 8.5. Flash SFC has a twofold increase in purification productivity and a greater than threefold decrease in solvent usage. Products from flash SFC were two to five times more concentrated than those from flash LC.

Chromatogram A

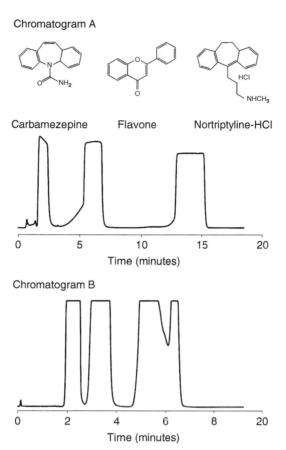

Carbamezepine   Flavone   Nortriptyline-HCl

Chromatogram B

Time (minutes)

Figure 8.23 Preparative separation of 1200 mg mixture of carbamezepine, flavone, and nortriptyline HCl. Chromatogram A: LC purification was conducted on Interchim 30 μm silica (150×19 mm i.d.) with a mobile phase gradient of methanol/dichloromethane/ammonium hydroxide (0.5/99.5/0.05 to 10/90/1) over 18.5 minutes. A flow rate of 40 mL/min was used. Chromatogram B: SFC purification was conducted on Interchim 30 μm silica (150×19 mm i.d.) with a mobile phase gradient of 5–55% methanol (w/0.2% diethylamine) in CO_2 over nine minutes, 100 bar. A flow rate of 80 mL/min was used. *Source:* Reprinted with permission from Miller and Mahoney [47]. Copyright 2012 Elsevier.

While the advantages of flash SFC are evident relative to flash LC there are still a number of technical issues that must be addressed before this technology becomes common place in the medicinal chemistry laboratory. The main issue involves the flash SFC equipment. Most of the work to date in this field has used standard preparative SFC systems with prepacked columns. Only recently has a beta version flash SFC instrument been developed. Current SFC equipment is too expensive to replace flash LC systems. A significant reduction in cost is

Table 8.5 Flash LC/SFC comparison of preparative separation of carbamezepine, flavone, and nortriptyline HCl.

Technique	Load (mg)	Peak 1 Volume (mL)	Peak 2 Volume (mL)	Peak 3 Volume (mL)	Solvent (L/g crude)	Productivity (g/hr)[a]
LC[b]	1200	43	100	108	0.62	2.52
SFC	1200	7.6	18	58	0.19	5.04

[a] Assume 10 minute re-equilibration for LC, 5 minutes for SFC.
[b] Higher loadings may have been possible but not investigated due to limited size of dry pack LC cartridge.
Source: Used with permission of [47].

needed to compete with the relatively low cost of flash LC equipment. Also, technology does not exist for flash SFC cartridges that is as easy to use, allows for a wide range of sizes, and is as versatile as flash LC cartridges. In addition, significant use of flash SFC requires a source and distribution network for CO_2. Finally, most preparative SFC collection uses cyclone type separators to separate carbon dioxide from organic solvent post purification. A design allowing collection with high purity and high recovery into test tubes is mandatory for a flash SFC system. It is the author's hope that SFC equipment vendors will begin to address these issues and introduce equipment suitable for flash SFC purification.

8.6 Best Practices for Successful SFC Purifications

Previous sections of this chapter presented theoretical information and a number of examples of SFC purifications. While knowledge of preparative SFC is important for successful purifications, at times this knowledge alone is not enough. It was military strategist Helmuth von Moltke who in the mid-nineteenth century noted "no battle plan survives contact with the enemy." The same can be said of SFC purifications; the best laid purification strategy may not survive contact with the sample. Each sample has its own characteristics which may only be evident once the purification has begun. This requires the chromatographer to change plans on the fly to deal with these peculiarities. The last section of this chapter presents a number of best practices that have been developed over the past 30 years which, if followed, can increase your purification success ratio. Some of these practices are relevant to all types of preparative chromatography, while others are specific to SFC purifications.

8.6.1 Sample Filtration and Inlet Filters

The first best practice relates to filtering of sample solutions. It is recommended to filter all samples prior to purification, even if the sample appears totally soluble. Small insoluble particles are difficult to visually detect. Inlet and outlet frits of preparative columns are designed to maintain the stationary phase inside the column. Columns packed with 5 µm particle size columns use frits with a porosity of 2 µm or less. The inlet frit can be easily clogged due to insoluble sample components. Clogging of the inlet frit can lead to increased column pressure drop. Material on the inlet frit will impact the mobile phase flow path across the frit, leading to reduction in column efficiency and in a preparative system, increased peak width which can lead to reduced purities, yields, and purification throughputs. If precipitation on the inlet frit is extensive, over time the pressure drop across the center of the frit is larger relative to the outer parts of the frit, which can lead to the frit deforming, resulting in the seal between the frit and column O-rings being compromised. This can cause packing material to pass around the inlet frit and escape from the column inlet, resulting in column failure. Even if frit deformation is not observed, collection of solids on the inlet frit will require the preparative column to be replaced, at a significant cost. Filtering of samples results in longer column lifetimes as well as preventing system shutdown due to high pressure.

Section 8.3.6 discussed the various options for sample introduction in SFC purifications. The majority of commercially available preparative SFC systems offer either mixed stream or modifier stream injection. These techniques require dissolution of the sample in organic solvent. It is only after injection that the sample has any interactions with CO_2. Depending on compound solubility in CO_2, it is possible that upon contact with CO_2 there may be sample precipitation. Sample precipitation upon injection can occur in any purification system where the sample dissolution solvent does not exactly match the mobile phase. As it is impossible to match sample dissolution with the mobile phase in preparative SFC (except with extraction as injection technique), this phenomenon is more prevalent in SFC purifications.

Because of potential sample precipitation, an in-line filter is absolutely required for SFC purifications. These filters contain a 2-µm disposable filter that helps to remove any precipitation that may occur upon contact with CO_2 after sample injection. These filters can also help remove any insoluble material present from sample dissolution. The use of in-line filters is cheap insurance for protection of expensive preparative SFC columns. In-line filters are available from a number of vendors. To avoid increased pressure drop it is necessary to use an in-line filter designed for preparative flow rates. The replacement frequency for in-line filters is difficult to predict. It will vary depending on amount of material purified, as well as sample components. As the inlet filter traps insoluble material, distribution across the filter will be modified, resulting in broader chromatographic peaks. This is shown in Figure 8.24. A clogged inlet filter leads to increased peak width, which can lead to decreased resolution between close eluting peaks. Replacement with a new filter restores the separation.

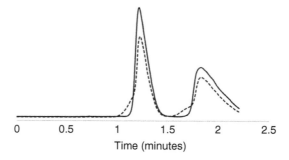

Figure 8.24 Effect of "dirty" inlet frit on SFC purification. Separation obtained with dirty filter (dashed line) and after installation of a new filter (solid line).

It is difficult to predict when sample precipitation will occur due to CO_2 in an SFC purification. Monitoring of column pressure drop is helpful for identifying precipitation. Always monitor system pressure, looking for pressure spikes upon injection. This can indicate material coming out of solution within the system. If the pressure quickly drops to standard operating pressures, this may not be a concern as whatever precipitated is quickly going back into solution. A steady increase in pressure with each injection is an indication that material is not being redissolved. When this occurs it must be immediately addressed. Reducing the amount of material applied to the column may eliminate the pressure increases. Another potential solution, if the chromatographic separation allows, is to increase the modifier percentage which may improve solubility.

8.6.2 Sample Purity

The goal of preparative SFC is to isolate pure material. One would think that the purity of the feedstock is not important. For small scale purification work, the type of impurities present in the sample is not as important, but it becomes more important as scale increases, and the need for high purification productivity becomes critical. For achiral purifications, a wide polarity range for sample components can lead to long cycle times or the need for gradients, or column washes to remove all components before the next injection. Gradients or column washes lead to longer purification times, increased solvent usage, and higher purification costs. Highly retained impurities may accumulate on the stationary phase, which reduces binding sites and limits loading capacity. Depending on the purification scale, a low efficiency prepurification (such as flash chromatography, crystallization, or extraction) can help remove components whose retention is far removed from the compound(s) of interest.

It is critical to remove residual metals prior to SFC purification. With the increased use of metal-based reactions, residual metals are often observed in samples submitted for SFC purification. Residual metals are often soluble in solvents used for sample dissolution (methanol) and are difficult to remove via filtration. Upon contact with CO_2, these metals often precipitate on the inlet frit, or in the column and may even bind to the stationary phase. Metal removal with metal scavengers or via flash chromatography is straightforward and highly recommended prior to SFC purification [142].

Sample purity is also important for chiral SFC purifications. Selectivity is lower in chiral separations relative to achiral separations; thus isocratic methods are used for nearly all chiral SFC purifications. Standard practice is to process racemates using multiple stacked injections. The presence of achiral impurities increases the required cycle time, leading to increased purification time and cost. The impact of achiral impurities on cycle time is well illustrated for the chiral separation in Figure 8.14 chromatogram B. The presence of an achiral impurity at approximately 9 minutes requires a cycle time of 3 minutes to achieve resolution of the enantiomers and removal of the achiral impurity. If the achiral impurity had been removed prior to purification, the cycle time could be reduced to 1.75 minutes, a purification increase of greater than 70%.

8.6.3 Salt vs. Free Base

It is often preferable to purify basic compounds as free bases rather than as salts when performing SFC purifications. Free bases often have higher solubility compared to salts under SFC conditions. This can lead to improved peak shape, higher loadings, and improved productivity. This effect is shown in Figure 8.25 for the preparative SFC separation of a proprietary racemate that was purified as a free base, and as a toluene sulfonic acid salt. The free base was soluble in methanol at 39 mg/mL, while the salt had a maximum solubility of 4.4 mg/mL. Twenty milligram of the free base was baseline resolved on a 2-cm-i.d. column. Using the same purification conditions, except for a larger 5-cm-i.d. column, baseline resolution was lost with a 9-mg injection of the salt.

SFC analysis and purification of basic compounds often requires the addition of a basic additive to reduce interactions with the silica and achieve good peak shape. Depending on the pKa of the racemate, as well as the strength of the base being used as an additive, it is possible to observe free base formation on the preparative column. As the compound converts from its salt form to the free base, multiple peaks (one for free base, one for salt) can be observed eluting from the column. Figure 8.26 shows this effect. Chromatogram A shows the resolution of 15 mg of the HCl salt of this proprietary basic racemate (Compound 7). The peak area for the two enantiomers did not increase as the load was increased, but a nonretained peak increased in size. Chromatogram B shows the resolution of 150 mg of the HCl salt. The majority of the peak area is the nonretained peak eluting at approximately 1 minute. Collection and analysis of the three peaks confirmed peaks 2 and 3 to be the product enantiomers and peak 1 to be the racemate.

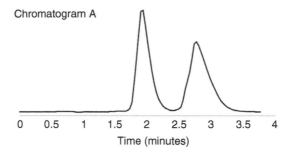

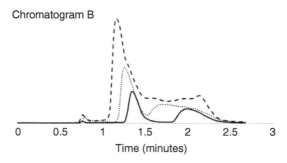

Figure 8.25 SFC purification of salt and free base forms of proprietary compound. Chromatogram A: SFC purification of free base was conducted on Chiralpak AD, 5 μm (25 × 2 cm i.d.) with a mobile phase of 40% methanol (w/0.2% diethylamine) in CO_2 with a flow rate of 80 mL/min. Detection at 254 nm. Injection: 20 mg in 0.5 mL 1:1 (v/v) methanol:dichloromethane. Chromatogram B: SFC purification of toluene sulfonic acid salt was conducted on Chiralpak AD, 5 μm (15 × 5 cm i.d.) with a mobile phase of 40% methanol (w/0.2% diethylamine) in CO_2 with a flow rate of 300 mL/min. Detection at 254 nm. Injection: 2, 9, and 22 mg dissolved at 4.4 mg/mL in 1:1 (v/v) methanol:dichloromethane (solid, small dash, and large dash lines, respectively).

The modifier for this separation contained 0.2% diethylamine. It was proposed that upon injection, this basic additive acted to convert the salt to the free base of the racemate. At 15 mg load there is sufficient diethylamine in the mobile phase to completely free base the injected racemate. As injection quantity increased, the diethylamine levels were insufficient to free base all of the injected racemate, leading to the HCl salt of the racemate eluting at approximately one minute. This theory was confirmed by the fact that past a certain level, peak areas for the two enantiomers did not increase, indicating all available diethylamine had been consumed and no additional conversion to the free base was possible.

Additional evidence for this theory was obtained after the racemic HCl salt was converted to the free base and then purified under identical conditions. Purification of the free base showed only two peaks, and no nonretained peaks were observed with increased loading. Ultimately 200 mg of racemate was injected with baseline resolution (Figure 8.26, chromatogram C).

Another option for purification when racemates are submitted as salts is to free base in solution prior to injection. This is accomplished by adding excess base to the sample solution. This technique has been used multiple times within Amgen laboratories and other pharmaceutical companies. When using this approach, it is important to select cycle times such that the salts present in

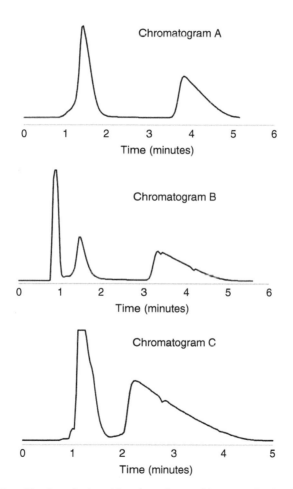

Figure 8.26 SFC purification of salt and free base forms of Compound 7. Purification performed on Chiralpak AD, 5 μm (15×2 cm i.d.) with a mobile phase of 40% methanol (w/0.2% diethylamine) in CO_2 with a flow rate of 80 mL/min. Detection at 220 nm. Chromatogram A: Injection of 15 mg in 0.1 mL 1 methanol. Chromatogram B: Injection of 150 mg in 1.0 mL methanol. Chromatogram C: Injection of 300 mg of free base in 1.5 mL 1:1 (v/v) methanol:dichloromethane.

the sample that are generated during the free basing process (which are usually nonretained) do not elute in a part of the chromatograms where products of interest are being collected. If salts do elute with product it will be necessary to remove them post purification via extraction or other separation technique. It is also important that the racemate not degrade due to higher levels of base used during dissolution/free base process.

8.6.4 Primary Amine Protection to Improve Enantiomer Resolution

It is known that basic functional groups, such as primary amines, on analytes can interact with residual silanols on all stationary phases and that these interactions lead to tailing peaks. For chiral separations this may cause nonspecific interactions that can reduce enantioseparation as well as cause tailing peaks. For analytical separations these interactions can be reduced via additives that serve to mask the residual silanols. Unfortunately, many of these additives are not suited for preparative separations. Another approach is to derivatize the primary amine function group of the analyte, removing the basic group that interacts with the residual silanols. This approach was investigated by Kraml et al. for the separation of amine enantiomers [143]. They showed that adding a carbobenzyloxy (cbz) protecting group offered enhanced chiral resolution compared to the nonprotected analogs for both HPLC and SFC. The enhanced chiral resolution was shown at both analytical and preparative scales. To be feasible addition and removal of the protecting group must occur rapidly and at high yields. Cbz derivatization hits on both of these requirements; Cbz derivatization is a simple chemical reaction that occurs at high yields and cbz removal is also possible at high yields using catalytic hydrogenolysis conditions. Besides cbz, tert-butyloxycarbonyl (boc) and fluorenylmethyloxycarbonyl (FMOC) have also been shown to improve chromatographic enantioseparations when added as protecting groups to primary amines.

8.6.5 Evaluation of Alternate Synthetic Intermediates to Improve SFC Purification Productivity

With the wide range of chiral stationary phases available these days it is possible to resolve most enantiomers at the milligram scale, even when low selectivity requires a brute force approach with long separation times and increased solvent consumption. However, these brute force approaches are not recommended as the amount of material to be purified increases. Minor changes in racemate structure can drastically influence important parameters such as solubility and selectivity that will impact purification productivity. Prior to any large scale separation it is advisable to examine a variety of intermediates in the synthetic scheme to identify the best resolution point.

When evaluating which intermediate to chromatographically resolve, there are a number of points to consider. These include the following:

- Compound solubility
- Chromatographic characteristics
  - Selectivity
  - Retention
- Solvent
  - Cost, availability, toxicity, viscosity, volatility, and purity
- Compound stability (in solution, during evaporation)
- Purity and physical state of compound
- Potential for racemization at subsequent steps (chiral resolution only)

It is preferable to perform the separation as early as possible in the synthetic route. This reduces the scale of subsequent reactions, eliminating waste, and reducing processing time. For large scale chiral separations, it is desirable to have racemate of high chemical purity. Achiral impurities can coelute with the enantiomers or elute such that collection using UV detection is complicated. Highly retained impurities can results in pollution of the CSP and ultimately in loss of separation. Ideally, a separation should be performed on a racemate that does not require the use of an acidic or basic additive. The presence of an additive complicates the mobile phase preparation step, and can complicate solvent recycling if it is being performed. Another point to consider is the physical state of the racemate. Is it a solid or an oil? At larger scale, an oil is difficult to weigh for dissolution and makes for a complicated isolation post chromatography. Upon chromatographic scale-up, the feed and resolved products are in solution for longer periods of time. Stability studies in the mobile phase should be performed and solvents with the potential to degrade the racemate should be avoided. Finally, for chiral separations, subsequent chemical steps should be evaluated for racemization potential. If the potential for racemization is identified, the enantioseparation should be placed after this synthetic step, or the chemistry modified to reduce the risk of racemization.

A standard approach is to obtain solubility data on the racemate and use this data to guide the method development process. A great analytical separation coupled with poor solubility is a recipe for a low productivity purification. This approach was used for a synthetic project requiring the generation of 100 g of final product. Evaluation of the synthetic scheme showed three potential molecules for resolution (Figure 8.27). All three compounds were screened by SFC. The best analytical methods for each racemate is shown in Figure 8.28. Resolution of the boc-protected amine was poor and no further work was performed on this compound. The free amine offered significantly improved resolution relative to the boc-protected compound. Lastly, the final compound was also resolved, although this compound suffered from poor solubility in methanol and was not chosen for subsequent work.

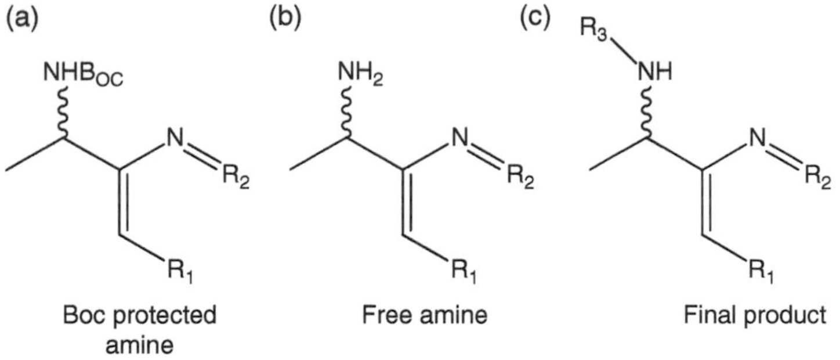

Figure 8.27 Chemical structures of resolution options (a) boc protected amine, (b) free amine, and (c) final product.

The free amine analog showed resolution on Chiralpak IC CSP with methanol, ethanol, and isopropanol as modifier. Solubility studies showed excellent solubility (~100 mg/mL) for all three solvents. The best enantioselectivity was observed with isopropanol. However, isopropanol was eliminated as a possibility for three reasons. The first was the long retention time for the second eluting enantiomer. This would lead to long cycle times and limit productivity. The second reason was the high viscosity of isopropanol would limit mobile phase flow rates. Finally the high boiling point of isopropanol would lead to longer distillation times. It was decided to resolve the free amine using Chiralpak IC and ethanol w/0.2% diethylamine additive.

The next step was to perform the separation on a small scale (~2 g). This work would confirm productivity as well as determine solvent requirements. Finally, the resolved enantiomers were taken through the final synthetic step to confirm racemization did not occur. The separation was scaled to a 2-cm-i.d. column and loading studies performed. As expected, based on excellent solubility and selectivity, high loading was possible. The separation of 500 mg racemate is shown in Figure 8.29, chromatogram A. Loading beyond 500 mg was possible but was not attempted due to lack of racemate. A loading of 500 mg/injection resulted in a productivity of 3.47 kkd racemate and a solvent usage of 0.32 L/g racemate.

A larger lot of 357 g racemate was then submitted for resolution. To accommodate the increased sample size, a larger column with 5-cm-i.d. was used. The resolution of 2 g of racemate is shown in Figure 8.29, chromatogram B. This separation resulted in a productivity of 3.6 kkd racemate and solvent usage of 0.37 L/g racemate. A subsequent 100 g lot of racemate was also submitted for chromatographic resolution. At the time of submission ethanol was not available in the required volumes and the separation was performed using methanol as a modifier. Separation of the primary amine intermediate had lower selectivity with methanol modifier relative to ethanol (2.27 vs. 5.77, see Figure 8.28) and a lower productivity was expected. Using methanol, a

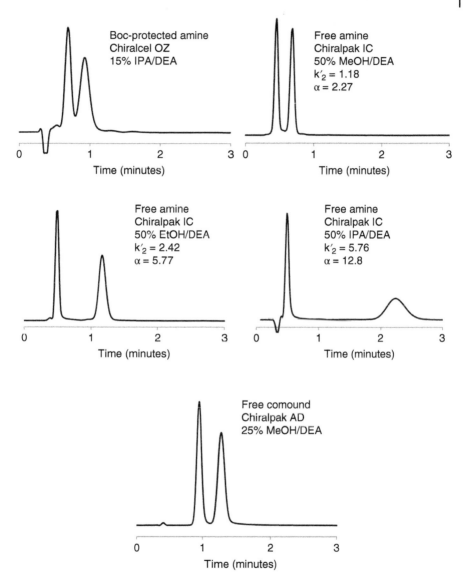

Figure 8.28 Analytical SFC separations of boc protected amine, free amine, and final product.

productivity of 3.18 kkd racemate and solvent usage of 0.37 L/g racemate was observed. Excellent scale-up from 2- to 5-cm-i.d. columns was observed for this separation. This example demonstrates the value of evaluating all synthetic steps in order to develop the most productive and cost-effective purification.

Chromatogram A

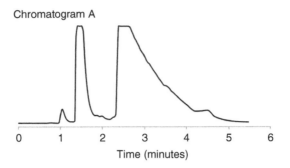

Chromatogram B

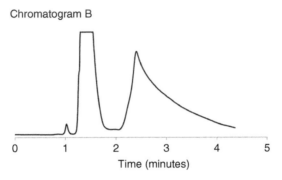

Figure 8.29 Preparative SFC separation of free amine. Chromatogram A: Resolution of 500 mg, dissolved in 5 mL ethanol was conducted on Chiralpak IC, 5 μm (2 cm i.d. ×25 cm), 100 bar, with a mobile phase of 50% ethanol w/0.2% diethylamine in CO_2, 100 bar, 265 nm. A flow-rate of 80 mL/min was used. Chromatogram B: Resolution of 2000 mg, dissolved in 20 mL ethanol, was conducted on Chiralpak IC, 5 μm (5 cm i.d. ×15 cm), 100 bar, with a mobile phase of 50% ethanol w/0.2% diethylamine in CO_2, 100 bar, 300 nm. A flow-rate of 325 mL/min was used.

8.7 Summary

Preparative SFC has many advantages over preparative LC purifications. These can include increased productivity, reduced solvent usage, and lower purification costs. For small scale purifications encountered in pharmaceutical discovery, SFC is the technique of choice for chiral separations and is rapidly becoming the technique of choice for achiral purifications. Currently large scale (>1 kg) SFC purifications are rarely performed due to lack of equipment and contract facilities, but as the advantages of SFC become better known it is hoped this situation will change.

References

1 Ettre, L.S. and Sakodynskii, K.I. (1993). M. S. Tswett and the discovery of chromatography II: completion of the development of chromatography (1903–1910). *Chromatographia* 35 (5): 329–338.

2 Ettre, L.S. and Sakodynskii, K.I. (1993). M. S. Tswett and the discovery of chromatography I: early work (1899–1903). *Chromatographia* 35 (3): 223–231.

3 Klesper, E., Corwin, A.H., and Turner, D.A. (1962). High pressure GAs chromatography above critical temperatures. *The Journal of Organic Chemistry* 27 (2): 700–701.

4 Jusforgues, P. (1995). Instrumental design and separation in large scale industrial supercritical fluid chromatography. In: *Process Scale Liquid Chromatography* (ed. G. Subramanian), 153. Weinheim: VCH.

5 Jusforgues, P., Shaimi, M., and Barth, D. (1998). Preparative supercritical fluid chromatography: Grams, Kilograms and Tons! In: *Supercritical Fluid Chromatography With Packed Columns* (eds. K. Anton and C. Berger), 403. New York, NY: Marcel Dekker.

6 Ruthven, D.M. (1984). *Principles of Adsorption and Adsorption Processes.* Wiley.

7 Ganetsos, G. and Barker, P.E. (1992). *Preparative and Production scale Chromatography.* CRC Press.

8 Guiochon, G., Felinger, A., and Shirazi, D.G. (2006). *Fundamentals of Preparative and Nonlinear Chromatography.* Academic Press.

9 Rodrigues, A.E., LeVan, M.D., and Tondeur, D. (2012). *Adsorption: Science and Technology.* Springer Science & Business Media.

10 White, C. and Burnett, J. (2005). Integration of supercritical fluid chromatography into drug discovery as a routine support tool. *Journal of Chromatography A* 1074 (1–2): 175–185.

11 Miller, L. (2014). Pharmaceutical purifications using Preparative supercritical fluid chromatography. *Chimica Oggi / Chemistry Today* 32 (2): 23–26.

12 Miller, L. and Potter, M. (2008). Preparative chromatographic resolution of racemates using HPLC and SFC in a pharmaceutical discovery environment. *Journal of Chromatography B, Analytical Technologies in the Biomedical and Life Sciences* 875 (1): 230–236.

13 Miller, L. and Potter, M. (2008). Preparative supercritical fluid chromatography (SFC) in drug discovery. *American Pharmaceutical Review* 11 (4): 112–117.

14 Francotte, E. (2016). Practical advances in SFC for the purification of pharmaceutical molecules. *LCGC Europe.* 29 (4): 194–204.

15 Ventura, M., Farrell, W., Aurigemma, C. et al. (2004). High-throughput preparative process utilizing three complementary chromatographic purification technologies. *Journal of Chromatography A* 1036 (1): 7–13.

16 Barnhart, W.W., Gahm, K.H., Thomas, S. et al. (2005). Supercritical fluid chromatography tandem-column method development in pharmaceutical sciences for a mixture of four stereoisomers. *Journal of Separation Science* 28 (7): 619–626.

17 Welch, C.J., Leonard, W.R. Jr., Dasilva, J.O. et al. (2005). Preparative chiral SFC as a green technology for rapid access to enantiopurity in pharmaceutical process research. *LCGC North America* 23 (1).

18 de la Puente, M.L., Soto-Yarritu, P.L., and Anta, C. (2012). Placing supercritical fluid chromatography one step ahead of reversed-phase high performance liquid chromatography in the achiral purification arena: a hydrophilic interaction chromatography cross-linked diol chemistry as a new generic stationary phase. *Journal of Chromatography A* 1250: 172–181.

19 Miller, L. (2012). Preparative enantioseparations using supercritical fluid chromatography. *Journal of Chromatography A* 1250: 250–255.

20 Riley, F., Zelesky, T., Marquez, B., and Brunelli, C. (2008). Packed column SFC: impacting project progression from drug discovery through development. *SFC*, Zurich, Switzerland (October 2008).

21 Francotte, E. (2015). SCF: a multipurpose approach to support drug discovery. *1st International Conference on Packed Column SFC China*, Shanghai, China (November 2015).

22 Caille, S., Boni, J., Cox, G.B. et al. (2010). *Organic Process Research & Development* 14 (1): 133–141.

23 Forssen, P., Samuelsson, J., and Fornstedt, T. (2013). Relative importance of column and adsorption parameters on the productivity in preparative liquid chromatography. I: investigation of a chiral separation system. *Journal of Chromatography A* 1299: 58–63.

24 Toribio, L., Alonso, C., Del Nozal, M. et al. (2006). Semipreparative enantiomeric separation of omeprazole by supercritical fluid chromatography. *Journal of Chromatography A* 1137 (1): 30–35.

25 Lazarescu, V., Mulvihill, M.J., and Ma, L. (2014). Achiral preparative supercritical fluid chromatography. In: *Supercritical Fluid Chromatography: Advances and Applications in Pharmaceutical Analysis* (ed. G.K. Webster), 97. Boca Raton, FL: CRC Press Taylor & Francis Group.

26 Ventura, M.C. (2014). Chiral preparative supercritical fluid chromatography. In: *Supercritical Fluid Chromatography: Advances and Applications in Pharmaceutical Analysis* (ed. G.K. Webster), 171. Boca Raton, FL: CRC Press Taylor & Francis Group.

27 Orihuela, C., Fronek, R., Miller, L. et al. (1998). Unique dynamic axial compression packing systems. *Journal of Chromatography A* 827: 193–196.

28 Pic Solution DAC Columns for SFC. http://www.pic-sfc.com/columns.html.

29 Novasep DAC Columns for SFC. http://www.novasep.com/technologies/supersep-supercritical-fluid-chromatography-systems.html.

30 Scott, R., Perrin, W.H., Ndzie, E. et al. (2007). Purification of difluoromethylornithine by global process optimization: coupling of chemistry and chromatography with enantioselective crystallization. *Organic Process Research and Development* 11 (817–824): 817.

31 Kaspereit, M., Lorenz, H., and Seidel-Morgenstern, A. (2001). Coupling of chromatography and crystallization for enantioseparation. *Chemie Ingenieur Technik* 73 (6): 720–721.

32 Seidel-Morgenstern, A. ed. (2008). Coupling of chromatography and crystallization to separate enantiomers. Talk presented at Verfahrenstechnisches Kolloquium. TU Hamburg-Harburg, Germany (6 November 2008).

33 Kaspereit, M., Lorenz, H., and Seidel-Morgenstern, A. (2002). Coupling of simulated moving bed technology and crystallization to separate enantiomers. *Fundamentals of Adsorption* 7: 101–108.

34 Newburger, J. and Guiochon, G. (1990). Utility of the displacement effect in the routine optimization of separations by preparative liquid chromatography. *Journal of Chromatography A* 523: 63–80.

35 Christopher, J., Welch, P.S., Spencer, G. et al. (2008). Microscale HPLC predicts preparative performance at millionfold scale. *Organic Process Research and Development* 12: 674–677.

36 Guillarme, D., Nguyen, D.T.T., Rudaz, S., and Veuthey, J.-L. (2007). Method transfer for fast liquid chromatography in pharmaceutical analysis: application to short columns packed with small particle. Part I: isocratic separation. *European Journal of Pharmaceutics and Biopharmaceutics.* 66 (3): 475–482.

37 Guillarme, D., Nguyen, D.T.T., Rudaz, S., and Veuthey, J.-L. (2008). Method transfer for fast liquid chromatography in pharmaceutical analysis: application to short columns packed with small particle. Part II: gradient experiments. *European Journal of Pharmaceutics and Biopharmaceutics.* 68 (2): 430–440.

38 Tarafder, A., Hudalla, C., Iraneta, P., and Fountain, K.J. (2014). A scaling rule in supercritical fluid chromatography. I. Theory for isocratic systems. *Journal of Chromatography A* 1362: 278–293.

39 Tarafder, A. and Hill, J.F. (2017). Scaling rule in SFC. II. A practical rule for isocratic systems. *Journal of Chromatography A* 1482: 65–75.

40 Cox, G.B. (2005). Introduction to preparative chromatography. In: *Preparative Enantioselective Chromatography* (ed. G.B. Cox), 19–47. Blackwell Publishing Ltd.

41 Dingenen, J. and Kinkel, J.N. (1994). Preparative chromatographic resolution of racemates on chiral stationary phases on laboratory and production scales by closed-loop recycling chromatography. *Journal of Chromatography A* 666 (1): 627–650.

42 Chin, G.J., Chee, Z.H., CHen, W., and Rajendran, A. (2010). Solubility of Flurbiprofen in CO_2 and CO_2 + Methanol. *Journal of Chemical & Engineering Data* 55: 1542–1546.

43 Gahm, K.H., Tan, H., Liu, J. et al. (2008). Purification method development for chiral separation in supercritical fluid chromatography with the solubilities in supercritical fluid chromatographic mobile phases. *Journal of Pharmaceutical and Biomedical Analysis* 46 (5): 831–838.

44 Gahm, K.H., Huang, K., Barnhart, W.W., and Goetzinger, W. (2011). Development of supercritical fluid extraction and supercritical fluid chromatography purification methods using rapid solubility screening with multiple solubility chambers. *Chirality* 23 (Suppl 1): E65–E73.

45 Miller, L. and Sebastian, I. (2012). Evaluation of injection conditions for preparative supercritical fluid chromatography. *Journal of Chromatography A* 1250: 256–263.

46 Blom, K.F. (2002). Two-pump at-column-dilution configuration for preparative liquid chromatography-mass spectrometry. *Journal of Combinatorial Chemistry* 4: 295–301.

47 Miller, L. and Mahoney, M. (2012). Evaluation of flash supercritical fluid chromatography and alternate sample loading techniques for pharmaceutical medicinal chemistry purifications. *Journal of Chromatography A* 1250: 264–273.

48 Shaimi, M. and Cox, G.B. (2014). Injection by Extraction: A Novel Sample Introduction Technique for Preparative SFC. *Chromatography Today* November/December: 42–45.

49 Mourier, P.A., Eliot, E., Caude, M.H. et al. (1985). Supercritical and subcritical fluid chromatography on a chiral stationary phase for the resolution of phosphine oxide enantiomers. *Analytical Chemistry* 57 (14): 2819–2823.

50 Yan, T.Q., Orihuela, C., and Swanson, D. (2008). The application of preparative batch HPLC, supercritical fluid chromatography, steady-state recycling, and simulated moving bed for the resolution of a racemic pharmaceutical intermediate. *Chirality* 20 (2): 139–146.

51 Speybrouck, D. and Lipka, E. (2016). Preparative supercritical fluid chromatography: a powerful tool for chiral separations. *Journal of Chromatography A* 1467: 33–55.

52 Okamoto, Y. and Yashima, E. (1998). Insight into the separation of enantiomers by HPLC on polysaccharide derivatives. *Angewandte Chemie (International Ed. in English)* 37: 1020–1043.

53 Okamoto, Y. and Yashima, E. (1998). Polysaccharide derivatives for chromatographic separation of enantiomers. *Angewandte Chemie International Edition* 37 (8): 1020–1043.

54 Gasparrini, F., Misiti, D., and Villani, C. (2001). High-performance liquid chromatography chiral stationary phases based on low-molecular-mass selectors. *Journal of Chromatography A* 906 (1–2): 35–50.

55 Welch, C.J. (1994). Evolution of chiral stationary phase design in the Pirkle laboratories. *Journal of Chromatography A* 666 (1): 3–26.

56 Haginaka, J. (2001). Protein-based chiral stationary phases for high-performance liquid chromatography enantioseparations. *Journal of Chromatography A* 906 (1–2): 253–273.

57 Cabrera, K. and Ludbda, D. (1992). Chemically-bonded β-cyclodextrin as chiral stationary phase for the separation of enantiomers of pharmaceutical drugs. *GIT Spezial Chromatographie* 12 (2): 77–79.

58 Menges, R.A. and Armstrong, D.W. (1991). Chiral separations using native and functionalized cyclodextrin-bonded stationary phases in high-pressure liquid chromatography. In: *Chiral Separations by Liquid Chromatography* (ed. S. Ahuju), 67–100. American Chemical Society.

59 Ward, T.J. and Farris, A.B. III (2001). Chiral separations using the macrocyclic antibiotics: a review. *Journal of Chromatography A* 906 (1): 73–89.

60 Stringham, R.W., Lynam, K.G., and Grasso, C.C. (1994). Application of subcritical fluid chromatography to rapid chiral method development. *Analytical Chemistry* 66: 1949–1954.

61 Zhao, Y., Woo, G., Thomas, S. et al. (2003). Rapid method development for chiral separation in drug discovery using sample pooling and supercritical fluid chromatography–mass spectrometry. *Journal of Chromatography A* 1003 (1–2): 157–166.

62 Welch, C.J. (2004). Chiral chromatography in support of pharmaceutical process research. In: *Preparative Enantioselective Chromatography* (ed. G.B. Cox), 1–18. Oxford, UK: Blackwell Publishing Ltd.

63 Maftouh, M., Granier-Loyaux, C., Chavana, E. et al. (2005). Screening approach for chiral separation of pharmaceuticals: Part III. Supercritical fluid chromatography for analysis and purification in drug discovery. *Journal of Chromatography A* 1088 (1): 67–81.

64 White, C. (2005). Integration of supercritical fluid chromatography into drug discovery as a routine support tool: Part I. Fast chiral screening and purification. *Journal of Chromatography A* 1074 (1–2): 163–173.

65 Laskar, D.B., Zeng, L., Xu, R., and Kassel, D.B. (2008). Parallel SFC/MS-MUX screening to assess enantiomeric purity. *Chirality* 20 (8): 885–895.

66 Mangelings, D. and Vander Heyden, Y. (2008). Chiral separations in sub- and supercritical fluid chromatography. *Journal of Separation Science* 31 (8): 1252–1273.

67 Hamman, C., Wong, M., Hayes, M., and Gibbons, P. (2011). A high throughput approach to purifying chiral molecules using 3mum analytical chiral stationary phases via supercritical fluid chromatography. *Journal of Chromatography A* 1218 (22): 3529–3536.

68 Zhang, Y. and Hicks, M.B. (2011). Advanced SFC method development with a multi-column supercritical fluid chromatography with gradient screening. *American Pharmaceutical Review* 14: 52–60.

69 Zeng, L., Xu, R., Zhang, Y., and Kassel, D.B. (2011). Two-dimensional supercritical fluid chromatography/mass spectrometry for the enantiomeric analysis and purification of pharmaceutical samples. *Journal of Chromatography A* 1218 (20): 3080–3088.

70 De Klerck, K., Mangelings, D., and Vander Heyden, Y. (2012). Supercritical fluid chromatography for the enantioseparation of pharmaceuticals. *Journal of Pharmaceutical and Biomedical Analysis* 69: 77–92.

71 De Klerck, K., Parewyck, G., Mangelings, D., and Vander Heyden, Y. (2012). Enantioselectivity of polysaccharide-based chiral stationary phases in supercritical fluid chromatography using methanol-containing carbon dioxide mobile phases. *Journal of Chromatography A* 1269: 336–345.

72 Hamman, C., Wong, M., Aliagas, I. et al. (2013). The evaluation of 25 chiral stationary phases and the utilization of sub-2.0mum coated polysaccharide chiral stationary phases via supercritical fluid chromatography. *Journal of Chromatography A* 1305: 310–319.

73 De Klerck, K., Vander Heyden, Y., and Mangelings, D. (2014). Generic chiral method development in supercritical fluid chromatography and ultra-performance supercritical fluid chromatography. *Journal of Chromatography A* 1363: 311–322.

74 Kalikova, K., Slechtova, T., Vozka, J., and Tesarova, E. (2014). Supercritical fluid chromatography as a tool for enantioselective separation; a review. *Analytica Chimica Acta* 821: 1–33.

75 West, C. (2014). Enantioselective Separations with supercritical fluids – review. *Current Analytical Chemistry* 10: 99–120.

76 Xia, B., Feng, M., Ding, L., and Zhou, Y. (2014). Fast separation method development for supercritical fluid chromatography using an autoblending protocol. *Chromatographia* 77: 781–791.

77 Dispas, A., Lebrun, P., Sacre, P.Y., and Hubert, P. (2016). Screening study of SFC critical method parameters for the determination of pharmaceutical compounds. *Journal of Pharmaceutical and Biomedical Analysis* 125: 339–354.

78 Miller, L. (2012). Evaluation of non-traditional modifiers for analytical and preparative enantioseparations using supercritical fluid chromatography. *Journal of Chromatography A* 1256: 261–266.

79 Miller, L. and Bush, H. (1989). Preparative resolution of enantiomers of prostaglandin precursors by liquid chromatography on a chiral stationary phase. *Journal of Chromatography* 484: 337–345.

80 Miller, L. and Weyker, C. (1990). Analytical and preparative resolution of enantiomers of prostaglandin precursors and prostaglandins by liquid chromatography on derivatized cellulose chiral stationary phases. *Journal of Chromatography* 511: 97–107.

81 Brocks, D.R., Pasutto, F.M., and Jamali, F. (1992). Analytical and semi-preparative high-performance liquid chromatographic separation and assay of hydroxychloroquine enantiomers. *Journal of Chromatography B: Biomedical Sciences and Applications* 581 (1): 83–92.

82 Miller, L. and Bergeron, R. (1993). Analytical and preparative resolution of enantiomers of verapamil and norverapamil using a cellulose-based chiral stationary phase in the reversed-phase mode. *Journal of Chromatography* 648: 381–388.

83 Miller, L., Honda, D., Fronek, R., and Howe, K. (1994). Examples of preparative chiral chromatography on an amylose-based chiral stationary phase in support of pharmaceutical research. *Journal of Chromatography A* 658: 429–435.

84 Miller, L., Orihuela, C., Fronek, R. et al. (1999). Chromatographic resolution of the enantiomers of a pharmaceutical intermediate from the milligram to the kilogram scale. *Journal of Chromatography A* 849: 309–317.

85 Nicoud, R.-M. (1999). The separation of optical isomers by simulated moving bed chromatography. *Pharmaceutical Technology Europe* 11 (3): 36.

86 Pflum, D.A., HSW, G.J.T., Kessler, D.W. et al. (2001). A large-scale synthesis of enantiomerically pure cetrizine dihydrochloride using preparative chiral HPLC. *Organic Process Research and Development* 5: 110–115.

87 Pflum, D.A., Wilkinson, H.S., Tanoury, G.J. et al. (2001). A large-scale synthesis of enantiomerically pure cetrizine dihydrochloride using preparative chiral HPLC. *Organic Process Research and Development* 5 (2): 110–115.

88 Andersson, S. (2007). Preparative chiral chromatography – a powerful and efficient tool in drug discovery. In: *Chiral Separation Techniques, A Practical Approach*, 3rde (ed. G. Subramanian), 585–600. Wiley-VCH.

89 Leonard, W.R. Jr., Henderson, D.W., Miller, R.A. et al. (2007). Strategic use of preparative chiral chromatography for the synthesis of a preclinical pharmaceutical candidate. *Chirality* 19 (9): 693–700.

90 Cox, G.B. (2008). *Preparative Enantioselective Chromatography*. Wiley.

91 Okamoto, Y., Aburatani, R., Miura, S.-I., and Hatada, K. (1987). Chiral stationary phases for HPLC: cellulose tris (3, 5-dimethylphenylcarbamate) and tris (3, 5-dichlorophenylcarbamate) chemically bonded to silica gel*. *Journal of Liquid Chromatography* 10 (8-9): 1613–1628.

92 Yashima, E., Fukaya, H., and Okamoto, Y. (1994). 3, 5-Dimethylphenylcarbamates of cellulose and amylose regioselectively bonded to silica gel as chiral stationary phases for high-performance liquid chr. *Journal of Chromatography A* 677 (1): 11–19.

93 Oliveros, L., Lopez, P., Minguillón, C., and Franco, P. (1995). Chiral chromatographic discrimination ability of a cellulose 3, 5-dimethylphenylcarbamate/10-undecenoate mixed derivative fixed on several chromatographic matrices. *Journal of Liquid Chromatography and Related Technologies* 18 (8): 1521–1532.

94 Franco, P., Minguillón, C., and Oliveros, L. (1998). Solvent versatility of bonded cellulose-derived chiral stationary phases for high-performance liquid chromatography and its consequences in column loadability. *Journal of Chromatography A* 793 (2): 239–247.

95 Franco, P., Senso, A., Oliveros, L., and Minguillón, C. (2001). Covalently bonded polysaccharide derivatives as chiral stationary phases in high-performance liquid chromatography. *Journal of Chromatography A* 906 (1): 155–170.

96 Francotte, E. (2000). Photochemically cross-linked polysaccharide derivatives as supports for the chromatographic separation of enantiomers. US Patent 6011149A.

97 Francotte, E. and Huynh, D. (2002). Immobilized halogenophenylcarbamate derivatives of cellulose as novel stationary phases for enantioselective drug analysis. *Journal of Pharmaceutical and Biomedical Analysis* 27 (3–4): 421–429.

98 Layton, C., Ma, S., Wu, L. et al. (2013). Study of enantioselectivity on an immobilized amylose carbamate stationary phase under subcritical fluid chromatography. *Journal of Separation Science* 36 (24): 3941–3948.

99 De Klerck, K., Vander Heyden, Y., and Mangelings, D. (2014). Pharmaceutical-enantiomers resolution using immobilized polysaccharide-based chiral stationary phases in supercritical fluid chromatography. *Journal of Chromatography A* 1328: 85–97.

100 Lee, J., Lee, J.T., Watts, W.L. et al. (2014). On the method development of immobilized polysaccharide chiral stationary phases in supercritical fluid chromatography using an extended range of modifiers. *Journal of Chromatography A* 1374: 238–246.

101 Dasilva, J.O., Coes, B., Frey, L. et al. (2014). Evaluation of non-conventional polar modifiers on immobilized chiral stationary phases for improved resolution of enantiomers by supercritical fluid chromatography. *Journal of Chromatography A* 1328: 98–103.

102 Miller, L. (2014). Use of dichloromethane for preparative supercritical fluid chromatographic enantioseparations. *Journal of Chromatography A*. 1363: 323–330.

103 Phinney, K.W., Sander, L.C., and Wise, S.A. (1998). Coupled achiral/chiral column techniques in subcritical fluid chromatography for the separation of chiral and nonchiral compounds. *Analytical Chemistry* 70 (11): 2331–2335.

104 Zhang, Y., Zeng, L., Pham, C., and Xu, R. (2014). Preparative two-dimensional liquid chromatography/mass spectrometry for the purification of complex pharmaceutical samples. *Journal of Chromatography A* 1324: 86–95.

105 Alexander, A.J. and Staab, A. (2006). Use of achiral/chiral SFC/MS for the profiling of isomeric cinnamonitrile/hydrocinnamonitrile products in chiral drug synthesis. *Analytical Chemistry* 78 (11): 3835–3838.

106 Ventura, M. (2013). Use of achiral columns coupled with chiral columns in SFC separations to simplify isolation of chemically pure enantiomer products. *American Pharmaceutical Review*. 16 (6): 90–95.

107 Weller, H.N., Ebinger, K., Bullock, W. et al. (2010). Orthogonality of SFC versus HPLC for small molecule library separation. *Journal of Combinatorial Chemistry* 12: 877–882.

108 Wang, Z., Zhang, H., Liu, O., and Donovan, B. (2011). Development of an orthogonal method for mometasone furoate impurity analysis using supercritical fluid chromatography. *Journal of Chromatography A* 1218 (16): 2311–2319.

109 West, C. and Lesellier, E. (2006). Characterization of stationary phases in subcritical fluid chromatography by the solvation parameter model. I.

Alkylsiloxane-bonded stationary phases. *Journal of Chromatography A* 1110 (1-2): 181–190.

110 West, C. and Lesellier, E. (2006). Characterisation of stationary phases in subcritical fluid chromatography by the solvation parameter model. II. Comparison tools. *Journal of Chromatography A* 1110 (1-2): 191–199.

111 West, C. and Lesellier, E. (2006). Characterisation of stationary phases in subcritical fluid chromatography with the solvation parameter model. III. Polar stationary phases. *Journal of Chromatography A* 1110 (1-2): 200–213.

112 West, C. and Lesellier, E. (2006). Characterisation of stationary phases in subcritical fluid chromatography with the solvation parameter model IV. Aromatic stationary phases. *Journal of Chromatography A* 1115 (1-2): 233–245.

113 Khater, S., West, C., and Lesellier, E. (2013). Characterization of five chemistries and three particle sizes of stationary phases used in supercritical fluid chromatography. *Journal of Chromatography A* 1319: 148–159.

114 Lesellier, E. and West, C. (2015). The many faces of packed column supercritical fluid chromatography – a critical review. *Journal of Chromatography A* 1382C: 2–46.

115 West, C., Khalikova, M.A., Lesellier, E., and Heberger, K. (2015). Sum of ranking differences to rank stationary phases used in packed column supercritical fluid chromatography. *Journal of Chromatography A* 1409: 241–250.

116 West, C., Lemasson, E., Bertin, S. et al. (2016). An improved classification of stationary phases for ultra-high performance supercritical fluid chromatography. *Journal of Chromatography A* 1440: 212–228.

117 Ray McClain, M.P. (2011). A systematic study of achiral stationary phases using analytes selected with a molecular diversity model. *LC/GC* 29 (10): 2–9.

118 McClain, R., Hyun, M.H., Li, Y., and Welch, C.J. (2013). Design, synthesis and evaluation of stationary phases for improved achiral supercritical fluid chromatography separations. *Journal of Chromatography A* 1302: 163–173.

119 Tarafder, A. (2016). Metamorphosis of supercritical fluid chromatography to SFC: an Overview. *TrAC Trends in Analytical Chemistry* 81: 3–10.

120 Aurigemma, C.M., Farrell, W.P., Simpkins, J. et al. (2012). Automated approach for the rapid identification of purification conditions using a unified, walk-up high performance liquid chromatography/supercritical fluid chromatography/mass spectrometry screening system. *Journal of Chromatography A* 1229: 260–267.

121 Rosse, G. (2016). Modern SFC-MS platform for achiral and chiral separations. *SFC Day San Diego,* San Diego (23 May 2016).

122 Miller, L. and Peterson, E.A. (2015). Chapter 4 greener solvent usage for discovery chemistry analysis and purification. In: *Green Chemistry Strategies for Drug Discovery: The Royal Society of Chemistry* (eds. E.A. Peterson and J.B. Manley), 66–93. Cambridge, UK: The Royal Society of Chemistry.

123 Isbell, J., Xu, R., Cai, Z., and Kassel, D. (2002). Realities of high-throughput liquid chromatography/mass spectrometry purification of large combinatorial libraries: a report on overall sample throughput using parallel purification. *Journal of Combinatorial Chemistry* 4 (6): 600–611.

124 Blom, K.F., Sparks, R., Doughty, J. et al. (2003). Optimizing preparative LC/MS configurations and methods for parallel synthesis purification. *Journal of Combinatorial Chemistry* 5: 670–683.

125 Shave, D., Brailsford, A., Potts, W., et al. (2004). High Throughput Mass-Directed Purification Of Drug Discovery Compounds. Application Note 720001011en. Waters Corporation.

126 Giger, R. (2000). High-throughput analysis, purification, and quantification of combinatorial libraries of single compounds. *CHIMIA International Journal for Chemistry* 54 (1-2): 37–40.

127 Kassel, D. (2001). Combinatorial chemistry and mass spectrometry in the 21st century drug discovery laboratory. *Chemical Reviews* 101 (2): 255–268.

128 Xu, R., Wang, T., Isbell, J. et al. (2002). High-throughput mass-directed parallel purification incorporating a multiplexed single quadrupole mass spectrometer. *Analytical Chemistry* 74 (13): 3055–3062.

129 Goetzinger, W., Zhang, X., Bi, G. et al. (2004). High throughput HPLC/MS purification in support of drug discovery. *International Journal of Mass Spectrometry* 238 (2): 153–162.

130 Jablonski, J. and Wheat, T. (2009). Optimized chromatography for mass directed purification of peptides. *Waters Application Note* 720000920EN.

131 Thomas, S., Notari, S., Semin, D. et al. (2006). Streamlined approach to the crude compound purification to assay process. *Journal of Liquid Chromatography and Related Technologies* 29 (5): 701–717.

132 Wang, T., Barber, M., Hardt, I., and Kassel, D.B. (2001). Mass-directed fractionation and isolation of pharmaceutical compounds by packed-column supercritical fluid chromatography/mas spectroscopy. *Rapid Communications in Mass Spectrometry* 15: 2067–2075.

133 Zhang, X., Towle, M.H., Felice, C.E. et al. (2006). Development of a mass-directed preparative supercritical fluid chromatography purification system. *Journal of Combinatorial Chemistry* 8: 705–714.

134 McClain, R.T., Dudkina, A., Barrow, J. et al. (2009). Evaluation and implementation of a commercially available mass-guided SFC purification platform in a high throughput purification laboratory in drug discovery. *Journal of Liquid Chromatography and Related Technologies* 32 (4): 483–499.

135 Miller, L. and Bergeron, R. (1994). Preparative liquid chromatographic isolation of unknown impurities in Arbidol and SI-5. *Journal of Chromatography A* 658: 489–496.

136 Roge, A.B., Firke, S.N., Kawade, R.M. et al. (2011). Brief review on: flash chromatography. *International Journal of Pharmaceutical Sciences and Research* 2 (8): 1930–1937.

137 Stevens, W.C. Jr. and Hill, D.C. (2009). General methods for flash chromatography using disposable columns. *Molecular Diversity* 13 (2): 247–252.

138 Still, W.C., Kahn, M., and Mitra, A. (1978). *LC GC North America* 43: 2923.

139 Taygerly, J.P., Miller, L.M., Yee, A., and Peterson, E.A. (2012). A convenient guide to help select replacement solvents for dichloromethane in chromatography. *Green Chemistry* 14 (11): 3020.

140 Ashraf-Khorassani, M., Yan, Q., Akin, A. et al. (2015). Feasibility of correlating separation of ternary mixtures of neutral analytes via thin layer chromatography with supercritical fluid chromatography in support of green flash separations. *Journal of Chromatography A* 1418: 210–217.

141 McClain, R., Rada, V., Nomland, A. et al. (2016). Greening flash chromatography. *ACS Sustainable Chemistry & Engineering* 4 (9): 4905–4912.

142 Christopher, J., Welch, J.A.-W., Leonard, W.R. et al. (2005). Adsorbent screening for metal impurity removal in pharmaceutical research. *Organic Process Research and Development* 9: 198–205.

143 Kraml, C.M., Zhou, D., Byrne, N., and McConnell, O. (2005). Enhanced chromatographic resolution of amine enantiomers as carbobenzyloxy derivatives in high-performance liquid chromatography and supercritical fluid chromatography. *Journal of Chromatography A* 1100 (1): 108–115.

9

Impact and Promise of SFC in the Pharmaceutical Industry

9.1 Introduction to Pharmaceutical Industry

Discovery and development of a pharmaceutical product is a complicated process, requiring at least 10 years with the average cost for research and development of a successful drug estimated to be 2.6 billion US dollars [1]. The phases of pharmaceutical discovery and development are displayed in Figure 9.1. Chromatographic analysis is an essential technique in all phases of pharmaceutical R&D. Liquid chromatography, as a more established technique, is the predominant chromatographic technique. SFC has been used in pharmaceutical discovery for the past 20 years, but only in the last 5 years have advances in theory and equipment that allowed SFC to move into pharmaceutical development and manufacturing. This chapter covers the use of supercritical

Modern Supercritical Fluid Chromatography: Carbon Dioxide Containing Mobile Phases,
First Edition. Larry M. Miller, J. David Pinkston, and Larry T. Taylor.
© 2020 John Wiley & Sons, Inc. Published 2020 by John Wiley & Sons, Inc.

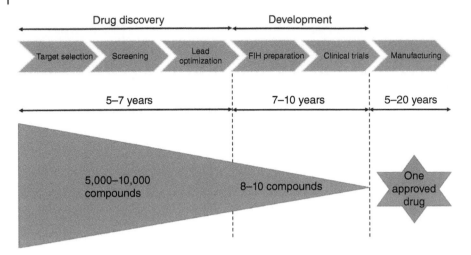

Figure 9.1 Phases of pharmaceutical discovery and development.

fluid chromatography (SFC) within the pharmaceutical industry from discovery to manufacturing. A number of different modalities including, but not limited to, small molecules, peptides, proteins, and monoclonal antibodies are used as pharmaceutical products. As discussed previously, SFC is not applicable for analysis and purification of larger molecules. This chapter discusses the use of SFC for small molecule analysis/purification within the pharmaceutical industry.

9.2 SFC in Pharmaceutical Discovery

9.2.1 Early Discovery Support

The first step in pharmaceutical discovery requires understanding of the workings of a disease including the biological origin of a disease, and the potential targets for intervention. After a potential target is identified, the process of drug discovery begins. Most pharmaceutical companies have in-house collections of hundreds of thousands of compounds. These collections are subjected to high throughput screening (HTS) to rapidly access biochemical activity against the biological target of interest. The "hits" from HTS identify molecules for the lead optimization step of the discovery process. Quality of the compound collection is critical to obtaining high quality hits. Samples are analyzed by LC/MS to confirm purity and mass prior to entering the compound collection. Often NMR is also used to confirm structure. Samples in a compound collection are acquired from a variety of sources including within the company and external organizations. It is estimated that 5–10% of samples in most large compound collections do not contain the reported compound. Others may contain the reported compound, but also additional impurities.

While LC/MS is the technique of choice for analyzing compound collections, SFC has been shown as a suitable alternative. Pinkston et al. compared LC/MS and SFC/MS for the analysis of a large and diverse library of pharmaceutically relevant compounds from the Procter & Gamble Pharmaceutical repository [2]. Their study evaluated 2266 small molecules with a wide variety of functional groups, including nonpolar aliphatics, aromatics, carotenoids, amine hydrohalides, quaternary ammonium salts, multicaroboxylate salts, sulfonates, sulfates, sulfamic acid salts, phosphates, phosphonates, multiphosphonate salts, polyhydroxy compounds, and nitro compounds. When including "hit's with a strong peak related to the reported compound," a total of 87% of the compounds were eluted and detected by SFC/MS and 89.9% of the compounds were eluted and detected by LC/MS. 3.7% of the compounds were detected only by SFC/MS and 8.1% detected only by LC/MS. The only class of compounds consistently detected by LC/MS and not by SFC/MS contained a phosphate, a phosphonate, or a bisphosphonate. This study showed that SFC/MS provided equivalent results to LC/MS for screening a large, diverse library of drug-like molecules with SFC displaying the advantages of speed, environmental friendliness, orthogonal selectivity, and reduced cost of operation.

In the early days of pharmaceutical research natural products were the major source of pharmaceutical products. By 1990, about 80% of drugs were either natural products or analogs of natural products, including antibiotics, antiparasitics, antimalarials, lipid control agents, immunosuppressants, and anticancer drugs [3]. Over the past 10 years the industry has moved away from natural products, although some companies and academic organization are still involved in this research [4]. Nothias et al. reported on the use of analytical and preparative SFC to discover potent antiviral compounds from Euphorbia semiperfoliata whole plant extract [5]. A number of unknown diterpene esters that displayed antiviral activity against Chikungunya virus as well as a potent and selective inhibitor of HIV-1 replication were isolated using SFC-MS/MS and semi-preparative SFC. The advantages of preparative SFC relative to HPLC were evident; SFC allowed isolation in one day compared to 15 days for HPLC and avoided the use of ~100 L of solvent.

9.2.2 SFC in Medicinal Chemistry

Once a "hit" is identified from a high throughput screen, the hard work of turning that molecule into a potential drug begins. This is the lead optimization stage of pharmaceutical discovery. For a typical small molecule program thousands of compounds are synthesized and tested to identify a molecule with the necessary attributes to move from discovery into development. The medicinal chemist must balance numerous properties during lead optimization. These properties fall into four categories: (i) target activity, (ii) safety, (iii) pharmacokinetics and drug metabolism, and (iv) physical properties. The relative importance of properties within each category varies depending on the target disease. It is not usually possible to design a perfect molecule that optimizes all attributes. Safety properties can include hepatatoxicity, cardiotoxicity, genotoxicity, reproductive toxicity, and carcinogenicity. Properties related to pharmacokinetics and drug metabolism can include reactive metabolites, in vivo clearance, bioavailability, and plasma protein binding. Physical properties to be optimized may include molecular weight, lipophilicity, polar surface area, permeability, and solubility. The discovery team draws on their expertise and experience to balance each attribute until a molecule with acceptable properties is identified and the compound can move into development.

Standard procedure in medicinal chemistry lead optimization is to design a molecule, synthesize it and then test it. The results from this cycle are used to design the next molecule, which hopefully will have fewer limitations than its precursor. The cycle is repeated hundreds of times with the knowledge from each cycle helping to design the next. During these cycles thousands of different compounds are synthesized and tested. The synthesis of each compound often involves multiple purification steps in addition to 10–30 analyses depending on the complexity of the molecule. Chromatographic analysis and purification is a critical step in small molecule synthesis.

9.2.2.1 Analytical SFC

While SFC is an acceptable analytical technique to support small molecule synthesis, only minor inroads have been made with SFC for achiral analysis. The majority of achiral analyses performed in pharmaceutical research use LC/MS for purity determination and mass confirmation. Early analytical SFC systems suffered from poor quantitative performance, as well as poor reproducibility and robustness [6]. By the time equipment vendors solved these problems, LC/MS was established as the gold standard for pharmaceutical analysis, a distinction that, even with the advantages of SFC, will be difficult to change. One area where analytical achiral SFC has shown its value is method development for purification. Prior to purification, an analytical method must be developed. The standard approach for analytical method development is to screen a number of stationary and mobile phases. The intrinsic advantages of SFC such as reduced viscosity, lower pressure drops, improved efficiencies at high flow rates, and reduced analysis times make SFC the first choice for high throughput method development.

A large number of stationary phases are available for analytical SFC. Even when the list is pared to include only phases available in preparative dimensions, there are still dozens of options. An efficient method development process must reduce the number of columns evaluated. A number of researchers have developed approaches for achiral SFC method development for scale up to preparative loadings. While there are few publications on analytical achiral SFC from pharma researchers, it is a commonly used technique within the pharmaceutical industry. In 2005, White et al. described the incorporation of achiral SFC into drug discovery in the UK laboratories of Eli Lilly [7]. They described a screening protocol consisting of seven stationary phases (2-ethylpyridine, cyano, premier, silica, diol, diol with high carbon load, and 2-cyano). In 2011, de la Puente et al. described the achiral screening protocol developed in Eli Lilly's Spain facility [8]. From an initial evaluation of 11 phases, it was shown that five of these columns (2-ethylpyridine, diethylaminopropyl, benzenesulfonamide, diol, and dinitrophenyl) were able to resolve a target compound from its impurities in greater than 85% of research mixtures. The following year the same laboratory reported on further improvement of the 5-column screen to a 2-column screen (HILIC cross-linked diol and 2-ethylpyridine), which afforded a success rate of 85–90% [9]. More recently Francotte reported on the achiral SFC screening strategy utilized at Novartis [10]. Their approach used five columns for the primary screen (4-ethylpyridine, propylphenylurea, HILIC silica, amino and di-amino). This screening approach is part of a strategy that has allowed Novartis to transition 75% of their achiral purifications from HPLC to SFC [10].

While achiral SFC has experienced tremendous growth in the past decade, its growth pales in comparison to chiral SFC. Since the first analytical chiral SFC separation was reported in 1985 [11], SFC rapidly became the first choice for enantioseparations in pharmaceutical discovery. The majority of analytical chiral SFC is performed with 5-µm stationary phases. Soon after introduction of improved analytical SFC equipment, there was a shift to 3-µm phases and more recently sub-2 µm columns [12]. Superficially porous chiral columns have also been recently introduced for analytical SFC [13]. Additional information on analytical chiral SFC is found in Chapter 5.

9.2.2.2 Preparative SFC

The major use of SFC in small molecule drug discovery is compound purification. Depending on whether an intermediate or final product is being purified, and the stage of drug discovery, purification scale can vary from 50 mg to greater than 1 kg. Chromatographic purification is a solvent and time intensive process. The increased speed, reduced solvent usage, decreased evaporation time, and increased environmental friendliness of SFC are the main reasons many pharmaceutical companies routinely use SFC for both chiral and achiral purifications [14].

Thousands of compounds are synthesized during small molecule lead optimization. Each of these molecule needs some type of purification to achieve acceptable purity for further evaluation. In the 1990s and early 2000s, high throughput purification platforms were developed using reversed phase chromatography and mass-based fraction collection [15–17]. While this technology is very powerful and was used, and is still used, for the purification of tens of thousands of compounds per year, it does suffer from some limitations. Many small molecules have poor aqueous solubility, leading to poor chromatography. Evaporation of aqueous mobile phases is a time consuming, energy intensive process. Finally, reversed phase purifications often use acidic modifiers that can generate desired products as salts, which may not be suitable for biological testing.

Many of the limitations of mass-directed reversed phase LC were eliminated with the introduction of mass directed SFC systems. McClain et al. reported on the evaluation of mass directed SFC as a replacement for mass directed LC in Merck research laboratories [18]. Besides offering a complimentary separation mechanism, evaporation time was reduced from greater than 8 to 1 hour. The advantage of generating free base products was observed by Searle et al. [19].

Individual enantiomers behave differently within a living system, thus the FDA requires that individual enantiomers be separated and tested. During drug discovery, small quantities (~25–100 mg) of thousands of different molecules must be synthesized. If the target molecules contains a chiral center, it is necessary to ensure high enantiomeric purity. At times an enantiomerically pure starting material can be sourced to provide chirality. Often enantiomerically pure starting materials are not available, and an alternate purification strategy must be employed. Due to the small scale, and wide variety of chemical structures, it is not time efficient to develop an asymmetric synthesis, an enzymatic process, or crystallization process to achieve high enantiomeric purity. The quickest route to enantiopure material during the drug discovery process is chromatographic resolution of a racemate. This approach has the added benefit of generating both enantiomers, especially important as the desired isomer is most often unknown at this stage of discovery.

The first preparative chiral separations were performed by HPLC in the 1980s with the introduction of polysaccharide-based chiral stationary phases (CSPs) [20, 21]. The transition from HPLC to SFC began in the mid-1990s with the introduction of preparative SFC equipment by Berger Instruments and other vendors. While this equipment was suitable for both achiral and chiral purifications, it found its main utility for preparative enantioseparations. Preparative resolution of racemates by LC is solvent intensive, requiring 1–5 L, or more liters of organic solvent per gram of racemate depending on the separation [22–24]. Preparative SFC requires lower solvent volumes, often less than 1 L/g of racemate, and sometimes as low as 200 mL/g [14, 25–28]. The reduction in flammable solvent volumes allowed more purification equipment to be located in one facility before reaching facility solvent limits. By the mid-2000s many pharmaceutical companies, especially in the United States, had made a near complete transition to SFC for preparative enantioseparations. The transition in Europe was slower, being completed in the early 2010s. Early uses of preparative SFC used 2- and 3-cm-i.d. columns and the separation of milligram to gram quantities of racemates. The introduction of 5-cm-i.d. and larger columns, as well as higher flow rate preparative SFC equipment now allows the routine resolution of 1 kg and larger quantities of racemate [28–30]. Additional information on both achiral and chiral preparative SFC is found in Chapter 8.

9.2.3 Physiochemical Measurement by SFC

During lead optimization, a medicinal chemist balances a number of physical properties of the molecules being synthesized and tested. A number of experimental techniques, some chromatography based, are used to measure these properties. Chromatographic measurements often have the advantage of being higher throughput than other techniques. HPLC has been used to measure lipophilicity [31–34], permeability [35], hydrophobicity [36], and log P [37]. Recently, researchers have begun to evaluate SFC for physiochemical measurements, although only a few examples have been published.

Goetz et al. reported on an SFC method to measure permeability of cyclic peptides [38]. This assay was developed for a project designing orally bioavailable peptides, with a goal to have a "reasonably high throughput method that can be reliably used to produce data related to the permeability of peptides." Goetz also developed an SFC method for the indirect detection of intramolecular hydrogen bonding [39]. It is known that increased intramolecular hydrogen bonding correlated with increased membrane permeability. The SFC method correlated retention with the exposed polarity of a molecule. Molecules that form an intramolecular hydrogen bond exhibits lower retention by "hiding their polarity." Figure 9.2 shows the SFC analysis of two isomers: compound 1

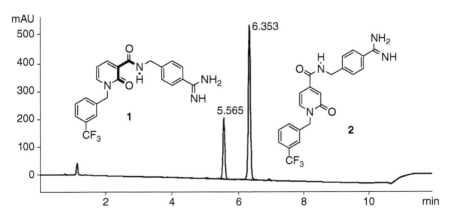

Figure 9.2 Superimposed chromatograms of compounds **1** capable of IMHB formation (5.565 minutes) and **2** incapable of IMHB formation (6.353 minutes). *Source:* Adapted from [39] with permission from American Chemical Society Publications.

is capable of intermolecular hydrogen bonding and has shorter retention relative to compound 2 which is not capable of intermolecular hydrogen bonding.

9.2.4 Use of SFC for Pharmacokinetic and Drug Metabolism Studies

As molecules with acceptable activity and physical properties are prepared in lead optimization, it is necessary to explore the pharmacokinetics and drug metabolism of the molecule. Pharmacokinetics is the study of the time course of drug adsorption, distribution, metabolism, and excretion. Metabolism is the metabolic breakdown of drugs by living systems. These studies involve administering the potential drug to a living system (e.g. mice) and measuring the level of drug as well as metabolites. While metabolism studies are also performed in animals, for early discovery work the typical test systems for metabolic stability studies are liver microsomes or hepatocytes. The introduction of advanced equipment with improved sensitivity has allowed SFC to be used to study bioavailability, metabolites, as well as "in vivo" interconversion during PKDM studies.

Bioavailability is a measure of the extent and rate at which a drug becomes available in general circulation [40]. When developing a potential pharmaceutical, high bioavailability is desired. Low bioavailability can result in varied exposure making it difficult to dose patients. High bioavailability results in administration of lower doses, which can help to reduce cost of drugs. Bioavailability is typically assessed in vivo in rats or mice in early discovery. In later stages of discovery bioavailability may be studied in dog or other species [41]. Geng et al. reported on the development of an SFC-tandem MS method for determination of lacidipine in beagle dog plasma for a bioavailability study [42]. This method was validated and used for quantitation studies. Compared to published LC/MS methods, the SFC method was simpler, used lower toxicity mobile phase, exhibited sharper peaks, and a faster separation time of less than 1.5 minutes/sample. SFC was used for compound measurement in dog plasma for ezetimibein [43] and 3-n-butylphthalide [44]. SFC-MS/MS was also used for measurement of components of Dengtaiye tablets after administration to rats [45].

Upon dosing compounds are chemically modified, or metabolized by various enzymes. Metabolism introduces polar groups into the compound, followed by conjugation to generate polar metabolites that are excreted. Metabolism studies generate a wide range of metabolites, all with unknown structures. These studies require sophisticated analytical technology to monitor metabolism. The gold standard for metabolism studies is LC coupled to triple quad MS (LC-MS/MS). SFC-MS/MS is now a viable option for metabolism studies of pharmaceutical products [46, 47] and has also been explored for metabolism of a fungicide on food products [48]. Yang reported on the use of SFC-MS/MS for simultaneous separation and quantitation of oxcarbazepine and its chiral metabolites in beagle dog plasma [49]. A previously developed LC-MS/MS method for this material had a run time of eight minutes and a LLOQ (lower limit of quantitation) of 50 ng/mL. The sensitivity was not adequate for analyte levels found in biological matrices and the long run time eliminated use in high throughput environments. SFC-MS/MS provide LLOQ of 5 ng/mL for oxcarbazepine and 0.5 ng/mL for the enantiomers with a run time of three minutes.

Standard practice within the pharmaceutical industry is development of individual enantiomers as opposed to racemates [50]. When performing in vivo pharmacokinetic studies on enantiomerically pure materials it is necessary to assess interconversion (i.e. conversion of one enantiomer to the other). Racemization or epimerization of labile chiral centers can be catalyzed by enzymes involved in metabolism [51]. Analysis of these types of studies are complicated by the need to separate and detect the drug product and metabolites, as well as their corresponding stereoisomers. In addition, when interconversion occurs, it is at low levels; resulting in the need to detect stereoisomers across a wider range of concentrations than in metabolic studies of nonchiral molecules. Most metabolic studies of interconversion use HPLC or GC for separation and quantitation; over the past three years a few papers have been published using SFC as the analytical technique.

Yan et al. reported on the use of SFC for biotransformation studies of atropisomers in pharmaceutical research [52]. Individual atropisomers were isolated using preparative SFC and the metabolic stability of the each isomer studied using various species microsomes, hepatocytes, and plasma. The major N-oxide metabolite in human plasma was formed as a racemic mixture of two atropoenantiomers at a concentration of 0.1–1 µg/mL. HPLC did not have adequate sensitivity to detect the metabolite isomers at this concentration. SFC had improved sensitivity over HPLC and was able to detect isomers at the 0.1–1 µg/mL level. The same SFC method was used to confirm that the individual atropisomers of the parent molecule did not racemize during the metabolic studies. Simeone et al. developed a method for separation of enantiomer of 9-hydroxyrisperidone metabolite of Risperidone [53]. This study showed the formation of the R isomer of the metabolite is favored.

As mentioned previously, development of analytical methods for interconversion metabolite studies is complicated by the need to resolve the drug and metabolites as well as corresponding stereoisomers. While the separation challenge can often be solved, high performing chromatographic techniques do not always have the resolving power to separate all peaks. Goel et al. reported the use of two-dimensional LC-SFC-MS [54]. The first dimension separation is reversed phase HPLC. A series of trapping columns is used to park the drug and metabolites from the initial separation. The trapped compounds from the first dimension are then sent to a second dimension chiral SFC method. An example of the separation potential of this configuration is shown in Figure 9.3.

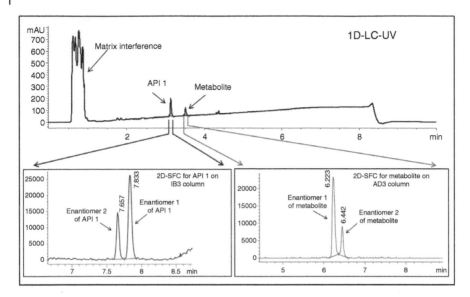

Figure 9.3 2D LC-SFC-MS analysis of API 1 and its metabolite. The chromatogram at the top is the RPLC separation, the bottom is the secondary SFC separation demonstrating resolution of the enantiomers for API 1 (left) and the metabolite (right). For experimental conditions see section 2.4.3 of [54]. *Source:* Reprinted from [54], with permission from Elsevier.

9.3 SFC in Development and Manufacturing

As a potential pharmaceutical moves from discovery into development, there is a large increase in the analytical burden. Discovery support requires development of generic methods suitable for many different compounds, or if developed for an individual molecule, are developed with minimal optimization. In development, analytical methods are fully optimized, and analyze not only final products and impurities, but also formulated products containing active pharmaceutical ingredient (API) and one or more excipients. Stability indicated methods, as well as bioanalytical methods for supporting clinical studies must also be developed. Finally, all analytical methods supporting clinical development and manufacturing must be validated, that is "proven it is suitable for its intended purpose." Validation of analytical methods is beyond the scope of this chapter. Readers are directed to the following for additional information [55–58].

9.3.1 Analytical SFC Analysis of Drug Substances and Drug Products

The first generation of analytical SFC instrumentation suffered from poor injection reproducibility, excessive detector noise, and inconsistent pump performance, especially under low modifier conditions. These limitations made

SFC impractical for use in regulated laboratory laboratories. Current generations of analytical SFC equipment have higher sensitivity, robustness, and reproducibility, and are routinely used for validated assays. Additional information on these equipment advances is found in Chapter 3.

Analytical characterization of drug substance in drug development is more intensive than characterization performed in drug discovery. Methods are required that resolve all impurities (many of them unknown) and degradation products present as low as 0.05%. The impurities are often closely related in structure to the drug substance and can be difficult to resolve. Methods must be developed for not only the drug substance (active pharmaceutical ingredient, API) but also for the drug product (formulated mixture of API and excipients in final dosage form). These requirements put a premium on effective method development. A number of generic SFC method development approaches are discussed throughout this book. Most of these approaches are not suitable when developing a method that requires validation and will be used for release of drug substance or drug product. With analytical SFC moving into validated analytical laboratories, a number of method development strategies suitable for validated methods have been published [59–67].

Galea et al. discussed an approach for selecting a dissimilar set of stationary phases for impurity profiling in SFC [59]. To maximize the possibility that all impurities are resolved from the API, method development should use columns and mobile phases that provide different selectivities [60, 61]. In this study, SFC retention of 64 drugs were measured on 27 columns. Correlation coefficients and principal component analysis were used to select of set of six dissimilar stationary phases (silica, C18, amino, Phenyl, HILIC, and cyano). The orthogonality of these six columns was demonstrated through the analysis of three drug mixtures. Lemasson et al. investigated selection of orthogonal columns [63] as well as optimization of mobile phase composition for impurity profiling of drug candidates [62]. Galea et al. investigated the effect of column temperature and back-pressure for drug impurity profiling [64].

Dispas et al. evaluated the impact of different critical method parameters (CMPs) including stationary phase, mobile phase, and injection solvent [65]. Design of experiments (DOE) and desirability function approaches enabled optimal chromatographic conditions to maximize peak capacity with acceptable values of symmetry factor. Interestingly, with the stationary phase used for this study, the accepted practice of "matching polarity between dissolution solvent (i.e. sample) and mobile phase" did not always hold true. Galea et al. also used DOE in their studies, this time to optimize column temperature, back-pressure, and gradient slope simultaneously [66]. Finally, Lemasson et al. compared high performance LC and SFC coupled to MS for impurity profiling of drug candidates [67]. Using a proprietary set of 140 pharmaceutical compounds they showed comparable quality between the best UHPLC and UHPSFC methods while also showing the two techniques to be highly orthogonal. It is their opinion that a combination of UHPLC and UHPSFC maximize the chance of resolving all impurities in a drug substance, and should be standard practice in pharmaceutical laboratories.

The first reported uses of SFC for drug substance and drug product analysis was in the early 2000s [68, 69]. The methods provided good sensitivity, but required high sample concentrations (10 and 12.5 mg/mL) to detect impurities at the 0.1% level. This is an indication of the poor sensitivity of early model analytical SFC systems. Current analytical systems use more typical sample concentrations of 0.5–1 mg/mL. Neither paper reported on validation of the SFC methods. In 2011, Wang et al. reported on an orthogonal method for impurity analysis of mometasone furoate [70]. The method was capable of trace level (0.05% of active) impurity analysis using an increased sample concentration and provided orthogonal selectivity to an existing HPLC method in one-third the analysis time. The method was partially validated and it was demonstrated with further validation that the method may be suitable for release and stability testing. Li et al. described the development of a sensitive and rapid method for the impurity analysis of the antibiotic rifampicin using SFC [71]. The SFC method had improved resolution of impurities with a reduced analysis time of 4 minutes compared to 50 minutes for the HPLC method. Figure 9.4 illustrates the orthogonality of SFC as well as the reduced analysis time relative to HPLC. Development and validation of analytical achiral SFC methods have also been reported for quinine sulfate [65] and salbutamol sulfate [72].

A number of synthetic impurities are formed during the synthesis of small molecules. The characterization and control of these impurities must be performed prior to filing with regulatory agencies. The origin and downstream fate of each impurity must be understood. Impurity fate and purge studies are often used to gain this knowledge [73]. Typically, these studies are performed using reversed phase LC-MS. Depending on separation complexity these studies may require two methods, one for percent area analysis and a different method for peak identification. Also, reverse phase HPLC is not suitable for intermediates or impurities that are not stable in aqueous environments. Pirrone et al. reported on the use of SFC-MS as an analytical technique for impurity fate mapping [74]. SFC showed improved separation for closely related components as well as stability for intermediates that reacted with water under HPLC conditions.

Release testing of compounds with a chiral center requires an enantioselective method to confirm enantiomeric purity. Marley et al. reported the development and validation of an SFC method to determine (R)-timolol in (S)-timolol [75]. The method was validated to meet the European Pharmacopeia requirements of a limit test for enantiomeric purity. Compared to the existing HPLC method listed in European Pharmacopoeia, SFC was 3 times faster and used 11 times less solvent. Hicks et al. reported on Merck Research Laboratories

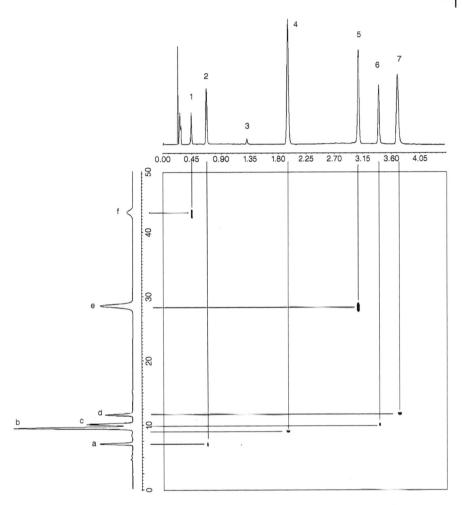

Figure 9.4 Spectral correlative map of rifampicin and its impurities obtained with SFC method RPLC method. Peak-2 and a, RQ; 4 and b, RF; 5 and e, 3-FR; 6 and c, RSV; 7 and d, RNO; 1, 3, and f, unknown impurities. *Source:* Reprinted from [71], with permission from Elsevier.

work to make chiral SFC a viable alternative to HPLC in support of pharmaceutical development and manufacturing activities [76]. Their findings illustrated that modern SFC equipment exhibits improved precision, reproducibility, accuracy, and robustness compared to earlier SFC equipment. SFC also provided superior resolution and peak capacity compared to HPLC. They recommend the use of SFC for GMP studies of stereochemistry in pharmaceutical development and manufacturing.

The lower mobile phase viscosity of carbon-dioxide-based mobile phases allows the use of longer column lengths, or coupling of different stationary phases to offer increased resolution and unique selectivities. Coupling of an achiral and a chiral column could theoretically allow both API assay, impurity determination, and enatiopurity to be measured in one analysis. Coupling of chiral and achiral columns has been reported, but we know of no publications on column coupling in a validated environment [77, 78]. Venkatramani et al. reported on a two-dimensional reversed phase LC-SFC method that would allow simultaneous achiral and chiral analysis of pharmaceutical compounds [79]. The peaks of interest from the achiral RPLC separation were concentrated on C-18 trapping columns and then injected onto the second dimension SFC system. The first dimension measured achiral purity and the second dimension provided chiral purity. While impurity detection in the first dimension was not discussed, one could envision this equipment configuration providing this capability.

Analytical SFC is also used for the analysis of formulated drug products. The first reported use of achiral SFC for dosage forms was reported in 2012 for the separation and quantitation of chlorzoxazone, Paracetomol, and Aceclofenac in their individual and combined dosage forms [80]. Alexander et al. reported on the use of SFC for analysis of the triple combination tablet of lamivudine, BMS-986001, and efavirenz as an antiretroviral human immunodeficiency virus type 1 (HIV-1) treatment [81]. The tablets contained three APIs and 13 possible impurity/degradation products. All 16 peaks were resolved by RPLC; by SFC 15 peaks were observed with one coeluting pair. A high degree of orthogonality and more even distribution of peaks across the separation space were observed by SFC. The authors also noted that the SFC method had advantages of a nonsloping baseline as well as fewer system peaks. SFC exhibited the sensitivity required for successful quantitation of impurities/degradation products at the 0.05–0.1 area percent level. Plachka et al. reported on SFC separation of agomelatine and its impurities [82]. The authors also developed a validated UHPLC method that was compared to UHPSFC. An informative comparison of the UHPSFC and UHPLC methods is shown in the spider diagram in Figure 9.5. Both methods demonstrated very good results in terms of repeatability, linearity, accuracy, and precision. UHPSFC provided slightly better results for method precision and resolution between the API and trailing impurity. All of the UHPSFC peaks were symmetrical and very sharp. UHPLC exhibited higher sensitivity as measured by lower limit of quantitation (LLOQ). The author noted the greatly increased method development time for UHPLC relative to UHPSFC.

An interested study was performed to evaluate SFC for determination of PEG adducts in pharmaceuticals [83]. Drug formulations containing polyethylene glycol (PEG) can generate reaction products between PEG and the API. Adduct formation was measured by incubating PEG with two different

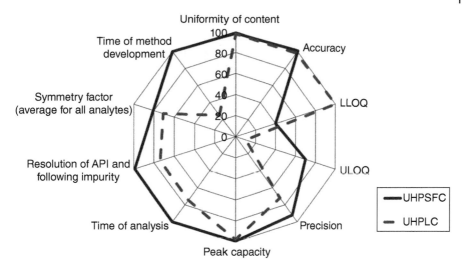

Figure 9.5 Spider diagram comparison of developed UHPSFC and UHPLC methods for the determination of agomelatine and its six impurities. *Source:* Reprinted from [82], with permission of Elsevier.

pharmaceutical compounds, cetirizine, and indomethacin, at 80 °C for 120 hours. The formation of polyethylene glycol esters was detected by a shift in the elution pattern using evaporative light scattering detector, and was confirmed by mass spectrometry. Scientists in France compared HPLC and SFC with ELSD detection for the determination of plasticizers in medical devices [84]. Plasticized-PVC is widely employed as tubing in medical devices used for blood transfusions, drug infusion, and dialysis. With some medical devices containing up to 40% w/w of plasticizers, regulatory agencies are interested in their identities, their levels, and their effect on human health. HPLC and SFC methods to resolve four plasticizers and the methods were validated. Validation studies showed HPLC to be more precise than SFC, but lower limits of quantitation were obtained with SFC. The HPLC method was validated in the ±10% acceptance limits but SFC lacked the accuracy to quantify the plasticizers. The authors felt the low accuracy in SFC may have been due to the instrument used, and different results may have been obtained with an SFC from a different manufacturer SFC can also be used for analysis of nontablet formulations. In 2007, the use of chiral SFC for assay of an aqueous formulation was reported [85]. Prior to this publication there was concern with direct injection of an aqueous sample in SFC. These concerns included sample freezing, precipitation, and distorted peak shape. This work showed that direct injection of a 100% aqueous formulation in SFC was feasible. The chiral method was validated, showing high degrees of selectivity, accuracy, precision, robustness, sensitivity, and linearity over a wide range of concentrations. SFC has also been explored for quality control of vitamin D3 oily formulations [86]. By SFC, it is possible to meet the specification established by the European Medicines Agency (EMA). These results illustrated the possibility of using SFC-MS for the QC of medicine and support the switch to greener analytical methods.

9.3.2 Preparative SFC in Development and Manufacturing

While preparative SFC is a widely used technique in pharmaceutical research, its use within pharmaceutical discovery and manufacturing is limited. Large scale purifications using liquid chromatography is a mature field with numerous examples in pharmaceutical manufacturing. The use of continuous chromatography (also called simulated moving bed, SMB), which offer high separation productivities and use little solvent due to recycling, can result in low purification costs [87–90]. Also, there exist numerous contract research and manufacturing companies with large scale liquid chromatography equipment (both batch and SMB) allowing separation of metric tons of material per year. To the authors' knowledge, there is only one contract facility with large scale preparative SFC equipment, and the size of the equipment limits production to less than one metric ton per year for most applications. Work on continuous purifications (SFC-SMB) has been reported in the literature, but equipment has not been manufactured beyond the proof of concept scale [91–94].

Preparative SFC has been used to isolate active ingredients from fish oil and palm oil. The first reports of use of SFC in fish oil purification were in the late 1980s. The use of the ethyl esters of eicosapentaenoic acid (EPA) and docosahexaenoic acid (DHA) has been promoted for numerous indications including psoriasis, bowel disease, mental illness, and cancer prevention [95]. Over the last 20 years there has been a shift to concentrated "omega-3 oils" containing higher levels of EPA and DHA with current over the counter products reaching EPA+DHA concentrations of 50–55% [96]. Pharmaceutical products containing fish oil were introduced in 2004 with the approval of Lovaza, containing EPA and DHA at concentrations of 85%. As an adjunct to diet, this mixture reduce triglyceride levels in patients with severe hypertriglyceridemia. The year 2012 saw FDA approval of Vascepa, a highly concentrated EPA (96%) formulation for reduction of triglyceride levels. The manufacturing processes for Lovaza and Vascepa are proprietary, but based on the number of recent patents and articles in the area of fish oil purification using SFC [97–100], it is probable that large scale SFC is being used in the manufacture of high purity EPA pharmaceutical products. The triglycerides of palm oil are used as a source of biodiesel, but also contain ~1% of phytonutrients such as tocols, carotenes andphytosterols, with a value of 970 USD per ton of biodiesel produced from palm oil [101]. The high value of the byproducts from biodiesel production made isolation economically feasible. While there have been additional publications on SFC purification of palm oil, it is unknown at this time if this process is being used at manufacturing scale [102, 103].

9.3.3 Metabolite/PKDM Studies in Development

Section 9.2.4 discussed the use of SFC for analysis of biological materials for pharmaceutical discovery support. This section centers on SFC for the analysis of pharmaceuticals in biological materials derived from humans. When a compound moves into development and clinical studies, the same pharmacokinetic, and metabolism studies performed on rat, dog, and other species, are also performed after human dosing. LC with triple quadrupole detection is the method of choice in this field for analysis of complex samples where the analytes of interest are present in the low ng/mL level or less [104]. SFC equipment advances now allow the required sensitivity for detection of these types of samples. Scientists in Norway developed an SFC method for enantiomeric separation and quantitation of citalopram in serum [105]. After validation, the method was implemented in routine use for both separation and quantitation of R/S-citalopram in greater than 250 serum samples as part of therapeutic drug monitoring project. On-line SFE-SFC-MS was used to quantify ketamine and its metabolites for metabolic profiling and pharmacological investigations of ketamine in humans [106].

An SFC method was developed to quantify 15 sulfonamides and their N4-acetylation metabolites in serum [107]. Separation of all 15 sulfonamides and their metabolites was achieved in seven minutes. Pilarova et al. developed an SFC-MS method for the high throughput determination of all isomeric forms of vitamin E in human serum [108]. Two methods to resolve all eight tocopherols and tocotrienols were developed, a high speed method with analysis time of 2.5 minutes, and a high resolution method with analysis time of 4.5 minutes. Readers with further interest are directed to the following review articles on the use of SFC for bioanalysis [109–111]. The use of SFC for metabolic phenotyping, also referred to as metabonomics or metabolomics, was recently explored [112]. A fast SFC-MS method for the analysis of medium and high polarity (clogP of −7 to 2) compounds was developed and its utility demonstrated for the separation of polar metabolites in human urine. A unique use of SFC was recently published, a high throughput method for the separation and quantitation of C_{19} and C_{21} steroids and their C11-oxy metabolites [113]. The unique selectivity of this method may hold promise in the identification of new steroid markers in steroid-linked endocrine diseases, in addition to profiling steroid metabolism and abnormal enzyme activity in patients.

9.3.4 SFC in Chemical Process Development

The goal of chemical process development is creation of a safe and cost effective synthetic route for generation of the larger quantities of active ingredient required for clinical studies and ultimately for manufacturing. Many of the types of analytical SFC support required for process development, such as achiral and chiral analysis, have already been discussed in this chapter. This section discusses a few specialized areas that do not fall in the previously discussed categories.

Development of a chemical process can involve evaluation of hundreds of reaction conditions to characterize the reaction, and determine optimum operating conditions. Many of these reactions are performed in a high throughput, parallel mode to minimize development time. Each of the reaction needs analysis to determine yield and purity. A number of analytical technique are in use, many based on LC separation [114–117]. SFC has been investigated as an analytical method to support high throughput experimentation. Scientists at Merck described the use of MISER chromatographic analysis (Multiple Injections in a Single Experimental Run) using SFC [118]. Using this technique, it is possible to analyze a full plate of 96 samples in less than 34 minutes. The introduction of sub-2 μm particles for chiral columns has expanded the potential of chiral chromatography for high-throughput screen of large compound libraries [12].

An additional use of SFC in chemical process development is impurity isolation using preparative SFC. It is necessary to confirm the structures of reaction byproducts in order to fully characterize a reaction step. The speed and reduced requirements for solvent removal, make preparative SFC an ideal technique to rapidly isolate milligram quantities of impurities for characterization by MS and/ or NMR. Due to the confidential nature of synthesis of compounds in the pharmaceutical industry, most isolation work is not published. Klobcar et al. reported on the isolation of oxidative degradation products of atrorvatatin [119]. Zhao discussed isolation of multiple impurities from a mother liquor sample [120].

9.4 SFC for Analysis of Illegal Drugs

Once seized by law enforcement, illegal drugs are analyzed by a combination of instrumental, immunoassay, microscopic, and wet chemical techniques. Often more than one separation technique is used to increase specificity. Acceptable separation techniques include TLC, GC, LC, and CE. Currently SFC is not an approved separation technique. A review of SFC-MS for forensic analysis can be found at [121]. Breitenbach et al. evaluated the use of UHPSFC for the analysis of synthetic cannabinoids [122]). UHPSFC was evaluated due to its superior resolution of positional isomers and diastereomers. SFC was able to resolve the synthetic cannabinoid and nine of its positional isomers in under 10 minutes. Principal component analysis was used to demonstrate the orthogonality of SFC, GC, and HPLC.

The S-enantiomer of amphetamine has a greater CNS stimulant activity than the R-enantiomer. The majority of prescribed amphetamine consists of the pure S-enantiomer. Illegal amphetamine is available mainly as a racemic mixture. To distinguish between legal and illegal amphetamine, chiral separation of R- and S- amphetamine in biological specimens is required. Hegstad et al.

presented the evaluation of SFC-MS for enantiomeric separation and quantitation of R/S-amphetamine in urine [123]. The method was used routinely and shown to be a reliable and useful tool for distinguishing R- and S-amphetamine in patient samples.

Anabolic steroid misuse improves body weight gain and feed-conversion efficiency in meat producing animals. Anabolic steroids are also used by athletes to increase sports performance. Analysis of glucuronide and sulfate steroids in urine is the accepted method to detect anabolic steroid misuse. Doue et al. compared SFC-MS to HPLC-MS for the analysis of glucuronide and sulfate steroids in urine [124]. They determined that SFC-MS-MS had improved sensitivity and reproducibility compared to HPLC-MS-MS. The technology was used to confirm doping of estradiol in two animals by detection and quantitation of conjugated steroids. Desfontaine et al. developed UHPLC and UHPSFC methods for screening of 43 anabolic agents in human urine [125]. These methods were compared to the reference GC-MS-MS method used for steroid analysis in anti-doping laboratories. The UHPLC and UHPSFC methods had numerous advantages relative to GC method, mainly sensitivity and matrix effects. Endogenous compounds from the sample matrix that elute at the same time as target compounds can negatively influence method accuracy, precision and sensitivity through ion suppression or enhancement [126]. The matrix effect of the three techniques (SFC, GC, and LC) is shown in Figure 9.6. These results are in agreement with previous studies of matrix effects in SFC [127].

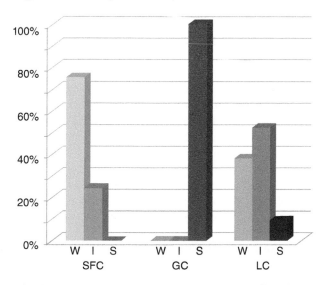

Figure 9.6 Graphical representation of the importance of matrix effects for the three methods. Weak matrix effects (W) correspond to $-20\% < ME_{rel} < 20$, intermediate matrix effects (I) to $20\% < |ME_{rel}| < 100\%$, and strong matrix effect (S) to ME_{rel} higher than 100% or lower than -100%. *Source:* Reprinted from [125], with permission from Elsevier.

9.5 Summary

Analytical and preparative SFC has been a core technology for pharmaceutical discovery for the past 20 years. Equipment and theoretical advances of the past five years have allowed SFC to move from the nonvalidated environment of pharmaceutical discovery to the validated environment of pharmaceutical development and manufacturing. SFC has also moved into formulated drug products, bioanalytical analysis, and analysis of illegal drugs; areas that have been typically served by HPLC or other analytical techniques. As the chromatographic and environmental advantages of SFC become more widely known, expansion into other areas is expected.

References

1 DiMasi, J.A., Grabowski, H.G., and Hansen, R.W. (2016). Innovation in the pharmaceutical industry: new estimates of R&D costs. *Journal of Health Economics* 47: 20–33.

2 Pinkston, J.D., Wen, D., Morand, K.L. et al. (2006). Comparison of LC/MS and SFC/MS for screening a large and diverse library of pharmaceutical relevant compounds. *Analytical Chemistry* 78 (7467–7472): 7467.

3 Li, J.W.-H. and Vederas, J.C. (2009). Drug discovery and natural products: end of an era or an endless frontier? *Science* 325 (5937): 161–165.

4 David, B., Wolfender, J.-L., and Dias, D.A. (2014). The pharmaceutical industry and natural products: historical status and new trends. *Phytochemistry Reviews* 14 (2): 299–315.

5 Nothias, L.-F., Boutet-Mercey, S., Cachet, X. et al. (2017). Environmentally friendly procedure based on supercritical fluid chromatography and tandem mass spectrometry molecular networking for the discovery of potent antiviral compounds from Euphorbia semiperfoliata. *Journal of Natural Products.*

6 Schafer, W., Chandrasekaran, T., Pirzada, Z. et al. (2013). Improved chiral SFC screening for analytical method development. *Chirality* 25 (11): 799–804.

7 White, C. and Burnett, J. (2005). Integration of supercritical fluid chromatography into drug discovery as a routine support tool: II. Investigation and evaluation of supercritical fluid chromatography for achiral batch purification. *Journal of Chromatography A* 1074 (1): 175–185.

8 de la Puente, M.L., Lopez Soto-Yarritu, P., and Burnett, J. (2011). Supercritical fluid chromatography in research laboratories: design, development and implementation of an efficient generic screening for exploiting this technique in the achiral environment. *Journal of Chromatography A* 1218 (47): 8551–8560.

9 de la Puente, M.L., Soto-Yarritu, P.L., and Anta, C. (2012). Placing supercritical fluid chromatography one step ahead of reversed-phase high performance liquid chromatography in the achiral purification arena: a hydrophilic

interaction chromatography cross-linked diol chemistry as a new generic stationary phase. *Journal of Chromatography A* 1250: 172–181.

10 Francotte, E. (2016). Practical advances in SFC for the purification of pharmaceutical molecules. *LC GC Europe* 29 (4): 194–204.

11 Mourier, P.A., Eliot, E., Caude, M.H. et al. (1985). Supercritical and subcritical fluid chromatography on a chiral stationary phase for the resolution of phosphine oxide enantiomers. *Analytical Chemistry* 57 (14): 2819–2823.

12 Sciascera, L., Ismail, O., Ciogli, A. et al. (2015). Expanding the potential of chiral chromatography for high-throughput screening of large compound libraries by means of sub-2mum Whelk-O 1 stationary phase in supercritical fluid conditions. *Journal of Chromatography A* 1383: 160–168.

13 Ciogli, A., Ismail, O.H., Mazzoccanti, G. et al. (2018). Enantioselective ultra-high performance liquid and supercritical fluid chromatography: the race to the shortest chromatogram. *Journal of Separation Science*.

14 Miller, L. (2012). Preparative enantioseparations using supercritical fluid chromatography. *Journal of Chromatography A* 1250: 250–255.

15 Xu, R., Wang, T., Isbell, J. et al. (2002). High-throughput mass-directed parallel purification incorporating a multiplexed single quadrupole mass spectrometer. *Analytical Chemistry* 74 (13): 3055–3062.

16 Zeng, L., Burton, L., Yung, K. et al. (1998). Automated analytical/preparative high-performance liquid chromatography–mass spectrometry system for the rapid characterization and purification of compound libraries. *Journal of Chromatography A* 794 (1): 3–13.

17 Kassel, D. (2001). Combinatorial chemistry and mass spectrometry in the 21st century drug discovery laboratory. *Chemical Reviews* 101 (2): 255–268.

18 McClain, R.T., Dudkina, A., Barrow, J. et al. (2009). Evaluation and implementation of a commercially available mass-guided SFC purification platform in a high throughput purification laboratory in drug discovery. *Journal of Liquid Chromatography and Related Technologies* 32 (4): 483–499.

19 Searle, P.A., Glass, K.A., and Hochlowski, J.E. (2004). Comparison of preparative HPLC/MS and preparative SFC techniques for the high-throughput purification of compound libraries. *Journal of Combinatorial Chemistry* 6: 175–180.

20 Miller, L. and Bush, H. (1989). Preparative resolution of enantiomers of prostaglandin precursors by liquid chromatography On a chiral stationary phase. *Journal of Chromatography* 484: 337–345.

21 Miller, L. and Weyker, C. (1990). Analytical and preparative resolution of enantiomers of prostaglandin precursors and prostaglandins by liquid chromatography on derivatized cellulose chiral stationary phases. *Journal of Chromatography* 511: 97–107.

22 Miller, L., Orihuela, C., Fronek, R. et al. (1999). Chromatographic resolution of the enantiomers of a pharmaceutical intermediate from the milligram to the kilogram scale. *Journal of Chromatography A* 849: 309–317.

23 Miller, L., Grill, C., Yan, T. et al. (2003). Batch and simulated moving bed chromatographic resolution of a pharmaceutical racemate. *Journal of Chromatography A* 1006 (1-2): 267–280.

24 Grill, C.M., Miller, L., and Yan, T.Q. (2004). Resolution of a racemic pharmaceutical intermediate. *Journal of Chromatography A* 1026 (1-2): 101–108.

25 Miller, L. and Potter, M. (2008). Preparative supercritical fluid chromatography (SFC) in drug discovery. *American Pharmaceutical Review* 11 (4): 112–117.

26 Miller, L. (2012). Evaluation of non-traditional modifiers for analytical and preparative enantioseparations using supercritical fluid chromatography. *Journal of Chromatography A* 1256: 261–266.

27 Miller, L. (2014). Pharmaceutical purifications using preparative supercritical fluid chromatography. *Chimica Oggi / Chemistry Today* 32 (2): 23–26.

28 Miller, L. (2014). Use of dichloromethane for preparative supercritical fluid chromatographic enantioseparations. *Journal of Chromatography A* 1363: 323–330.

29 Welch, C.J., Leonard, W.R. Jr., Dasilva, J.O. et al. (2005). Preparative chiral SFC as a green technology for rapid access to enantiopurity in pharmaceutical process research. *LCGC North America* 23 (1).

30 Leonard, W.R. Jr., Henderson, D.W., Miller, R.A. et al. (2007). Strategic use of preparative chiral chromatography for the synthesis of a preclinical pharmaceutical candidate. *Chirality* 19 (9): 693–700.

31 Du, C.M., Valko, K., Bevan, C. et al. (1998). Rapid gradient RP-HPLC method for lipophilicity determination: a solvation equation based comparison with isocratic methods. *Analytical Chemistry* 70 (20): 4228–4234.

32 Kaliszan, R., Haber, P., Bączek, T. et al. (2002). Lipophilicity and pKa estimates from gradient high-performance liquid chromatography. *Journal of Chromatography A* 965 (1-2): 117–127.

33 Valkó, K. (2004). Application of high-performance liquid chromatography based measurements of lipophilicity to model biological distribution. *Journal of Chromatography A* 1037 (1-2): 299–310.

34 Valkó, K.L. (2016). Lipophilicity and biomimetic properties measured by HPLC to support drug discovery. *Journal of Pharmaceutical and Biomedical Analysis* 130: 35–54.

35 Ermondi, G., Vallaro, M., and Caron, G. (2018). Learning how to use IAM chromatography for predicting permeability. *European Journal of Pharmaceutical Sciences* 114: 385–390.

36 Valkó, K., Bevan, C., and Reynolds, D. (1997). Chromatographic hydrophobicity index by fast-gradient RP-HPLC: a high-throughput alternative to log P/log D. *Analytical Chemistry* 69 (11): 2022–2029.

37 Valko, K., Du, C., Bevan, C. et al. (2001). Rapid method for the estimation of octanol/water partition coefficient (log Poct) from gradient RP-HPLC retention and a hydrogen bond acidity term (Sigma alpha2H). *Current Medicinal Chemistry* 8 (9): 1137–1146.

38 Goetz, G.H., Philippe, L., and Shapiro, M.J. (2014). EPSA: a novel supercritical fluid chromatography technique enabling the design of permeable cyclic peptides. *ACS Medicinal Chemistry Letters* 5 (10): 1167–1172.

39 Goetz, G.H., Farrell, W., Shalaeva, M. et al. (2014). High throughput method for the indirect detection of intramolecular hydrogen bonding. *Journal of Medicinal Chemistry* 57 (7): 2920–2929.

40 Hurst, S., Loi, C.-M., Brodfuehrer, J., and El-Kattan, A. (2007). Impact of physiological, physicochemical and biopharmaceutical factors in absorption and metabolism mechanisms on the drug oral bioavailability of rats and humans. *Expert Opinion on Drug Metabolism & Toxicology* 3 (4): 469–489.

41 van de Waterbeemd, H. and Testa, B. (2009). Introduction: the why and how of drug bioavailability research. In: *Drug Bioavailability: Estimation of Solubility, Permeability, Absorption and Bioavailability*, 2e (eds. H. van de Waterbeemd and B. Testa), 1–6. Germany: Wiley-VCH Verlag GmbH & Co. KgaA.

42 Geng, Y., Zhao, L., Zhao, J. et al. (2014). Development of a supercritical fluid chromatography-tandem mass spectrometry method for the determination of lacidipine in beagle dog plasma and its application to a bioavailability study. *Journal of Chromatography B, Analytical Technologies in the Biomedical and Life Sciences* 945–946: 121–126.

43 Di, M., Li, Z., Jiang, Q. et al. (2018). A rapid and sensitive supercritical fluid chromatography/tandem mass spectrometry method for detection of ezetimibein dog plasma and its application in pharmacokinetic studies. *Journal of Chromatography B, Analytical Technologies in the Biomedical and Life Sciences* 1073: 177–182.

44 Li, Y., Zhao, L., Li, X. et al. (2015). Quantification of 3-n-butylphthalide in beagle plasma samples by supercritical fluid chromatography with triple quadruple mass spectrometry and its application to an oral bioavailability study. *Journal of Separation Science* 38 (4): 697–702.

45 Yang, Z., Sun, L., Liang, C. et al. (2016). Simultaneous quantitation of the diastereoisomers of scholarisine and 19-epischolarisine, vallesamine, and picrinine in rat plasma by supercritical fluid chromatography with tandem mass spectrometry and its application to a pharmacokinetic study. *Journal of Separation Science* 39 (13): 2652–2660.

46 Matsubara, A., Fukusaki, E., and Bamba, T. (2010). Metabolite analysis by supercritical fluid chromatography. *Bioanalysis* 2 (1): 27–34.

47 Taguchi, K., Fukusaki, E., and Bamba, T. (2014). Supercritical fluid chromatography/mass spectrometry in metabolite analysis. *Bioanalysis* 6 (12): 1679–1689.

48 Tao, Y., Zheng, Z., Yu, Y. et al. (2018). Supercritical fluid chromatography-tandem mass spectrometry-assisted methodology for rapid enantiomeric analysis of fenbuconazole and its chiral metabolites in fruits, vegetables, cereals, and soil. *Food Chemistry* 241: 32–39.

49 Yang, Z., Xu, X., Sun, L. et al. (2016). Development and validation of an enantioselective SFC-MS/MS method for simultaneous separation and quantification of oxcarbazepine and its chiral metabolites in beagle dog plasma. *Journal of Chromatography B, Analytical Technologies in the Biomedical and Life Sciences* 1020: 36–42.

50 De Camp, W.H. (1989). The FDA perspective on the development of stereoisomers. *Chirality* 1 (1): 2–6.

51 Testa, B., Carrupt, P.A., and Gal, J. (1993). The so-called "interconversion" of stereoisomeric drugs: An attempt at clarification. *Chirality* 5 (3): 105–111.

52 Yan, T.Q., Riley, F., Philippe, L. et al. (2015). Chromatographic resolution of atropisomers for toxicity and biotransformation studies in pharmaceutical research. *Journal of Chromatography A* 1398: 108–120.

53 Simeone, J., Alelyunas, Y., Wrona, M., and Rainville, P. (2014). Enantiomeric Separation of the Active Metabolite of Risperidone and its Application for Monitoring Metabolic Stability using UPC2/MS/MS. Application note 720005181en. https://www.gimitec.com/file/720005181en.pdf.

54 Goel, M., Larson, E., Venkatramani, C.J., and Al-Sayah, M.A. (2018). Optimization of a two-dimensional liquid chromatography-supercritical fluid chromatography-mass spectrometry (2D-LC-SFS-MS) system to assess "in-vivo" inter-conversion of chiral drug molecules. *Journal of Chromatography B, Analytical Technologies in the Biomedical and Life Sciences* 1084: 89–95.

55 Dispas, A., Lebrun, P., and Hubert, P. (2017). Chapter 11 – validation of supercritical fluid chromatography methods. In: *Supercritical Fluid Chromatography*, 317–344. Elsevier.

56 Guideline IHT ed. (2005). Validation of analytical procedures: text and methodology Q2 (R1). *International Conference on Harmonization*, Geneva, Switzerland.

57 Huang, Y. (2014). Analytical SFC for impurities. In: *Supercritical Fluid Chromatography: Advances and Applications in Pharmaceutical Analysis* (ed. G.K. Webster), 225. Boca Raton, FL: CRC Press Taylor & Francis Group.

58 Dispas, A., Lebrun, P., Ziemons, E. et al. (2014). Evaluation of the quantitative performances of supercritical fluid chromatography: From method development to validation. *Journal of Chromatography A* 1353: 78–88.

59 Galea, C., Mangelings, D., and Heyden, Y.V. (2015). Method development for impurity profiling in SFC: The selection of a dissimilar set of stationary phases. *Journal of Pharmaceutical and Biomedical Analysis* 111: 333–343.

60 Van Gyseghem, E., Van Hemelryck, S., Daszykowski, M. et al. (2003). Vander Heyden Y. Determining orthogonal chromatographic systems prior to the development of methods to characterise impurities in drug substances. *Journal of Chromatography A* 988 (1): 77–93.

61 Van Gyseghem, E., Jimidar, M., Sneyers, R. et al. (2004). Selection of reversed-phase liquid chromatographic columns with diverse selectivity towards the potential separation of impurities in drugs. *Journal of Chromatography A* 1042 (1): 69–80.

62 Lemasson, E., Bertin, S., Hennig, P. et al. (2015). Development of an achiral supercritical fluid chromatography method with ultraviolet absorbance and mass spectrometric detection for impurity profiling of drug candidates. Part I: optimization of mobile phase composition. *Journal of Chromatography A* 1408: 217–226.

63 Lemasson, E., Bertin, S., Hennig, P. et al. (2015). Development of an achiral supercritical fluid chromatography method with ultraviolet absorbance and mass spectrometric detection for impurity profiling of drug candidates. Part II. Selection of an orthogonal set of stationary phases. *Journal of Chromatography A* 1408: 227–235.

64 Muscat Galea, C., Slosse, A., Mangelings, D., and Vander Heyden, Y. (2017). Investigation of the effect of column temperature and back-pressure in achiral supercritical fluid chromatography within the context of drug impurity profiling. *Journal of Chromatography A* 1518: 78–88.

65 Dispas, A., Lebrun, P., Sacre, P.Y., and Hubert, P. (2016). Screening study of SFC critical method parameters for the determination of pharmaceutical compounds. *Journal of Pharmaceutical and Biomedical Analysis* 125: 339–354.

66 Muscat Galea, C., Didion, D., Clicq, D. et al. (2017). Method optimization for drug impurity profiling in supercritical fluid chromatography: application to a pharmaceutical mixture. *Journal of Chromatography A* 1526: 128–136.

67 Lemasson, E., Bertin, S., Hennig, P. et al. (2016). Comparison of ultra-high performance methods in liquid and supercritical fluid chromatography coupled to electrospray ionization – mass spectrometry for impurity profiling of drug candidates. *Journal of Chromatography A* 1472: 117–128.

68 Gyllenhaal, O. and Karlsson, A. (2000). Packed-column supercritical fluid chromatography for the analysis of isosorbide-5-mononitrate and related compounds in bulk substance and tablets. *Journal of Biochemical and Biophysical Methods* 43 (135–146): 135.

69 Ashraf-Khorassani, M., Taylor, L.T., Williams, D.G. et al. (2001). Demonstrative validation study employing a packed column pressurized fluid chromatography method that provides assay, achiral impurities, chiral impurity, and IR identity testing for a drug substance. *Journal of Pharmaceutical and Biomedical Analysis* 26: 725–738.

70 Wang, Z., Zhang, H., Liu, O., and Donovan, B. (2011). Development of an orthogonal method for mometasone furoate impurity analysis using supercritical fluid chromatography. *Journal of Chromatography A* 1218 (16): 2311–2319.

71 Li, W., Wang, J., and Yan, Z.Y. (2015). Development of a sensitive and rapid method for rifampicin impurity analysis using supercritical fluid

chromatography. *Journal of Pharmaceutical and Biomedical Analysis* 114: 341–347.

72 Dispas, A., Desfontaine, V., Andri, B. et al. (2017). Quantitative determination of salbutamol sulfate impurities using achiral supercritical fluid chromatography. *Journal of Pharmaceutical and Biomedical Analysis* 134: 170–180.

73 Li, Y., Liu, D.Q., Yang, S. et al. (2010). Analytical control of process impurities in Pazopanib hydrochloride by impurity fate mapping. *Journal of Pharmaceutical and Biomedical Analysis* 52 (4): 493–507.

74 Pirrone, G.F., Mathew, R.M., Makarov, A.A. et al. (2018). Supercritical fluid chromatography-photodiode array detection-electrospray ionization mass spectrometry as a framework for impurity fate mapping in the development and manufacture of drug substances. *Journal of Chromatography B, Analytical Technologies in the Biomedical and Life Sciences* 1080: 42–49.

75 Marley, A. and Connolly, D. (2014). Determination of (R)-timolol in (S)-timolol maleate active pharmaceutical ingredient: validation of a new supercritical fluid chromatography method with an established normal phase liquid chromatography method. *Journal of Chromatography A* 1325: 213–220.

76 Hicks, M.B., Regalado, E.L., Tan, F. et al. (2016). Supercritical fluid chromatography for GMP analysis in support of pharmaceutical development and manufacturing activities. *Journal of Pharmaceutical and Biomedical Analysis* 117: 316–324.

77 Phinney, K.W., Sander, L.C., and Wise, S.A. (1998). Coupled achiral/chiral column techniques in subcritical fluid chromatography for the separation of chiral and nonchiral compounds. *Analytical Chemistry* 70 (11): 2331–2335.

78 Alexander, A.J. and Staab, A. (2006). Use of achiral/chiral SFC/MS for the profiling of isomeric cinnamonitrile/hydrocinnamonitrile products in chiral drug synthesis. *Analytical Chemistry* 78 (11): 3835–3838.

79 Venkatramani, C.J., Al-Sayah, M., Li, G. et al. (2016). Simultaneous achiral-chiral analysis of pharmaceutical compounds using two-dimensional reversed phase liquid chromatography-supercritical fluid chromatography. *Talanta* 148: 548–555.

80 Desai, P.P., Patel, N.R., Sherikar, O.D., and Mehta, P.J. (2012). Development and validation of packed column supercritical fluid chromatographic technique for quantification of chlorzoxazone, paracetamol and aceclofenac in their individual and combined dosage forms. *Journal of Chromatographic Science* 50 (9): 769–774.

81 Alexander, A.J., Zhang, L., Hooker, T.F., and Tomasella, F.P. (2013). Comparison of supercritical fluid chromatography and reverse phase liquid chromatography for the impurity profiling of the antiretroviral drugs lamivudine/BMS-986001/efavirenz in a combination tablet. *Journal of Pharmaceutical and Biomedical Analysis* 78-79: 243–251.

82 Plachka, K., Chrenkova, L., Dousa, M., and Novakova, L. (2016). Development, validation and comparison of UHPSFC and UHPLC methods for the determination of agomelatine and its impurities. *Journal of Pharmaceutical and Biomedical Analysis* 125: 376–384.

83 Schou-Pedersen, A.M., Ostergaard, J., Johansson, M. et al. (2014). Evaluation of supercritical fluid chromatography for testing of PEG adducts in pharmaceuticals. *Journal of Pharmaceutical and Biomedical Analysis* 88: 256–261.

84 Lecoeur, M., Decaudin, B., Guillotin, Y. et al. (2015). Comparison of high-performance liquid chromatography and supercritical fluid chromatography using evaporative light scattering detection for the determination of plasticizers in medical devices. *Journal of Chromatography A* 1417: 104–115.

85 Mukherjee, P.S. (2007). Validation of direct assay of an aqueous formulation of a drug compound AZY by chiral supercritical fluid chromatography (SFC). *Journal of Pharmaceutical and Biomedical Analysis* 43 (2): 464–470.

86 Andri, B., Dispas, A., Klinkenberg, R. et al. (2017). Is supercritical fluid chromatography hyphenated to mass spectrometry suitable for the quality control of vitamin D3 oily formulations? *Journal of Chromatography A* 1515: 209–217.

87 McCoy, M. (2000). SMB emerges as chiral technique. *Chemical and Engineering News* 78 (25): 17–19.

88 Houlton, S. (2001). SMB chromatography: a fast way forward? *Manufacturing Chemist* 72 (11): 23–26.

89 Quallich, G.J. (2005). Development of the commercial process for Zoloft/sertraline. *Chirality* 17 (Suppl): S120–S126.

90 Scott, R., Perrin, W.H., Ndzie, E. et al. (2007). Purification of difluoromethylornithine by global process optimization: coupling of chemistry and chromatography with enantioselective crystallization. *Organic Process Research and Development* 11 (817-824): 817.

91 Peper, S., Johannsen, M., and Brunner, G. (2007). Preparative chromatography with supercritical fluids. Comparison of simulated moving bed and batch processes. *Journal of Chromatography A* 1176 (1-2): 246–253.

92 Martínez Cristancho, C.A., Peper, S., and Johannsen, M. (2012). Supercritical fluid simulated moving bed chromatography for the separation of ethyl linoleate and ethyl oleate. *The Journal of Supercritical Fluids* 66: 129–136.

93 Liang, M.-T., Liang, R.-C., Huang, L.-R. et al. (2014). Supercritical fluids as the desorbent for simulated moving bed – application to the concentration of triterpenoids from Taiwanofugus camphorata. *Journal of the Taiwan Institute of Chemical Engineers* 45 (4): 1225–1232.

94 Johannsen, M. and Brunner, G. Supercritical fluid chromatographic separation on preparative scale and in continuous mode. *The Journal of Supercritical Fluids* 2017.

95 Rubio-Rodríguez, N., Beltrán, S., Jaime, I. et al. (2010). Production of omega-3 polyunsaturated fatty acid concentrates: a review. *Innovative Food Science and Emerging Technologies* 11 (1): 1–12.

96 Lembke, P. (2013). Production techniques for omega-3 concentrates. In: *Omega-6/3 Fatty Acids: Functions, Sustainability Strategies and Perspectives* (eds. F. De Meester, R.R. Watson and S. Zibadi), 353–364. Totowa, NJ: Humana Press.

97 Krumbholz, R., Lembke, P., and Schirra, N., inventors; Google Patents, assignee. Chromatography process for recovering a substance or a group of substances from a mixture 2011.

98 Adam, P., Valéry, E., and Bléhaut, J., inventors; Google Patents, assignee. Chromatographic process for the production of highly purified polyunsaturated fatty acids 2014.

99 Kelliher, A., Morrison, A., Oroskar, A., et al., inventors; Google Patents, assignee. SMB process for producing highly pure EPA from fish oil 2016.

100 Fiori, L., Volpe, M., Lucian, M. et al. (2017). From fish waste to omega-3 concentrates in a biorefinery concept. *Waste and Biomass Valorization* 8 (8): 2609–2620.

101 Han, N.M., May, C.Y., and Hawari, Y. (2008). Pilot scale supercritical fluid chromatography: an experience. *SFC*, Zurich, Switzerland (October 2008).

102 Han, N.M. and Choo, Y. (2015). Enhancing the separation and purification efficiency of palm oil carotenes using supercritical fluid chromatography. *Journal of Oil Palm Research* 27 (4): 387–392.

103 Ng, M.H. and Choo, Y.M. (2015). Packed supercritical fluid chromatography for the analyses and preparative separations of palm oil minor components. *American Journal of Analytical Chemistry* 6 (08): 645.

104 Nováková, L. (2013). Challenges in the development of bioanalytical liquid chromatography–mass spectrometry method with emphasis on fast analysis. *Journal of Chromatography A* 1292: 25–37.

105 Hegstad, S., Havnen, H., Helland, A. et al. (2017). Enantiomeric separation and quantification of citalopram in serum by ultra-high performance supercritical fluid chromatography-tandem mass spectrometry. *Journal of Chromatography B, Analytical Technologies in the Biomedical and Life Sciences* 1061-1062: 103–109.

106 Hofstetter, R., Fassauer, G.M., and Link, A. (2018). Supercritical fluid extraction (SFE) of ketamine metabolites from dried urine and on-line quantification by supercritical fluid chromatography and single mass detection (on-line SFE-SFC-MS). *Journal of Chromatography B, Analytical Technologies in the Biomedical and Life Sciences* 1076: 77–83.

107 Zhang, Y., Zhou, W.E., Li, S.H. et al. (2016). A simple, accurate, time-saving and green method for the determination of 15 sulfonamides and metabolites in serum samples by ultra-high performance supercritical fluid chromatography. *Journal of Chromatography A* 1432: 132–139.

108 Pilarova, V., Gottvald, T., Svoboda, P. et al. (2016). Development and optimization of ultra-high performance supercritical fluid chromatography mass spectrometry method for high-throughput determination of tocopherols and tocotrienols in human serum. *Analytica Chimica Acta* 934: 252–265.

109 Ríos, A., Zougagh, M., and de Andrés, F. (2010). Bioanalytical applications using supercritical fluid techniques. *Bioanalysis* 2 (1): 9–25.

110 Dispas, A., Jambo, H., André, S. et al. (2018). Supercritical fluid chromatography: a promising alternative to current bioanalytical techniques. *Bioanalysis* 10 (2): 107–124.

111 Lesellier, E. (2011). Supercritical fluid chromatography for bioanalysis: practical and theoretical considerations. *Bioanalysis* 3 (2): 125–131.

112 Sen, A., Knappy, C., Lewis, M.R. et al. (2016). Analysis of polar urinary metabolites for metabolic phenotyping using supercritical fluid chromatography and mass spectrometry. *Journal of Chromatography A* 1449: 141–155.

113 du Toit, T., Stander, M.A., and Swart, A.C. (2018). A high-throughput UPC(2)-MS/MS method for the separation and quantification of C19 and C21 steroids and their C11-oxy steroid metabolites in the classical, alternative, backdoor and 11OHA4 steroid pathways. *Journal of Chromatography B, Analytical Technologies in the Biomedical and Life Sciences* 1080: 71–81.

114 Robbins, D.W. and Hartwig, J.F. (2011). A simple, multidimensional approach to high-throughput discovery of catalytic reactions. *Science* 333 (6048): 1423–1427.

115 Collins, K.D., Gensch, T., and Glorius, F. (2014). Contemporary screening approaches to reaction discovery and development. *Nature Chemistry* 6 (10): 859.

116 Schafer, W., Bu, X., Gong, X. et al. (2014). 9.02 High-Throughput Analysis for High-Throughput Experimentation in Organic Chemistry A2. In: *Comprehensive Organic Synthesis II*, 2e (ed. P. Knochel), 28–53. Amsterdam: Elsevier.

117 Houben, C. and Lapkin, A.A. (2015). Automatic discovery and optimization of chemical processes. *Current Opinion in Chemical Engineering* 9: 1–7.

118 Zawatzky, K., Biba, M., Regalado, E.L., and Welch, C.J. (2016). MISER chiral supercritical fluid chromatography for high throughput analysis of enantiopurity. *Journal of Chromatography A* 1429: 374–379.

119 Klobčar, S. and Prosen, H. (2015). Isolation of oxidative degradation products of atorvastatin with supercritical fluid chromatography. *Biomedical Chromatography* 29 (12): 1901–1906.

120 Zhao, Y. (2014). SFC in process analytical chemistry. In: *Supercritical Fluid Chromatography: Advances and Applications in Pharmaceutical Analysis* (ed. G.K. Webster), 195. Boca Raton, FL: CRC Press Taylor & Francis Group.

121 Pauk, V. and Lemr, K. (2018). Forensic applications of supercritical fluid chromatography – mass spectrometry. *Journal of Chromatography B, Analytical Technologies in the Biomedical and Life Sciences* 1086: 184–196.

122 Breitenbach, S., Rowe, W.F., McCord, B., and Lurie, I.S. (2016). Assessment of ultra high performance supercritical fluid chromatography as a separation technique for the analysis of seized drugs: Applicability to synthetic cannabinoids. *Journal of Chromatography A* 1440: 201–211.

123 Hegstad, S., Havnen, H., Helland, A. et al. (2018). Enantiomeric separation and quantification of R/S-amphetamine in urine by ultra-high performance supercritical fluid chromatography tandem mass spectrometry. *Journal of Chromatography B, Analytical Technologies in the Biomedical and Life Sciences*: 1077–1078:7-12.

124 Doue, M., Dervilly-Pinel, G., Pouponneau, K. et al. (2015). Analysis of glucuronide and sulfate steroids in urine by ultra-high-performance supercritical-fluid chromatography hyphenated tandem mass spectrometry. *Analytical and Bioanalytical Chemistry* 407 (15): 4473–4484.

125 Desfontaine, V., Novakova, L., Ponzetto, F. et al. (2016). Liquid chromatography and supercritical fluid chromatography as alternative techniques to gas chromatography for the rapid screening of anabolic agents in urine. *Journal of Chromatography A* 1451: 145–155.

126 Cappiello, A., Famiglini, G., Palma, P. et al. (2008). Overcoming matrix effects in liquid chromatography – mass spectrometry. *Analytical Chemistry* 80 (23): 9343–9348.

127 Novakova, L., Rentsch, M., Grand-Guillaume Perrenoud, A. et al. (2015). Ultra high performance supercritical fluid chromatography coupled with tandem mass spectrometry for screening of doping agents. II: analysis of biological samples. *Analytica Chimica Acta* 853: 647–659.

10

Impact of SFC in the Petroleum Industry

10.1 Petroleum Chemistry

10.1.1 Crude Refining Processes

Diesel/petroleum is a naturally occurring, yellow-to-black liquid found in geological formations beneath Earth's surface, which is commonly refined into various types of fuels. Three conditions must be met for oil reservoirs to form:

Modern Supercritical Fluid Chromatography: Carbon Dioxide Containing Mobile Phases,
First Edition. Larry M. Miller, J. David Pinkston, and Larry T. Taylor.
© 2020 John Wiley & Sons, Inc. Published 2020 by John Wiley & Sons, Inc.

(i) a source rock rich in hydrocarbon material buried deep enough for subterranean heat to cook it into oil, (ii) a porous and permeable reservoir rock for it to accumulate in, and (iii) a cap rock or other mechanism that prevents heat from escaping to the surface. Within these reservoirs, fluids typically organize themselves like a three-layer cake with a layer of water below the oil layer and a layer of gas above it.

Crude oil is considered *light* if it has low density or *heavy* if it has high density. It is referred to as *sweet* if it contains relatively little sulfur or *sour* if it contains substantial amounts of sulfur. Light crude oil is more desirable than heavy crude oil since it produces a higher yield of gasoline; while sweet oil commands a higher price than sour oil because it has fewer environmental impurities and requires less refining to meet sulfur standards that are imposed on fuels in consuming countries. Petroleum includes only crude oil, but in common usage it includes all liquid, gaseous, and solid hydrocarbons. It is indeed a crude material, being messy, unstable, and dangerous to handle. An oil well produces predominantly crude oil, with some natural gas dissolved in it. The hydrocarbons in crude oil are mostly alkanes, various aromatic hydrocarbons, naphthenes, and asphaltics. The other organic compounds contain nitrogen, oxygen, sulfur, and trace amounts of metal such as iron, nickel, copper, and vanadium [1].

Components of petroleum are separated using fractional distillation. This process separates different components of crude oil so that they can be further refined. Fractional distillation begins when the crude oil is put into a high pressure steam boiler. The crude oil is heated to temperatures up to 1112 °F. After the oil vaporizes, it enters the bottom of the distillation column through a pipe. The distillation column is a tall tank that contains many trays or plates. The vapor rises in the column, cooling as it rises. The specific vapors cool at their boiling point and condense on the plates or trays in the column. Via this process the components are partially separated from one another into fractions.

Liquid fractions of crude oil are then placed into 10 main categories. These main products are further refined to create materials more common to everyday life. The processes employed in addition to fractional distillation include (i) vacuum flashing, (ii) thermal cracking, (iii) cat cracking, (iv) hydrocracking, (v) gas plant alkylation, (vi) catalytic reforming, and (vii) blending in that order. The yield of the previous process becomes the feedstock of the next process, with distillation being the first process and blending the final one [2]. The main *products* of petroleum are termed: asphalt, diesel, fuel Oil, gasoline, kerosene, liquefied petroleum gas, lubricating oil, paraffin wax, bitumen, and petrochemicals.

10.1.2 Petrochemical Processes

Petrochemicals are nonfuel compounds derived from crude oil and natural gas. The feed-stocks for petrochemical plants are provided largely by refineries and include gas, naphtha, kerosene, and light gas oil. Natural gas processing plants are also a source of feedstock, thereby providing natural gas, ethane, and

Figure 10.1 Petroleum refinery in Anacortes, Washington, USA. *Source:* https//wikipedia. org/wiki/oil refinery Anacortes, Wahington,U.S. picture.

liquefied petroleum gas. Petrochemical plants are typically built adjacent to refineries in order to be near the source of feedstock. Also, there often are byproduct streams created in the plants that can best be utilized by returning them to the refinery for processing. In fact, the boundary between the two types of plant often become blurred as progressive integration of both fluid streams and operating personnel takes place for increased efficiency. As a result several very large refining and petrochemical plants have developed around the world (see Figure 10.1). Petrochemical processes are, in general, much smaller than refining processes, but there are exceptions like very large ethylene plants [3].

10.2   Introduction to Petroleum Analysis

The true test of the usefulness of an analytical technique is its range of applicability in solving real-world problems. Supercritical fluid chromatography (SFC) and supercritical fluid extraction (SFE) have always been able to attract interest, owing to some to the intrigue stimulated by the use of the word "supercritical" [4]. Practical analytical applications of the two techniques have really only appeared in the last decade. The increasing demand for petroleum and the dwindling supply of the natural resource have encouraged the petroleum industry to process high boiling crude fractions and to consider alternate sources of the petroleum, such as tar sands, shale oil, and coal. With a greater diversity of fuels has come the need for new and better methods of hydrocarbon analysis, particularly for heavy hydrocarbons. SFC offers new opportunities for these hydrocarbon analyses that arise from (i) the availability of the flame ionization detector and (ii) the possibility of achieving high chromatographic resolution of heavy hydrocarbons that exceed the analysis range of gas chromatography.

Hydrocarbon type analysis primarily refers to the separation and quantitative measurement of saturates, olefins, aromatics, and polar components. Fractions may be further grouped into (i) straight-chain, branched, and cyclic paraffins; (ii) olefins according to the number of unsaturated groups, or aliphatic vs. alicyclic structures; and (iii) aromatics according to the number of rings. Polar compounds include heavy condensed aromatic hydrocarbons and compounds containing sulfur, nitrogen, and oxygen heteroatoms.

Broadly speaking, petroleum analyses concern (i) class-type (e.g. aliphatic vs. mono-aromatic vs. poly-aromatic) analyses and (ii) analyses of one or more target compounds. In the case of petroleum products, there are relatively few classes of molecules. The complexities of petroleum separations arise instead from the enormous number of components present within these classes [5]. Furthermore, petroleum is refined through fractional distillation based on the boiling points of compounds, and the number of components in a petroleum fraction increases exponentially with boiling point [6]. Thus, fractions that include low-boiling compounds are particularly complex. To add to the complexity, traditional fossil fuel sources are dwindling and new ways to isolate fuels from existing sources are being developed.

Given these enhanced oil extraction techniques and the inherent complexity of petroleum, the sheer number of compounds in a sample demand a variety of separation techniques for their analysis. In some cases, the number of components can quickly exceed the available peak capacity of single-chromatographic column techniques and thus require more advanced techniques, such as two-dimensional separations to resolve more of the components. This chapter looks at the history of petroleum separations with special attention concerning how SFC currently contributes to this field.

The application of SFC to the separation of hydrocarbons and related compounds started in the 1980s when it was discovered that SFC using carbon dioxide as the mobile phase combined many of the advantages associated with both gas chromatography (GC) and liquid chromatography (LG). For example, SFC combines GC-like detectors with a LC-like mobile phase.

This chapter also examines advanced techniques such as two-dimensional chromatography, which significantly increase the peak capacity of separations. Multidimensional separations are ideally suited for addressing the analytical challenges associated with measuring compounds in petroleum samples [7]. GC × GC can bring specific benefits to the entire range of petroleum analyses from profiling potential oilfields during exploration to assessing the presence of petroleum and its combustion products in the environment [8].

It is relatively easy to couple SFC with either GC or even comprehensive two-dimensional GC (i.e. SFC-GC × GC) because one obtains the sample as a gas after decompression of the supercritical carbon dioxide [9]. In addition, it seems reasonable to imagine a SFC × SFC system that could reach the performance of GC × GC because long, high resolution columns can be used in the first dimension (SFC) followed by a very fast separation which can be obtained in the second dimension (GC).

10.3 Historical Perspective

10.3.1 Hydrocarbon Analysis via FIA

Since its development first by Conrad [10] and later by Criddle and LeTourneau [11], the petroleum industry initially used a fluorescence indicator adsorption (FIA) method, ASTM D1319, for the determination of saturated, olefinic, and aromatic hydrocarbons in naphtha, jet fuel, and diesel oil. The method served as the industry standard for 30 years because it was the only direct method of measuring these groups of compounds. The method employed fluorescent dyes that were placed at the head of an open, packed silica gel column. A sample of fuel was added, followed by elution of the analytes through the column with isopropyl alcohol. Saturates, olefins, and aromatics traveled down the column at different speeds separating into discrete bands of different color. Under UV light, saturated hydrocarbons remained colorless, olefins turned yellow, aromatics turned blue, and the alcohol front turned red. The length of each band was measured wherein each length corresponded to the volume percent of the chemical group. The major problem associated with the FIA method was the operator's ability to consistently distinguish the sharp yellow, blue, and red fronts. The FIA method, however, was never useful for highly colored shale or coal-derived fractions since colored samples interfered with detection of the dyed zone boundaries.

In applying SFC to this separation, the use of high fluid densities, temperatures close to the critical point, and 1-mm-i.d. columns packed with small silica particles tended to minimize the volatility effects that caused band broadening. However, it was found that with CO_2, the separation of saturates from olefins was incomplete because of the low polarizability of CO_2 compared to hexane and perfluorinated hydrocarbons used as solvents in liquid chromatography. In other words, the gain in quantitation was offset by a loss in resolution. The use of sulfur hexafluoride as mobile phase that is reasonably compatible with FID is analogous to the use of perfluorinated hydrocarbons. In this case, the instrumentation was somewhat complex, however, and sulfur hexafluoride was an environmentally objectionable fluid.

10.3.2 SFC Replaces FIA

In 1984, Rawdon and Norris at Texaco were probably the first workers to propose SFC for fuel group separations [12] that retained the simplicity of a single silica column operated with CO_2 as the mobile phase. Their instrumentation was a modified liquid chromatograph (HP1082 LC) with independent control of pressure and flow. This arrangement was combined with a stand-alone Gow-Mac gas chromatograph and flame ionization detector (GC-FID). Rawdon's and Norris' results unfortunately could not be repeated owing to instability of the silver loaded GC columns which were incorporated into the method to separate olefins from paraffins [13, 14].

High performance liquid chromatography (HPLC) was another alternative to FIA because of its speed and resolution capability, but HPLC lacked a sensitive and a universal detector that could be used for the general analysis of petroleum fractions [15]. Nuclear magnetic resonance and mass spectrometry have been used, but neither of these techniques became operational for hydrocarbon analysis because of initial high capital cost and operational complexities.

Today, SFC with flame ionization detection has been shown to work well for hydrocarbon group analysis. It was determined that retention time reproducibility was greatly enhanced by controlling the temperature of the SFC pressure transducer on the pump. A UV, or other nondestructive detector, was often placed before the BPR. Pressure density programming was used to elute increasingly larger molecular weight analytes. In order for density-programmed SFC to be quantitative, however, the problem of variable split ratio injection accompanying pressure programming needed to be resolved [16].

Packed column SFC/FID generally splits the column effluent before the backpressure regulator, thus directing the lesser fraction to the FID, and the majority of the flow to the BPR. In other words, during a SFC pressure program run, with column effluent split between FID and UV detection, the total flow remained constant; while, the flow into the fixed restrictor associated with the FID increased. Under these conditions, the FID "response factor" ceased to be constant. Apparent molecular weight distributions were thus distorted, and indicated that larger amounts of heavier components were present than was actually the case.

Several temporary strategies to deal with this issue were subsequently published by Berger [14], but no long-term solution was apparent. Experimental problems persisted: (i) fixed restrictors were irreproducible, with no two restrictors being exactly the same, (ii) back pressure regulators tended to cause serious band broadening if used in front of an FID, and (iii) pressure programming caused major changes in column flow, dramatically changing peak widths and column efficiency, plus potentially changing the detector response factor. Nevertheless, with proper calibration, packed column SFC/FID remained a useful quantitative tool in petroleum analysis.

10.3.3 Hydrocarbon SFC Analysis via ASTM 5186-91

The first *standard analytical method* of any kind to use SFC was ASTM Method D 5186-91 for determination of aromatics in diesel fuel via the Western States Petroleum Association Round Robin [17]. With an SFC method in hand, aromatic content could be quantified below 1% in as little time as two minutes. The analysis was robust and repeatable on different instruments. The SFC method used (i) a silica column, (ii) pure CO_2 at constant pressure, and (iii) a flame ionization detector (FID). A sampling valve with a fixed loop was used to introduce the sample. The ASTM method utilized C_{22} and toluene as probes of resolution. Method specifications called for a resolution greater than R = 4 between the solutes. Resolution using a 25-cm column was as high as 13.5. By using higher flow rates, the analysis could be shortened to less than two minutes with minimal loss of resolution (R = 11). The ASTM method called for quantification using FID, but this method did not completely characterize actual fuels [16].

A better understanding of the composition of fuels was, however, realized by simultaneously using FID in conjunction with a UV detector. The method eliminated the need for the separation of olefins from saturates. As a result, the chromatographic equipment and methodology became extremely simple. Samples were injected neat, a single silica column and isopycnic conditions were used. The method was independent of sample size since calculations for quantifications were based solely on peak ratio area measurements. FID that acted as a general mass sensitive detector for all three carbon types provided quantitative information; while, UV detection at 190 nm provided high sensitive, uniform detector response that permitted semi-quantitation of the olefin fraction. Unlike the FIA method, the revised SFC method was suitable for colored samples as well as samples containing materials with mass lighter than hexane. Olefin quantitation by the SFC method correlated well with results by the standard FIA method for a wide range of commercial gasolines. Dual UV/FID chromatograms with a packed silica column are shown in Figure 10.2.

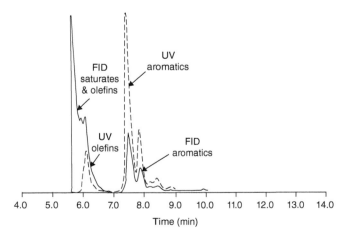

Figure 10.2 SFC of standard gasoline using dual UV/FID detection, Conditions: 50 cm × 1 mm i.d. packed column, silica particles; CO_2; 40 °C; 200 atm [18].

10.4 Early Petroleum Applications of SFC

10.4.1 Samples with Broad Polymer Distribution

Because the density of carbon dioxide employed with SFC strongly depends on the pressure of the system, pressure programming was used to process hydrocarbon mixtures that exhibited a wide molecular weight range. The chromatogram displayed in Figure 10.3, for example, has separate peaks for the oligomers

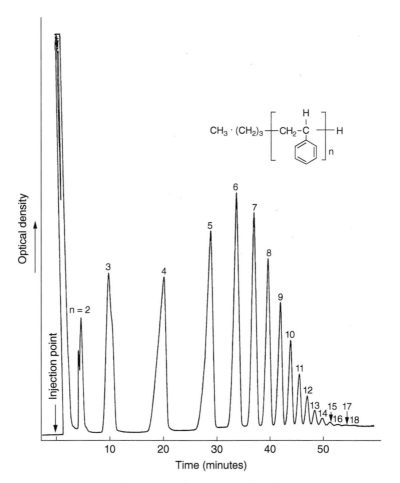

Figure 10.3 Chromatogram of a "monodisperse" polystyrene of 600 nominal molecular weight with dimethyl formamide followed by silica gel column chromatography. The B(a) A- and B(a)P-rich fraction is subsequently subjected to SFC for the final purification step. Even though a nonpolar solvent is used, the chromatograms from the two different runs are essentially similar albeit resolution is somewhat better on the right. *Source:* Altgelt and Gouw [19], figure 3.

of a polystyrene model mixture with molecular weights ranging from 266 to 1930 amu. Chromatography was performed with pentane/methanol (20:1) as the mobile phase on a stationary phase of Porasil C to which n-octyl groups were bonded. The molecular structure of each oligomer was described by $CH_3-(CH_2)_3-(CH_2CHC_6H_5)_n-H$, where n was the number of styrene molecules in the polymer [20]. Further work in this area was subsequently described by Klesper and Hartmann [21].

10.4.2   SFC Purification of Polycyclic Aromatic Hydrocarbons

Analysis of polycyclic aromatic hydrocarbons (PAHs) which are widely found in fossil fuels and combustion products are of interest because of their alleged carcinogenic properties. In 1976, Gouw and Jentoft reported near complete resolution. The goal of the study was the unambiguous determination of very pure fractions of benzo(a)pyrene [B(a)P] and benz(a)anthracene [B(a)A] from automobile exhaust. Concentration of the PAHs by extraction with dimethylformamide, followed by silica gel chromatography was carried-out [22]. The [B(a)A]- and [B(a)P]-rich fractions were subsequently subjected to SFC for the final purification step. In this step, the detector was first set at 280.5 nm to obtain maximum sensitivity for B(a)A. The wavelength was then shifted to 372 nm for optimum response to B(a)P. The analytes of interest were recovered in a high degree of purity by carefully evaporating off the mobile phase. Open tubular columns [23] containing bonded liquid crystalline poly(siloxane) stationary phases have achieved PAH resolution superior to that obtained by GC with the same phases.

10.4.3   Coal Tar Pitch

Sie and Rijnders [24] obtained the chromatogram shown on the left in Figure 10.4 with isopropanol as the supercritical fluid mobile phase and alumina as the substrate. The sample was a coal tar pitch commonly used in bitumen blending. The expected location of various condensed PNAHs can be seen in the figure. The chromatogram shows a series of distinctly separated peaks roughly corresponding to groups of increasing molecular weight and ring number. The chromatogram on the right in Figure 10.4, on the other hand, was obtained with n-hexane on alumina W-200 (neutral). Even though a nonpolar solvent was employed, it is interesting to note that the chromatograms from the two runs are essentially the same although the results shown by the Japanese workers showed improved resolution.

10.4.4   Enhanced SFC Performance

The performance of a modified SFC instrument that later passed the requirements of ASTM Method 5186-96 for accuracy, linearity, and repeatability proved to be supportive of the developmental work earlier described in

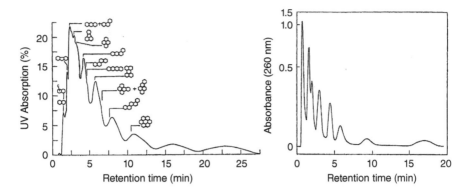

Figure 10.4 Chromatograms of coal tar. Left: isopropanol as the supercritical fluid and alumina as the stationary phase. The location of various condensed, polynuclear hydrocarbons are indicated. The last two broad peaks probably represent 7- and 8-ring aromatic compounds. Right: Chromatogram obtained with n-hexane on Almina W-200 (neutral). *Source:* Altgelt and Gouw [25], figure 5.

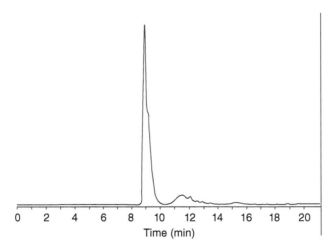

Figure 10.5 Chromatogram of jet fuel sample A. Chromatographic conditions: column, 75 cm × 1 mm i.d. (5 μm particles, 60 A pore width); fused silica restrictor, 15 μm × 12 cm; FID, 350°C; CO_2, 40 °C and 200 atm; H_2, 90 mL/min; column effluent flow, 120 mL/min as expanded gas. *Source:* Richter and Jones [16].

Reference 10. Up to a sixfold improvement in the precision was seen when the pressure transducer was temperature-controlled independent of the laboratory air temperature that was allowed to vary between 22 and 29 °C and the temperature of the transducer was controlled [16]. Figure 10.5 shows the chromatogram of diesel fuel whose results were typical of the results obtained for all samples when using the optimized experimental conditions. Table 10.1

Table 10.1 Effect of controlling the pressure transducer temperature.

	RSD (%) without temperature control	RSD (%) with temperature control
Diesel C		
Saturates	0.13	0.08
Monoaromatics	0.33	0.08
Polyaromatics	2.7	0.70
Total aromatics	0.25	0.14
Average	0.86	0.25
Jet fuel C		
Saturates	0.09	0.02
Monoaromatics	0.74	0.28
Polyaromatics	9.6	1.2
Total aromatics	0.48	0.09
Average	2.7	0.40
Jet fuel D		
Saturates	0.12	0.03
Monoaromatics	0.39	0.19
Polyaromatics	6.5	0.92
Total aromatics	0.41	0.09
Average	1.8	0.31

Source: Richter et al. [16]. Reproduced with permission of Oxford University Press.

shows that control of the pressure transducer temperature had a large impact on the assay reproducibility.

10.4.5   Sulfur Detection in a Petroleum Matrix

Shearer and Skelton in 1994 investigated the analysis of sulfur compounds in petroleum products via SFC using flameless sulfur chemiluminescence detection (SCD) [26]. At the time, this SCD was more sensitive and selective than any other sulfur-selective detector for SFC. It also produced a linear and nearly equimolar response to sulfur. A minimum detection limit of 0.3 pg of sulfur was measured, and response to sulfur in different sulfur species was nearly equimolar. Four years later, SFC with an open tubular column using 100% supercritical carbon dioxide as the mobile phase was used to successfully analyze diesel fuel and other heavy refinery samples [27]. Surprisingly, sulfur selective detection following chromatography of petroleum-derived products has not generated much excitement.

10.5 SFC Replacement for GC and LC

10.5.1 Simulated Distillation

Distillation is the primary separation process used to characterize petroleum products before processing. Distillation data can be obtained by using true distillation techniques or analytical techniques which simulate the distillation process. SIMDIS is a technique which is inexpensive and rapid compared to distillation techniques but it does not yield fractions for further characterization [28]. In the field of petroleum analysis, SFC has proven to be a very attractive substitute for the elution of hydrocarbons having more than 80–130 atoms of carbon which otherwise would have been difficult to elute via either GC or high temperature GC without cracking of the sample. Only one pump is required to deliver the carbon dioxide. The maximum operating pressure of the pump should be as high as possible (i.e. higher than 60 MPa) and pressure programming is mandatory at fixed temperature.

Implementation of micro or narrow bore packed-column SFC for SIMDIS application is quite straightforward because no split injection is required and the columns exhibit high load capacity. Using alkyl bonded phases, the longer the alkyl chain, the stronger the retention of hydrocarbons. They reported heaviest paraffin separated with capillary SFC is C_{126} on a coated 5% phenylpolydimethylsiloxane stationary phase using CO_2 at 160 °C and pressure programming starting 10 minutes from 100 to 550 bar at 13.3 bar/min [29, 30]. SFC can also be used for characterizing heavy crude components, with boiling points up to 760 °C. The effect of C_1 to C_{18} alkyl groups bonded to silica have been investigated and the oligomer peak resolution obtained with packed capillary columns approaches that obtained with open tubular columns. The true boiling and retention times of n-alkanes, alkylbenzenes, PAH, and thiophenes have also been correlated, and it has been found that the retention time differences do not exceed 1 minute for chemically different solutes with similar boiling points. Figure 10.6 shows the SFC of the SIMDIS calibration standard on a packed capillary hexylsilyl (C_6) column [28].

Data from SFC correlated well with data obtained by GC. Higher molecular weight hydrocarbons can easily be eluted at operating pressures below 415 bar (e.g. density of CO_2 mobile phase approximately 0.71 g/L). At this maximum pressure, a column packed with hexyl bonded silica elutes hydrocarbons boiling at more than 756 °C; whereas, a column packed with octadecyl silanol (ODS-2) (C_{18}); Figure 10.7 is more retentive and only elutes hydrocarbons boiling up to 686 °C. Hydrocarbons of even higher boiling points can be eluted if

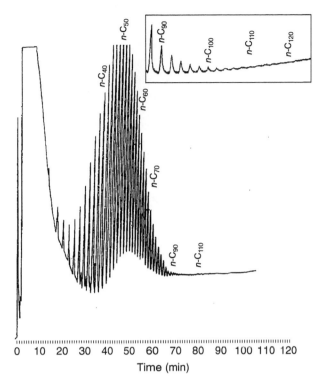

Figure 10.6 Simulated distillation chromatogram of Polywax 655 on a 0.25 m × 250 μm i.d. column packed with Waters Spherisorb SSW C6. Conditions: linear pressure programming from 100 to 415 bar (hold for 60 minutes) at 3.5 bar min, FID detection. *Source:* Wilson [31]. Reproduced with permission of Elsevier.

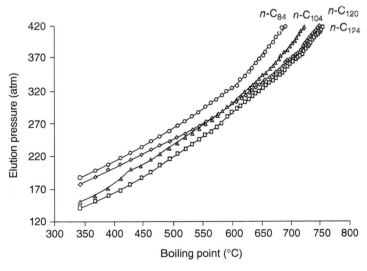

Figure 10.7 Graph of elution pressure versus the boiling point for packed columns. Diamonds, methyl; squares, hexyl; triangles, octyl; circles, ODS2. *Source:* Wilson [31]. Reproduced with permission of Elsevier.

the column length is changed. The significant deviation between the retention times of aromatics and straight chain alkanes of apparently similar boiling points which occurs when simulated distillation is performed by GC may be reduced using open tubular SFC and further minimized when packed capillary columns are used, especially for aromatic compounds with three or more rings. However, comparison may not be valid in view of the discrepancies between published values of PAH boiling points.

10.5.2 Hydrocarbon Group-Type Separations – PIONA Analysis

Another highly studied application in the petroleum industry is hydrocarbon group analysis. Group-type analysis refers to the separation and quantification of hydrocarbon groups. A classic example is the PIONA analysis that is used to profile the hydrocarbon content of a sample according to the concentrations of paraffins, isoparaffins, olefins, naphthenes, and aromatics. Heavier crude fractions are an unconventional petroleum source that has also been explored. Processes such as Fischer–Tropsch reactions are being used to generate liquid fuels from coal and shale. Liquid chromatography is commonly used but it suffers from (i) lack of resolution, (ii) lack of universal detection, and (iii) long analysis times. GC has also been used, but it is limited to the analysis of light distillates due to the column temperatures that are required [28].

HPLC-like pumps, on the other hand, that are available in most SFC systems can be implemented using a flow splitter prior to the column for the previous analyses. High pressure syringe pumps can also be used without splitting the flow rate before the column. In addition, packed column SFC can be an improvement over open tubular columns via the implementation of a variety of polar and nonpolar packed column stationary phases. Further experimentation protocol as it relates to GC-like and LC-like situations is available in reference [29]. Additional attempts to improve the resolution between groups or to separate more fractions in diesel fuels have involved the implementation of two different columns connected in series. A cyanopropyl-bonded silica connected in series to a bare silica column was shown to enhance selectivity between aromatic groups. The use of longer columns can also enhance the overall resolution between families and sub-groups. An increase in pressure also results in improved separation between saturates and aromatics. Separations using a combination of silica and cyano or amino columns or the combination of silica and 20% silver nitrate coated column are not as good as the separation obtained with a single 5 μm silica column at the same temperature.

The D5186 ASTM method approved in 1991 for a successful separation of nonaromatic and polyaromatic hydrocarbon groups required that temperatures in the range of 30–40 °C be used for this separation. At low temperature, the separation of saturates and mono-aromatics is easily achieved. However, low temperatures are not adequate for separation between mono- and polyaromatics. In ASTM D5186-03, a resolution of at least 4 between the non-aromatics (hexadecane) and mono-aromatics (toluene) must be obtained and the resolution between the mono-aromatics (tetraline) and polycyclic aromatics (naphthalene) must be at least 2. This means all the compounds eluted before the end of the peak corresponding to hexadecane are assigned to the non-aromatic hydrocarbons. The compounds eluting after hexadecane, but prior to the time corresponding to the start of the naphthalene peak, are assigned to mono-aromatics. All of the integrated area occurring after the start time of the naphthalene peak through the final return to baseline is assigned to polycyclic aromatic hydrocarbons [29].

More recently a different methodology has been reported, which involves the separation of petroleum distillate into aliphatic and aromatic fractions using a two-dimensional SFC × SFC system with a flow-switching interface. The columns used were a liquid crystal polysiloxane capillary column and a SB-biphenyl-30 capillary. The use of a liquid crystal column in the second dimension to provide shape selectivity allowed separation of various isomers, including chrysene, triphenylene, benz[a]anthracene, and benzofluoranthenes [28].

10.6 Biodiesel Purification

The proposed use of biodiesel esters derived from a variety of biological sources, such as canola, corn, fish, and other oils as diesel fuel blending components has led to the need within the petroleum industry to determine these compounds in the presence of the diesel fuel hydrocarbon matrix [32]. Fatty acid methyl and ethyl esters ranging in carbon number from C10 to C26 would be produced by trans-esterification of the mono-, di-, and tri-glycosides in the oils with methanol or ethanol. The product of this reaction would then be blended with conventional hydrocarbon-based diesel fuels.

As an energy source, biodiesel should have certain criteria. For example, the level of impurities such of triacylglycerol, diacylglycerol, monoacylglycerol, and free glycerol should be 0.2–0.8% for acylglycerols and 0.02% for glycerol. These impurities vary in polarity, solubility, and volatility. A single gas chromatographic method of analysis is not forthcoming unless one resorts to high temperature separation and analyte prederivatization. The standard method for determination of impurities in biodiesel currently involves nevertheless, these experimental conditions. Previously, a state of the art method that incorporates a single short column, sub-2 μm octadecyl bonded silica particles, and evaporative light scattering detection was developed for analysis of biodiesel and impurities in standardized samples. Employment of the method to analyze both a series of biodiesels prepared in house from tobacco seed oil and a commercially available B100 biodiesel have been described [33].

Figure 10.8 shows the separation of a mixture of C_{18} triacylglycerols, diacylglycerols, and monoacylglycerols, plus free glycerol spiked into model biodiesel which was a mixture of fatty acid ethyl esters FAEE's. Three different gradient elution schedules and flow rates were employed in order to obtain faster analysis with minimum lost in peak resolution and selectivity of all impurities. Initially, the modifier gradient percentage was increased from 80/20 to 50/50. Separation of all compounds was obtained in less than 10 minutes. Next, the gradient was changed from 98/2 to 90/10 and the initial gradient ramp time was changed from 10 to 2 minutes. Final modifier concentration was 50/50. The final trace in Figure 10.8 shows the separation of the same mixture using the faster gradient elution. All components were separated in less than five minutes. This final method was used for all analyses.

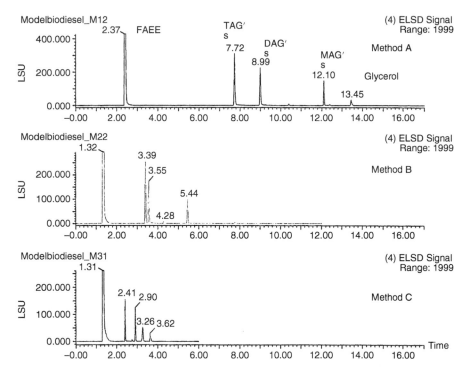

Figure 10.8 UPC^2 Single injection of a mixture of model biodiesel, glycerol, and C_{18} acylglycerols with different gradients, Oven temp.: 25°C, modifier 90/10 CH$_3$CN/MeOH. **Method A:** T = 0 minutes, 98/2 CO$_2$/modifier, T = 10 minutes, 80/20, T = 12 minutes, 50/50, T = 15.5 minutes, 98/2, Flow:1.0 mL/min. **Method B:** T = 0 minutes, 90/10 CO$_2$/modifier, T = 10 minutes, 50/50, T = 11.1 minutes, 90/10, **Method C:** T = 0 minutes, 90/10 CO$_2$/modifier, T = 2 minutes, 50/50 T = 6.1 minutes, 90/10. *Source:* Ashraf-Khorassani et al. [33], figure 3. Reproduced with permission of Elsevier.

In order to remove impurities from the synthesized biodiesel that was to serve as the matrix, a simple two-step column chromatography process was developed. In this process, tobacco seed oil biodiesel with associated impurities was passed through a bare silica column and eluted first using hexane and then by elution of the remaining analytes using ethanol. Solvent from each fraction was evaporated and the proper amount of each sample was dissolved in MeOH/DCM. Analysis of the fractions was performed using the SFC method developed previously. Figure 10.9 shows separation of the biodiesel matrix *before* purification, *after* purification, and *when* either hexane or ethanol was the eluting solvent. As can be observed, nearly pure synthetic biodiesel was obtained with the hexane elution. Monoacylgycerols were again easily separated and detected in these biodiesel samples using packed column SFC with post-column light scattering detection.

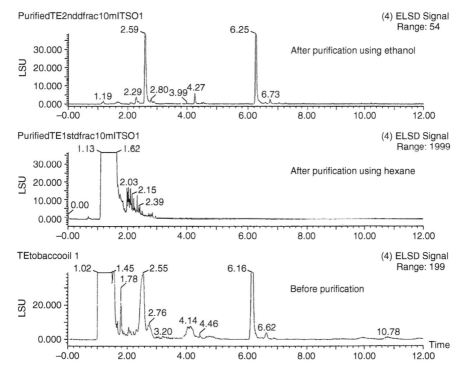

Figure 10.9 UHPSFC/ELSD of three different batches of synthetic biodiesel. Chromatography conditions: T = 0 minutes, 90/10 CO_2/modifier, T = 10 minutes, 50/50, T = 11.1 minutes, 90/10, T = 15.5 minutes, 98/2, Flow = 1.0 mL/min, 25 °C, Column: ACQUITY UPC^2 HSS C_{18} SB column (150 mm × 3.0 mm, 1.8 µm). *Source:* Ashraf-Khorassani et al. [33], figure 4. Reproduced with permission of Elsevier.

10.7 Multidimensional Separations

Petroleum is a sample that is inherently complex in terms of increasing molecular weight of the compounds in the sample. This poses a practical challenge for analytical chemists involved in petroleum research and production. To satisfy the increasing demand for petroleum-based fuels and related products, companies are turning more and more frequently to upgrading *heavier cuts* of crude oil. These samples are more complex than those from lighter oil resources. These heavy fractions also tend to include a higher proportion of nitrogen, sulfur, and oxygen containing species than the lighter fractions. These compounds cause problems both with refinery operations and finished products, thus making for interesting analytical targets [34].

The analysis problems presented by petroleum and other complex samples can be solved by using a high-resolution technique such as multidimensional chromatography which has been developed to overcome the limited peak capacity of one-dimensional chromatographic methods. In roughly the last 20 years, significant technological advances in areas such as column technology and column coupling technology, mobile phase flow control, and detector design have been coupled with improvements in electronics and computers resulting in a proliferation of multidimensional separation techniques. $GC \times GC$ was first demonstrated by Liu and Phillips in 1991 [35], and the petroleum community has played a very active role in supporting and promoting the technique.

Essentially all the applications of multidimensional techniques for the analysis of petroleum samples during this period have involved comprehensive two-dimensional gas chromatography ($GC \times GC$) or comprehensive gas chromatography–mass spectrometry ($GC \times MS$) with the bulk of the attention being paid to petroleum vacuum gas oils since GC is amenable to all but the heaviest petroleum samples. $GC \times GC$ is a dual-column technique wherein both separations are performed in the gas phase that are coupled by a modulator. The role of the modulator is to trap/collect effluent from the primary column and then periodically introduce the collected fraction to the secondary column. Several reviews have been published that trace recent developments in the application of comprehensive multidimensional techniques that involve gas chromatography [36, 37]. The following section will however deal exclusively with multidimensional techniques for analysis of petroleum products that incorporate SFC in the hyphenation process.

10.7.1 Comprehensive Two-Dimensional SFC

Two-dimensional gas chromatography is popular in many sectors such as oil, food, biology, and the environment because it improves the efficiency, the selectivity, and the resolution of separations to advantage for the analysis of very complex samples. It has become desirable to develop a comprehensive two-dimensional separation technique that allows the easy elution of low volatile compounds that are difficult to elute in GC × GC while preserving flame ionization detection. This technique is SFC when neat carbon dioxide is used as the mobile phase [38]. Given that the best SFC kinetic features that include (i) fast separations, and (ii) low mobile phase viscosity that allows (iii) long packed columns which exhibit (iv) c. 100 000 theoretical plates that under reasonable conditions could provide (v) a peak capacity in the same range as GC × GC can be feasible with SFC × SFC [32].

The design and implementation of comprehensive two-dimensional SFC using neat carbon dioxide as the mobile phase involved two conventional supercritical fluid chromatographs hyphenated via an online comprehensive 2D liquid "chromatography-like" interface. Two types of packed columns were used: bare silica and C_{18} bonded silica. It consisted of two "loop switching" valves that allowed the collection of the first dimension column effluent. Both dimensional separations were monitored via UV detection. The feasibility of comprehensive 2D SFC was demonstrated on synthetic mixtures of hydrocarbons, and its potential on real sample analysis was illustrated by the separation of heavier coal-derived distillates. The chromatogram of a vacuum distillate obtained from coal tar liquefaction demonstrated the effectiveness of the comprehensive 2D system. The separations reported here show that SFC × SFC can be a promising alternative to GC × GC for analyzing petroleum samples heavier than middle distillates. SFC might also be implemented as one dimension in a SFC × LC system or in an LC × SFC system.

10.7.2 SFC-GC × GC

Even though GC × GC has considerably improved the analysis of complex samples, it is still a hard task to fully characterize some diesels, and even extremely difficult when heavy fractions are concerned. These matrices are exceedingly complex due to their wide range of physicochemical properties preventing them from being separated by only two dimensions. Clearly, the addition of a separating mechanism similar to GC × GC is necessary. SFC is a valuable tool prior to comprehensive two-dimensional gas chromatographic (GC × GC) analysis to investigate the composition of petroleum vacuum oils (VGOs) and selectively analyze phenolic compounds in coal-derived middle distillates [39, 40]. In the first case, a multidimensional SFC system is used to separate saturates, aromatics, and polar compounds in VGO samples, and in the second case a single ethylpyridine stationary phase is sufficient to selectively separate phenols.

When considering online hyphenation of an LC system with a GC or GC × GC system, the key technical issue is the evaporation of the LC mobile phase that must induce a limited reinjection band for GC. SFC using CO_2

seems more compatible with a gas phase dimension than an LC dimension. The improved compatibility of SFC can be attributed to the expansion stage from the supercritical state to the gas state without any loss of the products of interest. Additionally, SFC separations can provide higher selectivity because of the variety of solvent, solutes and stationary phase interactions and because they are often faster than LC separations.

SFC-GC × GC enables the temporary storage of SFC fractions inside sampling loops for subsequent analysis by GC × GC. The interface was configured for the characterization of heavy petroleum fractions. The SFC dimension was adapted to perform a group type separation into saturated, unsaturated, and polar fractions. The GC × GC dimension was set in order to realize high temperature separations with two sets of columns.

10.7.3 Comprehensive – SFC-Twin-Two-Dimensional (GC × GC)

A new approach based on a three-dimensional chromatographic system has been proposed for PIONA analysis [41]. For the first time, SFC has been hyphenated to twin comprehensive two-dimensional gas chromatography resulting in a highly resolutive analytical tool. The valve diagram of a SFC system used for group type separation, and the valve diagram of SFC-Twin-GC ×

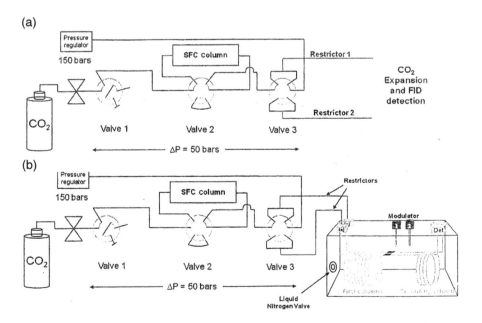

Figure 10.10 Valve diagram (a) of SFC system used for group type separation and valve diagram (b) of SFC-twin-GC × GC for extensive group type analysis of middle distillates. *Source:* Adam et al. [41]. Reproduced with permission of Elsevier.

GC for extensive group type analysis of middle distillates is shown in Figure 10.10. The comparison of results with reference methods (GC × GC) and (GC × MS) [34] has also been proposed and clearly illustrates the benefit of the SFC-twin-GC × GC approach. The additional separation dimension prior to GC × GC allowed unequaled quantitative possibilities and reduced the risk of misidentification. Saturated and unsaturated hydrocarbons that have been fractionated by SFC are transferred on two different GC × GC columns sets (twin GC × GC) placed in the same oven. The benefits of SFC-twin-GC × GC were demonstrated for the extended alkane, iso-alkane, alkene, naphthenes, aromatic analysis (so called PIONA) of diesel samples, which can be achieved in one single injection. For that purpose, saturated and unsaturated compounds were separated by SFC using a silver loaded column prior to GC × GC analysis. Significant discrepancies were observed between SFC-twin-GC × GC and reference methods. For instance, alkenes and naphthenes were quantitatively recovered in the saturated and unsaturated fractions respectively, allowing their identification in various diesel samples. Thus, resolution between each class of compounds was significantly improved compared to a single GC × GC run, and for the first time an extended PIONA analysis (i.e. the separation of hydrocarbons by chemical class and by carbon number) of diesel samples in a single injection can be reported [37].

References

1 Speight, J.G. (2002). *Handbook of Petroleum Product Analysis*. Hoboken, NJ: Wiley.
2 Altgelt, K.H.B. and Mieczyslaw, I. (1993). *Composition and Analysis of Heavy Petroleum Fractions*. New York, NY: Marcel Dekker, Inc.
3 Conaway, C.F. (1999). *The Petroleum Industry, A Nontechnical Guide*. Tulsa Oklahoma Publisher.
4 Lee, M.L. and Markides, K.E. (eds.) (1990). *Analytical Supercritical Fluid Chromatography and Extraction*. Provo, UT: Chromatography Conferences Inc.
5 Dallvge, J., Beens, J., and Brinkman, U.A.T. (2003). Comprehensive two-dimensional gas chromatography: a powerful and versatile analytical technique. *Journal of Chromatography. A* 1000: 69–108.
6 Nizio, K.D., McGinitie, T.M., and Harynuk, J.J. (2012). Comprehensive multidimensional separations for the analysis of petroleum. *Journal of Chromatography. A* 1255: 12–23.
7 Gorecki, T., Harynuk, J., and Panic, O. (2004). The evolution of comprehensive two dimensional gas chromatography. *Journal of Separation Science* 27: 359–379.
8 Adahchour, M., Breens, J., Vreuls, R.J.J., and Brinkman, U.A.T. (2006). Recent developments in comprehensive two-dimensional gas chromatography

(GC x GC) 1. Introduction and instrumental setup. *Trends in Analytical Chemistry* 25: 438–454.

9 Phillips, J.B. and Xu, J.Z. (1995). Comprehensive two-dimensional gas chromatography: a powerful and versatile technique. *Journal of Chromatography A* 703: 327.

10 Conrad, A.L. (1948). Aromatic compounds in petroleum products. *Analytical Chemistry* 20: 725–726.

11 Criddle, D.W. and LeTourneau, R.L. (1951). Fluorescent indicator adsorption method for hydrocarbon-type analysis. *Analytical Chemistry* 23: 1620–1624.

12 Norris, T.A. and Rawdon, M.G. (1984). Determination of hydrocarbon types in petroleum liquids by supercritical fluid chromatography with flame ionization detection. *Analytical Chemistry* 56: 1767–1769.

13 Rawdon, M.G. (1984). Modified flame ionization detector for supercritical fluid chromatography. *Analytical Chemistry* 56: 831.

14 Berger, T.A. (ed.) (1995). Packed column SFC. In: *SFC and the Petroleum Industry*, RSC Chromatography Monographs, 212–226. Cambridge: The Royal Society of Chemistry.

15 Hayes, P.C. and Anderson, S.D. (1988). Quantitative determination of hydrocarbons by structural group type via HPLC with dielectric constant detection – a review. *Journal of Chromatographic Science* 26: 210.

16 Richter, B.E., Jones, B.A., and Porter, N.L. (1998). Optimized supercritical fluid chromatographic instrumentation for the analysis of petroleum fractions. *Journal of Chromatographic Science* 36 (9): 444–448.

17 M.S. Klee, M.Z. Wang, Hewlett Packard Application Note 228-226, (1993), M.S. Klee, M.Z. Wang, HP Application Note 228-167 (1992), V. Giarrocco, M. Klee, Hewlett Packard Application Note, 228-231 (1993)

18 Schulz, W.W. and Genowitz, M.W. (1990). *Pittsburgh Conference on Analytical Chemistry*, New York, NY, (5–9 March 1990), Paper No. 133.

19 Altgelt, K.H. and Gouw, T.H. (eds.). Chromatography in Petroleum Analysis, 324. https://www.osti.gov/biblio/5268159.

20 Gouw, T.H. and Jentoft, R.E. (1979). Chromatography in petroleum analysis. In: *Supercritical Fluid Chromatography*, 314–327. New York: Marcel Dekker, Inc.

21 Klesper, E. and Hartman, W. (1977). Preparative supercritical fluid chromatography of styrene oligomers. *Journal of Polymer Science, Polymer Letters Edition* 15: 713–719.

22 Jentoft, R.E. and Gouw, T.H. (1969). Supercritical fluid chromatography of a "polydispersed" polystyrene. *Journal of Polymer Science. Part B* 7: 811–813.

23 Chang, H.C.K., Markides, K.E., Bradshaw, J.S., and Lee, M.L. (1988). Selectivity enhancement for petroleum hydrocarbons using a smectic liquid crystalline stationary phase in supercritical fluid chromatography. *Journal of Chromatographic Science* 26: 280–289.

24 Sie, S.T. and Rijnders, G.W.A. (1967). High pressure gas chromatography and chromatography with supercritical fluids IV, fluid-solid chromatography. *Separation Science* 2: 755.

25 Altgelt, K.H. and Gouw, T.H. (eds.). *Chromatography in Petroleum Analysis*, 321. https://www.osti.gov/biblio/5268159.

26 Shearer, R.L. and Skelton, R.J. (1994). Supercritical fluid chromatography of petroleum products using flameless sulfur chemiluminescence detection. *Journal of High Resolution Chromatography* 17: 251–254.

27 Heng, S., Taylor, L.T., Fujiari, E.M., and Yan, X. (1998). Sulfur determination in heavy diesel fuel with open-tubular column supercritical fluid chromatography. *LC GC* 16: 276–280.

28 Robson, M.B. (2000). Fuels and lubricants: supercritical fluid chromatography. In: *Encyclopedia of Separation Science* (ed. I. Wilson), 2894–2901. Academic Press.

29 Thiebaut, D. (2012). Separations of petroleum products involving supercritical fluid chromatography. *Journal of Chromatography. A* 1252: 177–188.

30 Dulaurent, A., Dahan, L., Thiebaut, T., and Espinat, F.B.D. (2007). Extended simulated distillation by capillary supercritical fluid chromatography. *Oil & Gas Science and Technology* 62: 33–43.

31 Wilson, I. (ed.) (2000). *Encyclopedia of Separation Science*, 1e, 2900. Academic Press.

32 Diehl, J.W. and DiSanzo, F.P. (2007). Determination of total biodiesel fatty acid methyl ester, ethyl esters, and hydrocarbon types in diesel fuels by supercritical fluid chromatography-flame ionization detection. *Journal of Chromatographic Science* 45: 690–693.

33 Ashraf-Khorassani, M., Yang, J., Rainville, P. et al. (2015). Ultrahigh performance supercritical fluid chromatography of lipophilic compounds with application to synthetic and commercial biodiesel. *Journal of Chromatography B* 983–984: 94–100.

34 Qian, K. and Di Sanzo, F.P. (2016). Detailed analysis of olefins in processed petroleum streams by combined multi-dimensional supercritical fluid chromatography and field ionization time-of-flight mass spectrometry. *Energy & Fuels* 30: 98–103.

35 Liu, Z. and Phillips, J.B. (1991). Comprehensive two-dimensional gas chromatography using an on-column thermal modulator interface. *Journal of Chromatographic Science* 29: 227–231.

36 Cortes, H.J., Winniford, B., Luong, J., and Pursch, M. (2009). Comprehensive two-dimensional gas chromatography review. *Journal of Separation Science* 1120: 883–904.

37 Adahchour, M., Breens, J., Vreuls, R.J.J., and Brinkman, U.A.T. (2006). Recent developments in comprehensive two-dimensional gas chromatography (GC x GC) 1. Introduction and instrumental setup. *Trends in Analytical Chemistry* 25: 438–454.

38 Guibal, P., Thiebaut, D., Sassiat, P., and Vial, J. (2012). Feasibility of neat carbon dioxide packed column comprehensive two dimensional supercritical fluid chromatography. *Journal of Chromatography. A* 1255: 252–258.

39 Omais, B., Dutriez, T., Courtiade, M. et al. (2011). SFC-GC × GC to analyze matrices, from petroleum and coal. *LCGC* 24 (7): 10–20.

40 Durtiez, T., Thiebaut, D., Courtiade, M. et al. (2013). Application to SFC-GC × GC to heavy petroleum fractions analysis. *Fuel* 104: 583–592.

41 Adam, F., Thiebaut, D., Bertoncini, F. et al. (2010). Supercritical fluid chromatography hyphenated with twin comprehensive dimensional gas chromatography for ultimate analysis of middle distillates. *Journal of Chromatography. A* 1217: 1386–1394.

11

Selected SFC Applications in the Food, Polymer, and Personal Care Industries

11.1 Introduction

SFC has had its greatest economic impact in recent years in the pharmaceutical industry, as illustrated by the applications described in Chapter 9. Yet some of the earliest applications of SFC were in the foods, polymer, and personal care industries [1]. The impact of SFC in these areas continues to this day, fueled by the widening applicability of SFC to more polar mixtures than in the early years of SFC. This chapter illustrates some of the more prominent applications of SFC in these areas.

Modern Supercritical Fluid Chromatography: Carbon Dioxide Containing Mobile Phases,
First Edition. Larry M. Miller, J. David Pinkston, and Larry T. Taylor.
© 2020 John Wiley & Sons, Inc. Published 2020 by John Wiley & Sons, Inc.

11.2   Selected Applications in the Foods Industry

11.2.1   Fats, Oils, and Fatty Acids

Fats and oils are important, high value components of most foods. They are also complex, nonpolar mixtures, and were thus logical targets for early practitioners of SFC, given the nonpolar nature of pure CO_2 [2]. Refined oils and fats typically contain neutral lipids, consisting of triglycerides, with lower levels of mono- and diglycerides, free fatty acids, and trace levels of many other natural components. The latter include carotenoids, tocopherols and tocotrienols, phytosterols, phosphatides, waxes, alkanes, and alkenes such as squalene. Figure 11.1 provides structures of some of these major and minor components. Some oils may contain ultra-trace levels of process-derived species, such as chlorinated propanediol esters, dichloropropanol esters, and glycidyl esters. As a result of oxidative and hydrolytic stress during cooking, oils and fats produce a complex mixture of oxidation products including hydroperoxides, aldehydes, alcohols (both free and incorporated into triglyceride structures), alkanes and alkenes, dimers and trimers, and even species as polar as glycerin [3]. Natural oils and fats present in foods also contain polar, complex lipids. Many of these contain a polar head group, linked to a glyceride structure, as with phospholipids, glycolipids, sphingolipids, ceramides, and gangliosides. These are incorporated into cellular structures or have distinct bioactive roles. Even the relatively "simple" triglyceride has dozens of distinct structures with a wide variety of properties related to the length and degree of unsaturation of its fatty acid components, not to speak of stereoisomers. So it's little surprise that separation scientists have worked long and hard to provide separations and improved understanding of the world of fats and oils.

Chemical derivatization of polar groups, and the use of short, open tubular columns coated with high-temperature-stable, cross-linked, thin-film stationary phases provided successful separations of fats and oils by gas chromatography. Yet the very high temperatures required for the elution of the least volatile triglycerides (>350 °C) results in structural changes in, and even degradation of, a fraction of the "high boilers." HPLC has also been used for fats and oils analysis. Detection of lipids can be challenging in traditional HPLC. In contrast, some of the first applications of SFC to fats and oils used water-saturated CO_2 as the mobile phase and took advantage of the universal and sensitive flame ionization detector [4]. A wide range of applications of SFC to lipids has been described in more recent years. Laboureur et al. [5], Donato et al. [6], and Yamada and Bamba [7] have provided recent reviews of the use of SFC and SFC/MS in the field of "lipidomics."

Sandra et al. illustrated the potential of SFC for the characterization of fats and oils [8]. They made good use of argentation chromatography to resolve the various saturated and unsaturated triglycerides, and used both pressure and

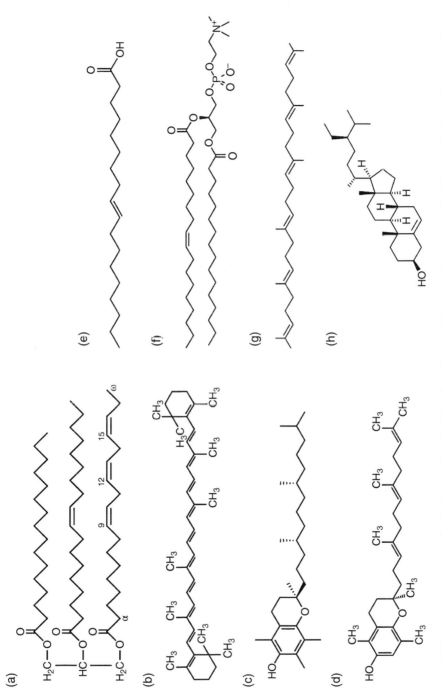

Figure 11.1 Some components of refined vegetable oils: (a) triglyceride containing palmitic (top), oleic (middle), and linolenic (bottom) acids. A diglyceride contains two fatty acids and an unesterified hydroxyl group on the glycerol backbone, while a monoglyceride contains a single fatty acid and two unesterified hydroxyl groups; (b) all-trans beta-carotene (a source of vitamin A); (c) alpha-tocopherol (a source of vitamin E); (d) beta-tocotrienol; (e) elaidic acid, a trans fatty acid; (f) phosphatidylcholine, a phosphatide; (g) squalene; and (h) sitosterol.

modifier gradients to effect the separation. Not only did they use silver impregnated SFC columns for better selectivity for degree of unsaturation, but they also added silver nitrate in methanol as a make-up fluid post-column, just before pneumatically assisted electrospray ionization. The triglycerides formed intense silver adduct ions. Atmospheric pressure chemical ionization (APCI) provided comparable results. Figure 11.2 shows an example of the triglyceride separations they achieved. Both APCI and silver-ion-adduct ESI in SFC/MS provided far better response than did UV absorbance detection, as detailed in the caption. Note the excellent separation according to the number of double bonds in the triglyceride structures.

While conceptually simpler than actual oils and fats, chromatographers have devoted much effort to the separation of fatty acids. The fatty acid distribution

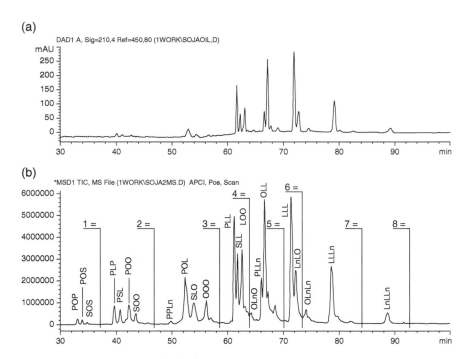

Figure 11.2 SFC separation of refined soybean oil with UV absorbance detection at 210 nm (a) and APCI-MS detection (b). Triglyceride composition and isomer distribution key: P = palmitic moiety, O = oleic moiety, S = stearyl moiety, L = linolyl moiety, Ln = linolenyl moiety. Chromatographic conditions: sulphonic silica based strong cation-exchanger column (Nucleosil 100-5 SA, 25 cm × 4.6 mm i.d. × 5 μm) impregnated with silver ions; column temperature: 65 °C; flow rate: 1 mL/min; modifier: 6/4 acetonitrile/2-propanol; pressure gradient: 150 bar (2 minutes) to 300 bar at 1.5 bar/min; modifier gradient: 1.2% (2 minutes) to 7.2% (28 minutes) at 0.3%/min, then to 12.2% at 0.54%/min. Injection volume: 5 μL of a 10% solution for UV and of a 0.1% solution for MS, both in $CHCl_3$. *Source:* Reprinted with permission from reference [8]. Copyright 2002, Elsevier.

of an oil has been related to its health effects and to the authenticity of high value oils. For example, many believe that trans fatty acids, and perhaps saturated fatty acids, to have negative health effects, while polyunsaturated fatty acids, notably those associated with fish oils, are reported to have positive health effects [9]. Another example is tracing the origin and authenticity of extra-virgin olive oil. Gas chromatography is the mainstay of fatty acid analysis. But longer chain fatty acids must be derivatized (most commonly by methylation) to be eluted by GC. Even with derivatization, structural changes (double bond migration) are known to occur for the most unsaturated long-chain fatty acids during GC analysis. In contrast, SFC has been used to separate underivatized fatty acids at relatively low temperatures, avoiding this issue, as described below.

Qu et al. used the solvating mobile phase of SFC to separate underivatized fatty acids [10]. Figure 11.3 shows the SFC/MS separation of a mixture of fatty acid mixtures on two very different columns: 2-ethylpyridine (2-EP) and octadecylsilane (C18). The C18 column appears to provide better selectivity. Also, note that the two columns show quite different retention for saturated and unsaturated fatty acids (more highly unsaturated eluting later on the 2-EP, but earlier on the C18). Conditions are listed in the figure legend.

Similarly, Ashraf-Khorassani et al. demonstrated the SFC separation of underivatized fatty acids, as well as complex oil samples using ELSD, UV, and mass spectrometric detection [11]. Figure 11.4 shows an impressive separation of all the components of a commercial fish oil sample, ranging from triglycerides and fatty acid methyl esters, to free fatty acids and glycerol, without derivatization.

Clearly SFC is widely applicable to fats and oils, but there are many other lipophilic, and some not so lipophilic, mixtures of importance in foods where SFC has shown its utility. The first example we'll discuss is the widely present tocopherols.

11.2.2 Tocopherols

Tocopherols are the subject of increasing interest in the food industry. They are primarily known for their vitamin E activity, and are widely present in seed and nut oils. Many foods, such as breakfast cereals, are fortified with tocopherols for their vitamin E activity. But their use as natural antioxidants has become increasingly common as food producers remove synthetic antioxidants (such as BHT, BHA, and TBHQ) from their foods. Therefore, separation and quantification of tocopherols has become even more important to food scientists as they study the stability and antioxidant effects of these congeners. Tocopherols are nonpolar, lipophilic species, so have long been the subject of separation using SFC [12–15]. More recent research has focused on the ability of SFC to separate these mixtures much more rapidly than practical using

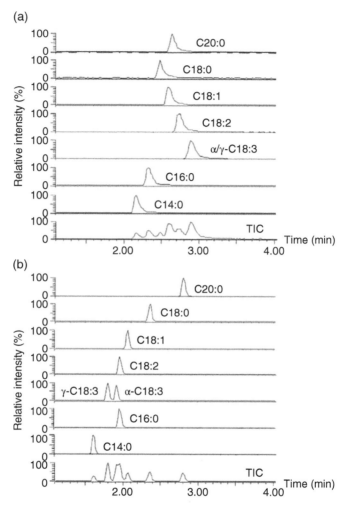

Figure 11.3 Total ion chromatogram (TIC) and extracted ion chromatograms of standard free fatty acid mixture (10.0 μg/mL each) from (a) BEH 2-EP column (1.7 μm, 3.0 mm × 100 mm i.d.) and (b) HSS C18 SB (1.8 μm, 3.0 mm × 100 mm i.d.). Peak identifications: myristic acid (C14:0, m/z 227.20), palmitic acid (C16:0, m/z 255.23), α-linolenic acid (α-C18:3, m/z 277.22), γ-linolenic acid (γ-C18:3, m/z 277.22), linoleic acid (C18:2, m/z 279.23), oleic acid (C18:1, m/z 281.25), stearic acid (C18:0, m/z 283.26), and arachidic acid (C20:0, m/z 311.30). Backpressure: 1500 psi; eluent A: CO_2, eluent B: 50/50 methanol/acetonitrile with 0.1% formic acid; flow rate: 1.6 mL/min; column temperature: 40 °C. The mobile phase gradient was held at 97% A for 1 minute, and was ramped to 93% A at 6 minutes. *Source:* Reprinted with permission from reference [10]. Copyright 2015, Elsevier.

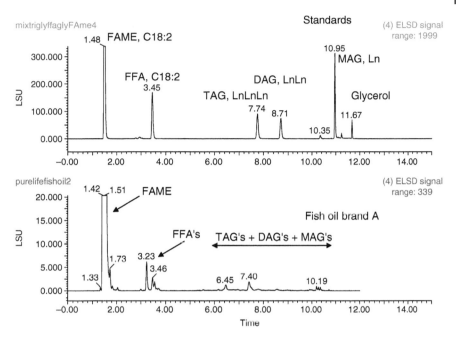

Figure 11.4 UHPSFC/ELSD separation of fish oil Brand A and a mixture of standards containing fatty acid methyl ester (FAME), free fatty acid (FFA), triacylglycerol (TAG), diacylglycerol (DAG), monoacylglycerol (MAG), and glycerol. Ln = lenolenic acid. Chromatographic conditions: initial mobile phase mixture: 98/2 (A/B), 98/2 at 0.5 minutes, 96/4 at 8 minutes, 80/20 at 9 minutes, 80/20 at 11 minutes, 50/50 at 11.1 minutes, 50/50 at 13 minutes, 98/2 at 13.1 minutes, and 98/2 at 15.0 minutes. A = CO_2, B = MeOH with 0.1% (v/v) formic acid, flow: 1 mL/min, column temp.: 25 °C, back pressure: 1500 psi, column HSS C18 150 × 3.0 mm, 1.8 μm, ELSD with makeup flow of 2-propanol at 0.2 mL/min. Fish oil Brand A conc.: 5%. *Source:* Reprinted with permission from reference [11]. Copyright 2015, Elsevier.

HPLC [16, 17]. For example, Mejean et al. compared a number of columns, modifiers, and other chromatographic conditions, and came up with an optimized separation of the tocopherols, as shown in Figure 11.5 [16], with a total analysis time of less than five minutes.

11.2.3 Other Vitamins

Tocopherols are somewhat unusual in that they are valued both for their role as vitamin E, and for their natural antioxidant activity. But SFC has been useful in the separation of other vitamins as well. Many of the vitamins are fat soluble, so were natural targets for separation by SFC in its early development [18, 19], and the literature is particularly rich in descriptions of the use of SFC for the characterization of mixtures containing carotenoids and vitamin A.

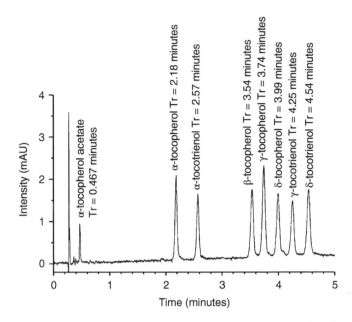

Figure 11.5 Chromatogram of tocopherol and tocotrienol standards produced using a NH_2 column after optimization, with a gradient of CO_2/EtOH + formic acid (0.1%). Flow rate of 1.5 mL/min, at a column temperature of 30 °C, 1 μL injected with a concentration of 1 mmol/L for each tocopherol and tocotrienol. UV detection at 295 nm. *Source:* Reprinted with permission from reference [16]. Copyright 2015, Springer Nature.

A remarkable example of taking advantage of the uniqueness of SFC was the coupling of SFC with NMR for the characterization of vitamin A acetate by Braumann et al. [20]. They separated five cis/trans-isomers of vitamin A acetate, and noted that when using pure CO_2 as the mobile phase, "...no solvent signal suppression is necessary and the unrestricted observation of the whole spectral range is possible, contrary to the HPLC separation in n-heptane, where in the aliphatic region almost 2 ppm of the ~ H NMR spectrum are affected by the suppression technique" [20]. Lesellier, West, and coworkers have used carotenoids as probe analytes in their extensive work to compare and classify columns for HPLC and SFC [21–26]. Bamba and coworkers demonstrated impressive separations of carotenoids as well [27]. Figure 11.6 shows an example of their work with carotenoids in which advantage is taken of the low viscosity of the CO_2-based mobile phase to couple three 100-mm monolithic columns while maintaining a flow rate of 3 mL/min [27]. The longer column clearly provides a better separation of the isomers. Note the retention-time inversion of the lutein and beta-carotene when moving from one to three coupled columns.

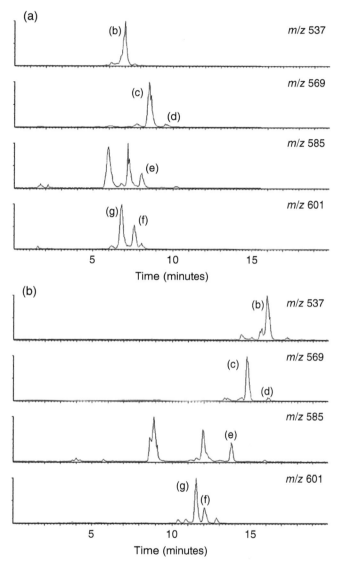

Figure 11.6 Analysis of the carotenoids in *C. reinhardtii* by SFC-MS using monolithic ODS columns. Analysis conditions: flow rate, 3 mL/min; outlet pressure, 10 MPa; column temperature, 35 °C; mobile phase, CO_2 and methanol (+0.1% ammonium formate) 10–30% 20 minutes. Column: (A) one monolithic ODS column and (B) three connected monolithic ODS columns; The plotted signals are for mass chromatograms of (b) b-carotene, (c) lutein, (d) zeaxanthin, (e) antheraxanthin, (f) neoxanthin, and (g) violaxanthin. *Source:* Reprinted with permission from reference [27]. Copyright 2009, Wiley.

11.2.4 Food Preservatives (Other Antioxidants and Antimicrobials)

Tocopherols belong to both the important classes of vitamins and natural antioxidants, and were discussed in Section 11.2.2. But SFC has been used for the characterization of other up-and-coming natural food antioxidants as well. For example, Ramirez et al. used SFC to separate rosemary extract actives [28]. The remarkable aspect of this work is that the use of a stationary phase consisting of silica particles coated with what are generally considered gas-chromatography phases (5% phenyl, 95% methyl silicone and poly(ethylene glycol)) allowed the elution of the relatively polar actives (carnosol, carnosic acid, and rosmarinic acid) using pure CO_2, with its wide ranging compatibility with detection options. Ramirez et al. used pressure programming at a relatively high column temperature of 100 °C. Their successful approach was in sharp contrast to most contemporary packed column SFC, which uses bonded stationary phases, mobile phase composition gradients, and relatively low temperatures, as described in Chapter 6.

Antimicrobials are an important class of food preservatives. Berger et al. described a surprisingly simple and very rapid method for the determination of two widely used antimicrobials, sorbate and benzoate salts, as well as the corresponding acids [29]. Twenty-five beverages and semi-solid foods (such as mustard and mayonnaise) were simply diluted with acidified methanol (to ensure that the salts were converted to the protonated form), and injected. The SFC/UV method was isocratic, operating at 3.5 mL/min of a mobile phase of CO_2 with 8.5% methanol containing 0.3% acetic acid. The column was a 4×250 mm (in total 1 m in length), 4.6-mm-i.d., 5-μm diol column operated at 50 °C and 150 bar column outlet pressure. While the authors found all the beverages correctly labeled, they found some of the semi-solid foods incorrectly labeled. This was indeed an unusual method for SFC, in which aqueous samples are simply diluted and injected. Not only was the method simple and highly efficient, but it was also very fast – ~7-fold more rapid than the published HPLC method.

11.2.5 Coloring Agents

Producers of food, both small and large companies, are moving rapidly to the adoption of natural coloring agents for their foods. These are most often extracts of fruits, vegetables, and other plant materials, and can be quite complex. Determining the important colorants in these extracts, with the goal of standardizing the potency of the colorants, is useful in assuring reliable and reproducible extracts, and, ultimately, consumer acceptance. Berger and Berger explored the use of a variety of 1.8-μm stationary phases, including three C18 phases, a cyano, a silica, and a diol for the separation of pigments in paprika oleoresin using a CO_2/2-propanol mobile phase [30]. The C18 phases were best for unsaponified pigments, while the silica column provided the best separation for saponified mixtures. The separations required less than 11 minutes, far faster than similar HPLC separations. Interestingly, one of the commercial pepper products examined appeared to contain artificial dyes.

Many natural pigments are apocarotentoids, C_{40} isoprenoids such as bixin (a coloring agent present in annatto) and crocetin. Giuffrida et al. developed an impressive separation of apocarotenoids from red habanero peppers [31]. They used SFC/MS/MS with APCI, combined with a fused-core C30 column and a simple CO_2/methanol mobile phase to determine 25 different apocarotenoids. The fused-core column and the selectivity of their selected reaction monitoring (SRM) detection approach allowed a very rapid analysis – less than five minutes. The same group has explored the advantages of online SFE/SFC/MS for the extraction and determination of carotenoids [32].

Natural pigments are often more expensive and less stable (especially across the wide range of pHs encountered in foods) than synthetic food colorants. There is therefore economic incentive for the rare unscrupulous supplier to supplement their extracts with synthetic legal, or even illegal, dyes. Not surprisingly, researchers have used SFC for rapid screening of foods and natural pigment extracts for these unwanted dyes. For example, Dolak et al. developed a rapid (less than eight minutes) SFC/UV/MS separation of the Sudan I–IV dyes, which are classified as Class 3 carcinogens, in foods [33]. Khalikova et al. developed and validated an impressive method for the extraction and determination of 11 illegal dyes in chili spices [34]. The SFC/UV separation was completed in less than five minutes using a fluoro-phenyl stationary phase operated at 70 °C. The mobile phase was CO_2 with a methanol:acetonitrile (1,1) modifier containing 2.5% formic acid. Limits of detection were in the sub-ppm range, while recoveries and reproducibilities were comparable to those obtained by HPLC/UV with much longer analysis times.

11.2.6 Sugars

Sugars are of great importance in all kinds of foods, and are among the most polar of nonionic organic molecules. As such, they would not be obvious candidates for SFC separation, and, in fact, they are not common solutes in SFC. However, Herbreteau et al. [35] and Morin-Allory and Herbreteau [36] published impressive separations of mono-, di-, and trisaccharides using SFC/ELSD. The columns were polar phases (cyano-, diol- and nitro-bonded silicas), and the mobile phase was CO_2 with a gradient of methanol. While the retention mechanism in SFC was shown to be similar to that of HPLC [36], the selectivities for various saccharides was different for the two methods [35].

A subsequent study of the SFC separation of saccharides by Salvador and the group referenced above expanded the mobile phases investigated to CO_2/

methanol, CO_2/methanol + water additive, and CO_2/methanol + water and trimethylamine additives, and compared "bare" silica to C1-silica columns [37]. A somewhat elevated column temperature (60 °C) and a higher flow rate (5 mL/min for a 4-mm-i.d. column) allowed the elution of a range of saccharides in <10 minutes.

11.3 Selected Applications in the Field of Synthetic Polymers

Natural polymers are the materials of life, while synthetic polymers have become the materials of everyday life. It is widely accepted that synthetic polymers have made great contributions to the quality of modern life, and what we consider modern living would be difficult without them. Polymers are complex structures. These structures enable their utility, and vary greatly in molecular weight distribution, terminal groups, blocks, crosslinking, branching, etc. Chemists use a variety of separation methods to characterize polymers, as they seek to establish structure–function relationships. This section focuses on selected applications of SFC in the characterization of synthetic polymers.

11.3.1 Molecular Weight Distribution

The most common application of SFC in the characterization of polymers has been the determination of the molecular weight distribution of relatively low molecular weight polymers/oligomers [38]. Many low to medium polarity oligomers have good solubility in pure CO_2, and SFC has long been used to determine molecular weight distributions. Examples include polyethylenes, polyethylene glycols [39, 40], polysiloxanes [41–43], and polyurethanes [38]). Polysiloxanes have great solubility in pure CO_2, and polysiloxanes ranging up to average molecular weights of ~20 000 Da can be characterized. Figure 11.7 shows the SFC separation of a polysiloxane by Berger and Todd [44]. These authors demonstrate the advantages of CO_2 as a mobile phase, using both flame ionization detection and UV detection at 191 nm. The siloxanes don't absorb at all, and are therefore not detectable, at 205 nm, typically the lowest detection wavelength used in reversed phase HPLC. The authors state that UV absorbance of the siloxanes drops off precipitously as wavelength rises, with little absorbance at 195 nm. In the same publication, the authors show the advantage of using hexane as a modifier for polysiloxanes. Hexane provides an even greater elution range.

Takahashi et al. have gone one step further in the application of SFC as a separation tool for oligomers according to degree of polymerization (i.e. molecular weight) [45, 46]. They used preparative-scale SFC to isolate single oligomers from a distribution, each with a particular degree of polymerization.

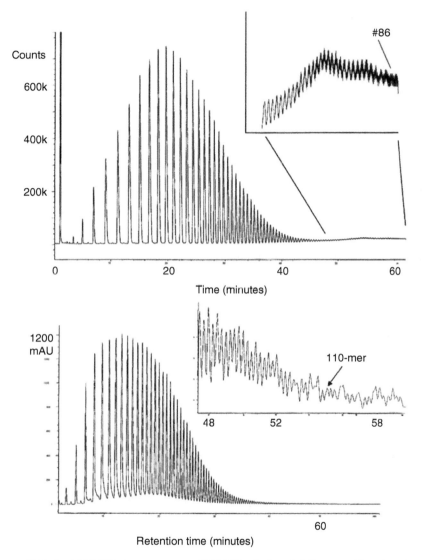

Figure 11.7 (top) SFC/FID chromatogram of DC-200 polydimethylsiloxane with a viscosity of 10 centistokes. The tail of the chromatogram is presented at an expanded scale to show more detail. Column: 4.6 × 200 mm, 5 μm Hypersil ODS, 2 mL/min, 150 °C, initial outlet pressure 80 bar, 5 bar/min to 370 bar, hold. 0.5 μL injection of 1:5 sample in hexane. FID detection, split flow ~10–40 mL/min expanded gas (pressure dependent). (bottom) Same sample separated using a 2.1 × 250 mm, 5-μm Deltabond Octyl column, at 0.4 mL/min, with an asymptotic density program. Injection size 10 μL neat. UV detection at 191 nm. *Source:* Reprinted with permission from reference [44]. Copyright 2001, Springer Nature.

They then used these pure oligomers to characterize the response of a variety of detectors (ultraviolet absorbance, evaporative light scattering, and matrix-assisted laser desorption/ionization mass spectrometry) as a function of mass.

11.3.2 Structural Characterization

Polymers are often not simple repetitions of a single monomer. They may have a variety of blocks or branches, each with its own distribution, as well as a variety of terminal groups. These structural differences can have dramatic effects on function, so understanding these differences is helpful. For example, Pretorius et al. used analytical and preparative-scale SFC, size-exclusion chromatography, and electrospray ionization mass spectrometry (ESI-MS) to characterize polyesters synthesized from phthalic acid and propylene glycol [47]. Important parameters were degree of polymerization, end-group functionality, and molecular topology (i.e. linear, cyclic, and branched). The most successful approach, providing detailed information about the nature of the polymer, involved fractionation by preparative-scale SFC followed by ESI-MS of the fractions.

Another example of the use of SFC for the separation and characterization of copolymers is shown in Figure 11.8. Pinkston et al., used SFC/MS to explore the structures of ethoxylated-propoxylated block copolymers [48]. The complexity of the polymer was such that multiple coupled columns (dimensions up to 2 m × 4.6 mm) operated at relatively high temperature (100 °C) were used for the separation, and qualitative and quantitative characterization was performed using an image-analysis-based approach. The low-m/z background ions (visualized as horizontal streaks in the image) were removed by the image analysis software to allow for easier identification and integration of low-m/z peaks. Separation based on retention time and m/z allowed identification and quantification of the individual oligomers in the blocks as well as the terminal groups. These analyses were used to help guide the polymer chemists as they optimized their synthetic approach. This is a great example of the power of coupled columns to provide high efficiency separations in SFC.

11.3.3 "Critical Condition" Group/Block Separations of Complex Polymers Using CO_2-containing Mobile Phases

An interesting application of CO_2-containing "enhanced fluidity" mobile phases is the separation of complex polymers at the "critical condition" [49] (not to be confused with "supercritical"). In this approach using traditional liquid mobile phases, the mobile phase composition is tuned such that the separation is "blind" to at least one oligomeric block, but separates the oligomers of another block or separates the polymers based upon terminal groups. Finding the critical condition by tuning the mobile phase composition can be

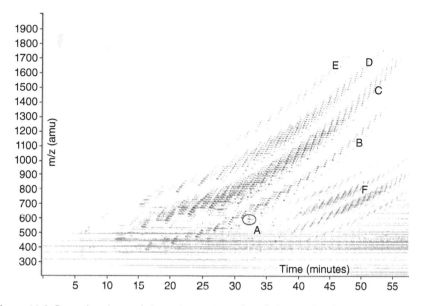

Figure 11.8 Retention time-m/z-intensity contour plot of a low-molecular-weight alkoxylated polymer. A: C_8E_{10} (octyl decaoxyethylene – single component) internal standard; B: unreacted starting material with an alkyl capping group on only one end of the molecule; C: reaction product with alkyl capping groups on both ends of the molecule; D: reaction product with a dimerized alkyl capping group on one end of the molecule; E: reaction product with a trimerized alkyl capping group on one end of the molecule; and F: doubly-charged species. *Source:* Reprinted with permission from reference [48]. Copyright 2002, Springer Nature.

challenging in traditional adsorption chromatography or gel permeation chromatography (GPC) with liquid mobile phases. But the addition of CO_2 to the mobile phase allows one to use composition, pressure, and temperature to affect the retention of various components of the polymer, and to more easily achieve the critical condition. Olesik and her colleagues expanded on this approach with examples of critical condition separations of poly(styrene-methyl acrylate) copolymers [50, 51], and of di-block and tri-block telechelic polymers [52]. ("Telechelic" polymers are polymers that are designed to react further to form even larger or more complex polymers.)

11.3.4 Polymer Additives

Additives are often what turn polymers and polymer blends into useful, resilient products. Additives are used to make polymers more rigid or more supple, to extend the life of products by resisting oxidation and photodegradation, or to help provide novel properties by allowing the blending or layering of dissimilar polymers. Additives are often lipophilic and polyfunctional, so SFC is a great tool for their separation and characterization.

Carrott and Davidson [53] explored the use of SFC for the characterization of polymer additives with both positive and negative APCI. For example, they found limits of detection in the tens of pg for Tinuvin 327 (a light stabilizer) and Irganox 1010 (an antioxidant) using either positive or negative-mode ionization. Figure 11.9 shows the rapid (<14 minutes) separation of six common polymer additives.

Performance improvements in the intervening years were illustrated by Zhang and Du in 2014 [54]. So called "ultra-performance SFC" was used to separate seven common polymer additives in less than five minutes. Microwave-assisted extraction (MAE) was used to extract additives from real samples. The combination of MAE with SFC/UV was able to provide recoveries of the seven common additives from 70 to 119% with RSDs of less than 10%.

Online supercritical fluid extraction/SFC of additives in polyethylene minimized sample handling and the use of organic solvents [55]. Zhou et al. used a simple glass-wool-packed cryogenic trap as an interface between the SFE and SFC steps. Results were comparable to those obtained with off-line SFE/HPLC and off-line liquid extraction/HPLC, with the exception that precision was lower in the SFE/SFC analysis. The poorer precision was ascribed to the much smaller sample size used in the online approach and sample inhomogeneity.

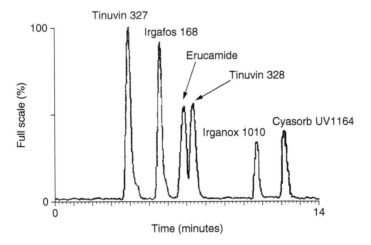

Figure 11.9 Positive ion atmospheric pressure chemical ionization total-ion-current chromatogram of six common polymer additives [53]. Column: 250×4.6 mm C_{18}; Mobile phase: 2% methanol in CO_2 for 1 minutes, then ramped to 10% methanol over 10 minutes, then held at 10%. Flow rate of 2 mL/min and a column inlet pressure (note – rather than outlet pressure) of 200 bar. *Source:* Reprinted with permission from reference [53]. Copyright 1998, the Royal Society of Chemistry.

11.4 Selected Applications in the Personal Care Industry

11.4.1 Lipophilic Components of Cosmetics

Many cosmetic products contain lipophilic mixtures that are critical to their function. These mixtures protect the skin, stabilize coloring agents, provide adhesion, enhance or diminish light reflection, etc. The lipophilic nature of these mixtures made them natural candidates for separation by SFC. An example is the SFC separation of natural and synthetic waxes [56, 57]. Li used supercritical fluid extraction to extract complex mixtures of waxes from mascaras, followed by characterization by SFC and multivariate analysis [57].

11.4.2 Surfactants in Cleaning Mixtures

A great number of personal care cleaning products contain surfactants. Nonionic surfactants are mild, effective, and rarely display adverse effects, so they are popular choices among surfactants. SFC and SFC/MS have long been used for their characterization [58]. One example is shown in Figure 11.10, where Hoffman et al. used 1,3-diphenyl-1,1,3,3-tetramethyldisilazane to derivatize the terminal hydroxyl group of a $C_{18}(EO)_{10}$ polyethoxylate surfactant [59]. This derivatizing agent converted the terminal hydroxyl to a $-O-Si(CH_3)_2-$ phenyl group, which provided much improved chromatographic behavior and UV detection sensitivity. The figure shows a comparison of separations using various single and coupled columns. The authors explored the use of other derivatizing agents, as well as high temperature elution of the underivatized surfactant with low UV detection (195 nm). Best results were obtained by coupling two "polar embedded" C18 columns (a Discovery C18 and a Discovery RP-AmideC16 column), as shown in Figure 11.10.

11.4.3 Emulsifiers in Personal Care Products

Emulsifiers are used to stabilize personal care products containing multiphase mixtures of polar and nonpolar ingredients, such as creams and lotions. Polysorbate 80 is one emulsifier commonly used in this way. It's quite a complex mixture, including polyethylene glycol (PEG), and PEG coupled to fatty alcohols, to saccharides, and to saccharides which are themselves esterified to one or more of a variety of fatty acids. The complexity is compounded by the oligomeric distribution of the PEG chain. Variation in the relative proportions of these components provides different properties and better, or worse, performance in various product applications. So characterizing Polysorbate 80 can be helpful in correlating structure to performance. Pan et al. used "ultra performance" SFC, coupled with tandem mass spectrometry, to characterize

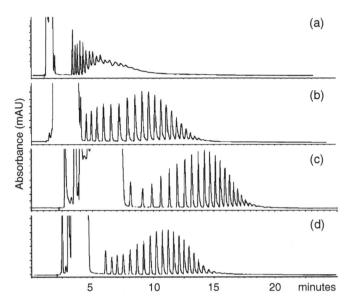

Figure 11.10 SFC/UV of SFC of $C_{18}EO_{10}$ surfactant derivatized with 1,3-diphenyl-1,1,3,3-tetramethyldisilazane; mobile phase: CO_2 modified with acetonitrile. Oven temperature: 40 °C; mobile phase flow rate: 2.4 mL/min; outlet pressure: 120 bar; detection at: 215 nm. Linear modifier gradient: 1% modifier held for 5 minutes increased to 20% at 1%/min, held at 20% for 5 minutes. (a) Discovery C18 column; (b) Discovery RP-AmideC16 column; (c) Two Discovery RP-AmideC16 columns; and (d) Discovery C18 column+Discovery RP-AmideC16 column. All columns were 4.6 mm × 250 mm, 5 µm. *Source:* Reprinted with permission from reference [59]. Copyright 2004, Elsevier.

polysorbate 80 [60]. They optimized flow rate, column temperature, and pressure and obtained a fast (~8 minutes) separation of the various oligomeric series present in the mixture. The analysis was suitable to help develop structure vs. performance relationships.

11.4.4 Preservatives

Preserving a personal care product from microbial degradation is of critical importance to its effectiveness and reputation. One last set of applications from the personal care industry includes preservatives, and illustrates the versatility of "SFC" instruments. Kapalavavi et al. [61], and Yang et al. [62] used high temperature HPLC and subcritical water chromatography to separate sunscreens in skincare creams [61] and preservatives in skincare products [62]. Figure 11.11 provides an example of the results with a series of paraben preservatives. No organic solvent was used in the mobile phase, making subcritical water chromatography (SCWC) one of the "greenest" condensed-phase chromatographic methods available. The authors recommend the use of a

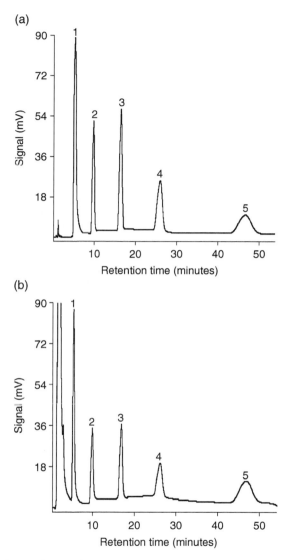

Figure 11.11 Subcritical water chromatography chromatograms obtained using 100% water on XBridge C18 column with programmed temperature at 1.0 mL/min. (a) Preservatives standard mixture; (b) SC-EC3 skincare cream sample. Best chromatogram mode for UV detection: first two peaks (benzyl alcohol and methyl paraben) were detected at 210 nm, whereas the other parabens were detected at 256 nm. Programmed temperature: initial temperature of 100 °C was increased to 150 °C at 15 °C/min and then maintained at 150 °C. Peak identification: 1, benzyl alcohol; 2, methyl paraben; 3, ethyl paraben; 4, propyl paraben; and 5, butyl paraben. *Source:* Reprinted with permission from reference [62]. Copyright 2012, Wiley.

commercial SFC instrument for SCWC, as most fit the requirements for SCWC as long as a column oven with a high (150–250 °C) upper temperature limit is available. The solvating power of subcritical water for relatively nonpolar analytes, like sunscreens and preservatives, increases as the temperature of the water rises. Yang et al. therefore used temperature programming to strengthen the mobile phase in the experiment shown in Figure 11.11. SCWC is an environmentally friendly tool for thermally stable analytes.

11.5 Conclusions

The health care and pharmaceutical arenas have benefited from widespread application of SFC. But SFC has also been a useful tool in many other areas. Here, we've illustrated selected applications of SFC in the foods, polymers, and personal care worlds. SFC provides speed and/or higher efficiencies, coupled with a normal-phase separation mode, usually orthogonal to the selectivity of reversed-phase HPLC.

References

1 Chester, T.L., Pinkston, J.D., and Raynie, D.E. (1994). Supercritical fluid chromatography and extraction. *Analytical Chemistry* 66 (12): 106–130.

2 Hannan, R.M. and Hill, H.H. Jr. (1991). Analysis of lipids in aging seed using capillary supercritical fluid chromatography. *Journal of Chromatography A* 547 (0): 393–401.

3 Porter, N.A., Caldwell, S.E., and Mills, K.A. (1995). Mechanisms of free radical oxidation of unsaturated lipids. *Lipids* 30 (4): 277–290.

4 France, J.E., Snyder, J.M., and King, J.W. (1991). Packed-microbe supercritical fluid chromatography with flame ionization detection of abused vegetable oils. *Journal of Chromatography A* 540 (0): 271–278.

5 Laboureur, L., Ollero, M., and Touboul, D. (2015). Lipidomics by supercritical fluid chromatography. *International Journal of Molecular Sciences* 16 (6): 13868–13884.

6 Donato, P., Inferrera, V., Sciarrone, D., and Mondello, L. (2017). Supercritical fluid chromatography for lipid analysis in foodstuffs. *Journal of Separation Science* 40 (1): 361–382.

7 Yamada, T. and Bamba, T. (2017). Lipid profiling by supercritical fluid chromatography/mass spectrometry. In: *Lipidomics* (ed. P. Wood), 109–131. New York, NY: Springer New York.

8 Sandra, P., Medvedovici, A., Zhao, Y., and David, F. (2002). Characterization of triglycerides in vegetable oils by silver-ion packed-column supercritical fluid chromatography coupled to mass spectroscopy with atmospheric pressure

chemical ionization and coordination ion spray. *Journal of Chromatography A* 974 (1–2): 231–241.

9 Kasim-Karaks, S. (2016). Omega-3 fish oils and insulin resistance. In: *Handbook of Nutraceuticals and Functional Foods*, 2e (ed. R. Wildman), 155–163. Boca Raton, FL: CRC Press.

10 Qu, S., Du, Z., and Zhang, Y. (2015). Direct detection of free fatty acids in edible oils using supercritical fluid chromatography coupled with mass spectrometry. *Food Chemistry* 170 (0): 463–469.

11 Ashraf-Khorassani, M., Isaac, G., Rainville, P. et al. (2015). Study of ultraHigh performance supercritical fluid chromatography to measure free fatty acids without fatty acid ester preparation. *Journal of Chromatography B, Analytical Technologies in the Biomedical and Life Sciences* 997: 45–55.

12 Saito, M. and Yamauchi, Y. (1990). Isolation of tocopherols from wheat germ oil by recycle semi-preparative supercritical fluid chromatography. *Journal of Chromatography A* 505 (1): 257–271.

13 Señoráns, F.J., Markides, K.E., and Nyholm, L. (1999). Determination of tocopherols and vitamin A in vegetable oils using packed capillary column supercritical fluid chromatography with electrochemical detection. *Journal of Microcolumn Separations* 11 (5): 385–391.

14 Turner, C., King, J.W., and Mathiasson, L. (2001). Supercritical fluid extraction and chromatography for fat-soluble vitamin analysis. *Journal of Chromatography A* 936 (1–2): 215–237.

15 Jiang, C., Ren, Q., and Wu, P. (2003). Study on retention factor and resolution of tocopherols by supercritical fluid chromatography. *Journal of Chromatography A* 1005 (1–2): 155–164.

16 Méjean, M., Brunelle, A., and Touboul, D. (2015). Quantification of tocopherols and tocotrienols in soybean oil by supercritical-fluid chromatography coupled to high-resolution mass spectrometry. *Analytical and Bioanalytical Chemistry* 407 (17): 5133–5142.

17 Pilařová, V., Gottvald, T., Svoboda, P. et al. (2016). Development and optimization of ultra-high performance supercritical fluid chromatography mass spectrometry method for high-throughput determination of tocopherols and tocotrienols in human serum. *Analytica Chimica Acta* 934: 252–265.

18 Lesellier, E., Tchapla, A., Péchard, M.R. et al. (1991). Separation of trans/cis α- and β-carotenes by supercritical fluid chromatography: II. Effect of the type of octadecyl-bonded stationary phase on retention and selectivity of carotenes. *Journal of Chromatography A* 557 (0): 59–67.

19 Aubert, M.C., Lee, C.R., Krstuloviác, A.M. et al. (1991). Separation of trans/cis- α- and β-carotenes by supercritical fluid chromatography: I. Effects of temperature, pressure and organic modifiers on the retention of carotenes. *Journal of Chromatography A* 557 (0): 47–58.

20 Braumann, U., Händel, H., Strohschein, S. et al. (1997). Separation and identification of vitamin A acetate isomers by supercritical fluid

chromatography – 1H NMR coupling. *Journal of Chromatography A* 761 (1–2): 336–340.

21 Lesellier, E. and Tchapla, A. (2005). A simple subcritical chromatographic test for an extended ODS high performance liquid chromatography column classification. *Journal of Chromatography A* 1100 (1): 45–59.

22 West, C. and Lesellier, E. (2006). Characterisation of stationary phases in subcritical fluid chromatography with the solvation parameter model IV. Aromatic stationary phases. *Journal of Chromatography A* 1115 (1–2): 233–245.

23 Lesellier, E., West, C., and Tchapla, A. (2006). Classification of special octadecyl-bonded phases by the carotenoid test. *Journal of Chromatography A* 1111 (1): 62–70.

24 West, C. and Lesellier, E. (2007). Characterisation of stationary phases in supercritical fluid chromatography with the solvation parameter model - V. Elaboration of a reduced set of test solutes for rapid evaluation. *Journal of Chromatography A* 1169 (1–2): 205–219.

25 West, C., Fougere, L., and Lesellier, E. (2008). Combined supercritical fluid chromatographic tests to improve the classification of numerous stationary phases used in reversed-phase liquid chromatography. *Journal of Chromatography A* 1189 (1–2): 227–244.

26 Lesellier, E. (2010). Extension of the C18 stationary phase knowledge by using the carotenoid test. *Journal of Separation Science* 33 (19): 3097–3105.

27 Matsubara, A., Bamba, T., Ishida, H. et al. (2009). Highly sensitive and accurate profiling of carotenoids by supercritical fluid chromatography coupled with mass spectrometry. *Journal of Separation Science* 32 (9): 1459–1464.

28 Ramírez, P., Señoráns, F.J., Ibañez, E., and Reglero, G. (2004). Separation of rosemary antioxidant compounds by supercritical fluid chromatography on coated packed capillary columns. *Journal of Chromatography A* 1057 (1–2): 241–245.

29 Berger, T.A. and Berger, B.K. (2013). Rapid, direct quantitation of the preservatives benzoic and sorbic acid (and salts) plus caffeine in foods and aqueous beverages using supercritical fluid chromatography. *Chromatographia* 76 (7–8): 393–399.

30 Berger, T.A. and Berger, B.K. (2013). Separation of natural food pigments in saponified and un-saponified paprika oleoresin by ultra high performance supercritical fluid chromatography (UHPSFC). *Chromatographia* 76 (11–12): 591–601.

31 Giuffrida, D., Zoccali, M., Giofrè, S.V. et al. (2017). Apocarotenoids determination in Capsicum chinense Jacq. cv. Habanero, by supercritical fluid chromatography-triple-quadrupole/mass spectrometry. *Food Chemistry* 231: 316–323.

32 Zoccali, M., Giuffrida, D., Dugo, P., and Mondello, L. (2017). Direct online extraction and determination by supercritical fluid extraction with chromatography and mass spectrometry of targeted carotenoids from red Habanero peppers (Capsicum chinense Jacq.). *Journal of Separation Science* 40 (19): 3905–3913.

33 Dolak, L.A., Cole, J., and Lefler, J.L. (2007). Resolution and identification of sudan dyes I-IV via supercritical fluid chromatography. *LCGC Europe* 2 (Suppl S): 22–23.

34 Khalikova, M.A., Satinsky, D., Solich, P., and Novakova, L. (2015). Development and validation of ultra-high performance supercritical fluid chromatography method for determination of illegal dyes and comparison to ultra-high performance liquid chromatography method. *Analytica Chimica Acta* 874: 84–96.

35 Herbreteau, B., Lafosse, M., Morin-Allory, L., and Dreux, M. (1990). Analysis of sugars by supercritical fluid chromatography using polar packed columns and light-scattering detection. *Journal of Chromatography A* 505 (1): 299–305.

36 Morin-Allory, L. and Herbreteau, B. (1992). High-performance liquid chromatography and supercritical fluid chromatography of monosaccharides and polyols using light-scattering detection: chemometric studies of the retentions. *Journal of Chromatography A* 590 (2): 203–213.

37 Salvador, A., Herbreteau, B., Lafosse, M., and Dreux, M. (1997). Subcritical fluid chromatography of monosaccharides and polyols using silica and trimethylsilyl columns. *Journal of Chromatography A* 785 (1): 195–204.

38 Sirrine, J.M., Ashraf-Khorassani, M., Moon, N.G. et al. (2016). Supercritical fluid chromatography with evaporative light scattering detection (SFC-ELSD) for determination of oligomer molecular weight distributions. *Chromatographia* 79 (15): 977–984.

39 Pyo, D. (2008). Supercritical fluid chromatographic separation of polyethylene glycol polymer. *Bulletin of the Korean Chemical Society* 29 (1): 231–233.

40 Brossard, S., Lafosse, M., and Dreux, M. (1992). Comparison of ethoxylated alcohols and polyethylene glycols by high-performance liquid chromatography and supercritical fluid chromatography using evaporative light-scattering detection. *Journal of Chromatography A* 591 (1–2): 149–157.

41 Baker, T.R. and Pinkston, J.D. (1998). Development and application of packed-column supercritical fluid chromatography/pneumatically assisted electrospray mass spectrometry. *Journal of the American Society for Mass Spectrometry* 9 (5): 498–509.

42 Pyo, D. and Lim, C. (2005). Supercritical fluid chromatographic separation of dimethylpolysiloxane polymer. *Bulletin of the Korean Chemical Society* 26 (2): 312–314.

43 Wu, N., Yee, R., and Lee, M.L. (2000). Fast supercritical fluid chromatography of polymers using packed capillary columns. *Chromatographia* 53 (3): 197–200.

44 Berger, T.A. and Todd, B.S. (2001). Packed column supercritical fluid chromatography of polysiloxanes using pure and hexane modified carbon dioxide with flame ionization and ultraviolet detection. *Chromatographia* 54 (11): 771–775.

45 Takahashi, K., Matsuyama, S., Saito, T., and Kinugasa, S. (2011). Development of the Certified Reference Materials for Molecular-Weight Distributions of Synthetic Polymers. *Bunseki Kagaku* 60 (3): 229–237.

46 Takahashi, K. (2013). Polymer analysis by supercritical fluid chromatography. *Journal of Bioscience and Bioengineering* 116 (2): 133–140.

47 Pretorius, N.O., Willemse, C.M., de Villiers, A., and Pasch, H. (2014). Combined size exclusion chromatography, supercritical fluid chromatography and electrospray ionization mass spectrometry for the analysis of complex aliphatic polyesters. *Journal of Chromatography A* 1330: 74–81.

48 Pinkston, J.D., Marapane, S.B., Jordan, G.T., and Clair, B.D. (2002). Characterization of low molecular weight alkoxylated polymers using long column SFC/MS and an image analysis based quantitation approach. *Journal of the American Society for Mass Spectrometry* 13 (10): 1195–1208.

49 Just, U. and Much, H. (1996). Characterization of polymers using supercritical fluid chromatography: application of adsorption chromatography, size exclusion chromatography and adsorption chromatography at critical conditions. *International Journal of Polymer Analysis and Characterization* 2 (2): 173–184.

50 Souvignet, I. and Olesik, S.V. (1997). Liquid chromatography at the critical condition using enhanced-fluidity liquid mobile phases. *Analytical Chemistry* 69 (1): 66–71.

51 Yun, H., Olesik, S.V., and Marti, E.H. (1998). Improvements in polymer characterization by size-exclusion chromatography and liquid chromatography at the critical condition by using enhanced-fluidity liquid mobile phases with packed capillary columns. *Analytical Chemistry* 70 (15): 3298–3303.

52 Phillips, S. and Olesik, S.V. (2002). Fundamental studies of liquid chromatography at the critical condition using enhanced-fluidity liquids. *Analytical Chemistry* 74 (4): 799–808.

53 Carrott, M.J. and Davidson, G. (1998). Identification and analysis of polymer additives using packed-column supercritical fluid chromatography with APCI mass spectrometric detection. *The Analyst* 123 (9): 1827–1833.

54 Zhang, Y. and Du, Z. (2014). Determination of seven additives in polymer products by ultra performance supercritical fluid chromatography. *Se Pu* 32 (1): 52–56.

55 Zhou, L.Y., Ashraf-Khorassani, M., and Taylor, L.T. (1999). Comparison of methods for quantitative analysis of additives in low-density polyethylene using supercritical fluid and enhanced solvent extraction. *Journal of Chromatography A* 858 (2): 209–218.

56 Brossard, S., Lafosse, M., and Dreux, M. (1992). Analysis of synthetic mixtures of waxes by supercritical fluid chromatography with packed columns using evaporative light-scattering detection. *Journal of Chromatography A* 623 (2): 323–328.

57 Li, J. (1999). Quantitative analysis of cosmetics waxes by using supercritical fluid extraction (SFE)/supercritical fluid chromatography (SFC) and multivariate data analysis. *Chemometrics and Intelligent Laboratory Systems* 45 (1–2): 385–395.

58 Chester, T.L. and Pinkston, J.D. (2002). Supercritical fluid and unified chromatography. *Analytical Chemistry* 74 (12): 2801–2811.

59 Hoffman, B.J., Taylor, L.T., Rumbelow, S. et al. (2004). Separation of derivatized alcohol ethoxylates and propoxylates by low temperature packed column supercritical fluid chromatography using ultraviolet absorbance detection. *Journal of Chromatography A* 1034 (1–2): 207–212.

60 Pan, J., Ji, Y., Du, Z., and Zhang, J. (2016). Rapid characterization of commercial polysorbate 80 by ultra-high performance supercritical fluid chromatography combined with quadrupole time-of-flight mass spectrometry. *Journal of Chromatography A* 1465: 190–196.

61 Kapalavavi, B., Marple, R., Gamsky, C., and Yang, Y. (2012). Separation of sunscreens in skincare creams using greener high-temperature liquid chromatography and subcritical water chromatography. *International Journal of Cosmetic Science* 34 (2): 169–175.

62 Yang, Y., Kapalavavi, B., Gujjar, L. et al. (2012). Industrial application of green chromatography – II. Separation and analysis of preservatives in skincare products using subcritical water chromatography. *International Journal of Cosmetic Science* 34 (5): 466–476.

12

Analysis of Cannabis Products by Supercritical Fluid Chromatography

12.1   Introduction

The medical marijuana industry is in a state of rapid expansion and acceptance across the world [1, 2]. While still classified in the United States as a schedule 1 drug under the Drug Enforcement Administration (DEA) Controlled Substances act, at the time of writing the medical use of cannabis is legal in 33 states and the recreational use of cannabis legal in 10 states. Each state is preparing regulations to guide the creation of safe products for consumers. Supercritical fluid chromatography (SFC) has become an increasingly important tool in the characterization of *Cannabis* and *Cannabis*-derived products. As these natural products and derivatives take on increasing financial and regulatory importance, the prominence of SFC in the field has grown. While there is nothing unique in the relationship between SFC and *Cannabis*-derived products, studying this relationship provides a good example of the power of SFC in the characterization of natural products in general. This chapter begins with a short glimpse into the history of *Cannabis* and *Cannabis*-derived

Modern Supercritical Fluid Chromatography: Carbon Dioxide Containing Mobile Phases,
First Edition. Larry M. Miller, J. David Pinkston, and Larry T. Taylor.
© 2020 John Wiley & Sons, Inc. Published 2020 by John Wiley & Sons, Inc.

products from a regulatory, pharmaceutical, and financial perspective, and then discuss the use of analytical and preparative SFC in the field.

12.1.1 *Cannabis* History

In the past 50 years, *Cannabis sativa (C. sativa)* and *indica (C. indica)* has grown from a substance that is essentially prohibited worldwide to one that has gained acceptance culturally and legally for both medicinal and recreational use. Cannabis was brought to America in 1619, mainly for applications in the textile industry. In 1937, the Marijuana Tax Act effectively banned *Cannabis* use and sales in the United States. This Tax Act was replaced by the Controlled Substances Act in the 1970s which again established cannabis as a schedule I substance and not accepted for medicinal use by the Drug Enforcement Agency (DEA). In 1996, California became the first state to legalize the use of *Cannabis* for medical purposes. As jurisdictions legalize cannabis products and the complexity of these products begins to surpass that of the classically dried plant material, appropriate methods for measuring biologically active constituents are paramount to ensure regulatory compliance. It is generally agreed that *C. indica and C. sativa* comprise a significant part of the forensic drug laboratory's case load [3]. Furthermore, illicit preparations of marijuana and hashish contain more than 500 compounds of differing polarities. These, along with their metabolites, make definitive analysis a difficult task. In addition to trace amounts of many minor cannabinoids, major cannabinoids tested in forensic laboratories have included cannabinol (CBN), delta-9-tetrahydrocannabinol (THC), and cannabidiol (CBD) (Figure 12.1).

Standard methodology to extract and analyze *Cannabis* has become an urgent matter to ensure both high product purity and compliance with existing laws. Product labels list cannabis plant parts or extracts as ingredients and/or make claims related to the product content of specific *cannabis* and cannabinoids. Cannabinoids have been proposed to have therapeutic effect for a number of diseases (Figure 12.2) [4]. In 2018, the U.S. Food and Drug Administration approved Epidiolex, an oral solution of cannabidiol for the treatment of seizures associated with two rare and severe forms of epilepsy (e.g. Lennox-Gastaut syndrome and Dravet syndrome) in patients two years of age and older. Epidiolex became the first FDA-approved drug that contained a purified substance derived from marijuana [5, 6]. It made headlines for two reasons: the strawberry-flavored syrup is designed to be palatable to young children and its active pharmaceutical ingredient is one of the cannabinoids with the greatest potential for therapeutic use.

Cannabis sativa L. has an extremely complex composition. Greater than 500 unique molecules are identified in cannabis to date [7, 8]. These molecules comprises many chemical classes, such as mono- and sesquiterpenes, sugars, steroids, and flavonoids including terpenophenolic compounds whose most

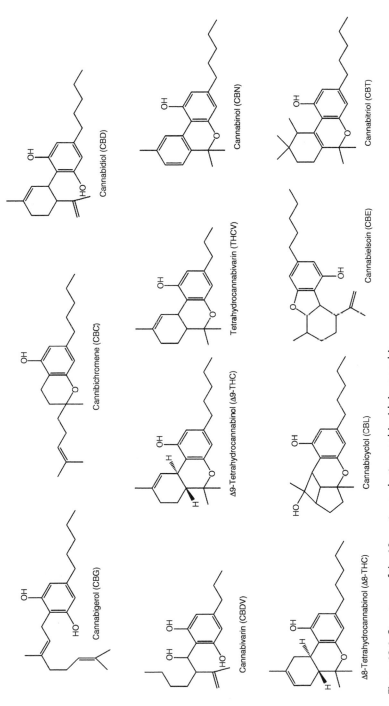

Figure 12.1 Structures of the 10 most prevalent cannabinoids in cannabis.

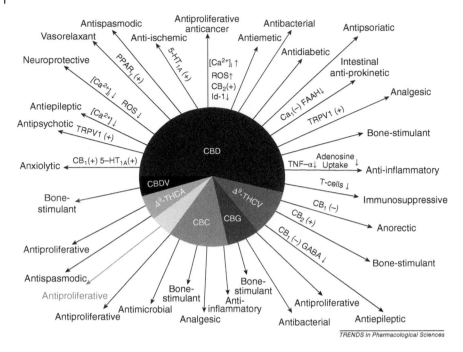

Figure 12.2 Pharmacological actions of nonpsychotropic cannabinoids (with the indication of the proposed mechanism of actions). *Source:* Reprinted from Izzo et al. [4] with permission from Elsevier.

prominent constituent is delta-9-trans-tetrahydrocannabinol. The complexity of cannabis demands high resolution analytical methods with excellent repeatability and relatively low limits of detection. The orthogonality of SFC and SFC/MS relative to GC and HPLC provides a forensic laboratory with a powerful combination for comprehensive characterization [9].

The cannabis industry is now becoming well organized. With legitimization comes regulation to ensure consumer safety. At the heart of this regulation are chromatographic methods to (i) test cannabis products for potency and (ii) confirm the absence of harmful contaminants. Because marijuana is illegal under federal law, there is no federal oversight to monitor quality or safety of cannabis and cannabis-containing products. As a result, each state has had to take responsibility for setting its own testing guidelines. In this regard, Colorado quickly embraced SFC as one of the primary chromatographic method for THC potency testing. In addition, manufacturing processes can benefit from the use of preparative SFE and SFC for the isolation and purification of individual cannabinoids for custom formulation preparation.

12.2 Analytical SFC

12.2.1 Introduction

Cannabis analyses are generally carried out using combinations of presumptive tests such as thin-layer chromatography (TLC), high performance liquid chromatography (HPLC), gas chromatography (GC), and gas chromatography-mass spectroscopy (GC/MS). While TLC is rapid and inexpensive, it is neither definitive nor quantitative. HPLC with nonspecific detectors, such as UV absorbance detection, offers greater resolution than TLC, but suffers from long analysis times, short analytical column life, and is not definitive. HPLC/MS provides improved specificity and sensitivity, yet still provides relatively long analysis times. GC and GC/MS offer the greatest resolution of the cannabis components, but derivatization of the sample is required before complete comparison of the samples can be made [9]. Chromatographic methodologies with suggested associated analytical tasks for cannabis analysis as described are outlined in Table 12.1.

When SFC was introduced in 1962, it focused primarily on petroleum and small polymer analysis, as well as testing the decaffeination of coffee and tea. As demonstrated in the previous chapters of this book, the focus has recently shifted to pharmaceuticals and personal care products, plus flavors, fragrances, pesticides and, of course, small polymers and petroleum products. Since 2012, there has been increasing interest in using SFC to meet the demand for cannabinoid testing in states that have legalized cannabis for medicinal or recreational use [10]. As described in Chapter 2, SFC's popularity is attributed to a number of factors including:

- low viscosity of supercritical fluids and higher analyte diffusion coefficients, yielding higher efficiency per unit time (compared to HPLC)
- orthogonal selectively (compared to RP-HPLC)

Table 12.1 Chromatographic methodologies to the budding cannabis industry.

- **Supercritical Fluid Chromatography**: Facilitates analysis of the complex interactions and degradation pathways of the cannabinoids through the cultivation and manufacturing process – critical for patient dosing and product shelf life
- **Gas Chromatography**: Identification of the phytochemical terpene constituents which have been indicated in having therapeutic synergy with the cannabinoids
- **Headspace Gas Chromatography**: Identify the residual solvents that may result from the many extraction and processing methods
- **Liquid Chromatography with Mass Spectrometric Detection**: Plays a role in pesticide testing
- **Preparative Supercritical Fluid Chromatography**: Purification of individual cannabinoids for the preparation of custom formulations to target specific cannabinoid therapies.

- limited analyte thermal degradation (compare to GC)
- low cost, recyclable, ecologically friendly solvents
- introduction of well-integrated, reliable instrumentation by major vendors
- incorporation of sub-3-µm particle columns
- lower limits of detection than earlier SFC instruments

This surge in popularity gave rise to an increased demand for SFC equipment that offers higher resolution for testing cannabinoids [11]. Numerous scientists have noted the importance of both analytical and preparative-scale chromatography in the analysis and accpetance of cannabis products. Identification of both achiral and chiral components and the establishment of product purity have been the focus. Technologies with carbon dioxide-based mobile phases for supercritical fluid chromatography and supercritical fluid extraction will play pivotal roles in cannabinoid characterization by facilitating analytical testing, plant extractions, and purification for custom formulations.

12.2.2 Early SFC of Cannabis Products

As described in Chapter 2, open-tubular SFC was popular in the 1980s and early-to-mid-1990s. Some of the early applications of SFC to cannabinoids used open-tubular SFC. Pure supercritical CO_2 was used as the mobile phase in conjunction with a wall-coated methyl-polysiloxane stationary phase and a flame ionization detector [12]. The authors demonstrated the open-tubular capillary SFC separation of a mixture of equal amounts of tetrahydrocannabinol and six cannabinoid metabolites. Tetrahydrocannabinol, the major psychoactive component in marijuana, was clearly separated from the other metabolites.

Near the turn of the century, cannabinoids were separated with packed-column SFC using a cyanopropyl silica column and a methanol gradient [13]. The SFC was interfaced with a mass spectrometer fitted with an atmospheric pressure chemical ionization (APCI) source. Cannabidiol plus three isomers of tetrahydrocannabinol were all separated in ~7 minutes and were identified with mass spectroscopy. While this technique offered attractive advantages over HPLC and GC/MS, its principal difficulty was blocking of the fused-silica interface as the CO_2 became subcritical (losing its solvating power) prior to entry into the mass spectrometer.

Subsequent to the previous report, a simple method for SFC-UV employing analysis of cannabis was described [14], which provided an alternate means for identification and comparison of cannabis samples. Authenticated cannabinoid standards (delta 9-THC, delta 8-THC, and cannabidiol) at 1 mg/mL in ethanol were studied.

As analysis of cannabinoids became more complicated, a novel, simple, and rapid SFC method was needed as a screening tool for both natural and synthetic cannabinoids and their metabolites in biological samples. "Ultra high performance" SFC instrumentation has filed this need, and is described in the following section.

12.2.3 Achiral SFC

SFC using achiral stationary phases can be applied in many areas of cannabis analysis including potency, identity, and impurity profiles of pure as well as formulated products. Wang et al. developed an SFC method for the analysis of cannabis plant extracts and preparation [15]. Nine of the most abundant cannabinoids were quantitively determined and unlike GC, no derivatization or decarboxylation was required prior to analysis. The SFC method showed high orthogonality compared to existing HPLC methods.

Cannabis contains both neutral and carboxylic acid forms of many cannabinoids. These acidic cannabinoids are thermally unstable and can be decarboxylated when exposed to light or heat via smoking, refluxing, or baking. As part of a study to understand kinetics of decarboxylation reactions that occur with phytocannabinoids, Wang et al. developed SFC/MS methods that provided consistent and sensitive analysis of phytocannabinoids and their decarboxylation products and degradants [16]. This study showed a first-order decarboxylation reaction, with different cannabinoids showing different rate constants.

The practice of synthesizing novel drugs with slight chemical structure modifications is commonplace for controlled substances. These substances were initially developed and studied for therapeutic use but were later misused [17, 18]. These "designer drugs" are made with the intent of circumventing controlled substance laws, and they represent a major challenge to both separation scientists and law enforcement laboratories charged with investigating the nature of seized materials. Synthetic cannabinoids represent one of more than 20 classes of designer drugs under federal control in the United States [19]. Synthetic cannabinoids possess (i) wide structural variability, (ii) potent cannabimimetic pharmacological activity, and (iii) binding capability to the same cannabinoid receptors as THC. Modifications made to these compounds result in structural analogues, structural homologues, positional isomers, and stereoisomers. Synthetic cannabinoids therefore present unique challenges for the separation, determination, and purification of these new analytes. A new question therefore arises. Will SFC be just as effective with synthetic cannabinoids as with traditional cannabinoids, and can it complement or replace UHPLC/MS and GC/MS, which are currently the standard analytical techniques for the analysis of synthetic cannabinoids in seized drugs [20]?

An answer to the above question is provided by Breitenbach et al. who investigated SFC as a separation technique for analysis of synthetic cannabinoids [21]. The authors demonstrated the orthogonality of SFC relative to GC and

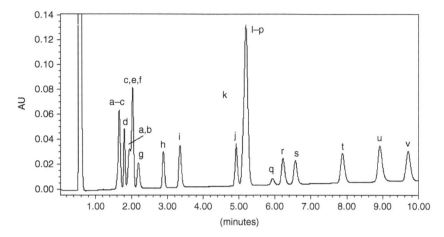

Figure 12.3 UHPSFC separation of controlled synthetic cannabinoids. Conditions: injection size, 1 µL; column 15 cm × 3.0 mm 2.5 µm Acquity UPC2 Trefoil CEL1. Initial conditions: 20% isopropanol, 80% carbon dioxide. Final conditions: 31% isopropanol, 69% carbon dioxide, 10.3 min linear gradient, 1.0 minutes gradient re-equilibration; flow rate 1.25 mL/min; temperature 55 °C, ABPR 2200 psi; UV detection 215 nm. Peaks 5 µg/mL. See [21] for compound identification. *Source:* Reprinted from Breitenbach et al. [21], with permission from Elsevier.

HPLC using principal component analysis. The separation of 22 controlled synthetic cannabinoids is shown in Figure 12.3. In summary, the authors "felt this new technique should prove useful in the analysis and detection of seized drug samples and will be a useful addition to the compendium of methods for drug analysis." A more recent use of synthetic cannabinoids involves spraying them on plants that are marketed as "legal highs" and sold on the internet and other shops. Jambo et al. developed a generic SFC/MS method for the quality control of cannabis plants that could be adulterated with synthetic cannabinoids [22]. Using design-of-experiments and design-space methodology (DoE-DS), an SFC method that can separate natural cannabinoids from the synthetics was assessed. Mass spectrometric detection was used to provide sensitivity and specificity. The method was validated with a LOD value as low as 14.4 ng/mL.

12.2.4 Chiral SFC

Depending on how it is derived, THC exists in many isomeric forms. Four (*trans*) isomers were predominant including (+)-Δ8-THC, (−)-Δ8-THC, (+)-Δ9-THC, and (−)-Δ9-THC. (−)trans-D9-THC is the main plant derived stereoisomer, but many synthetic preparations produce the more stable Δ8-THC isomer or a mixture of positional and stereoisomers. Under acidic conditions, cannabidiol is converted to delta-9-THC and other THC isomers [23]. FDA requires that stereoisomeric composition be quantified for active chiral pharmaceutical compounds. Likewise, other regulatory agencies are requiring that consumable products be monitored for mixtures of positional and stereoisomers that often give rise to changes in potency, pharmacological activity, or toxicity.

Runco et al. developed a method to resolve the stereoisomers of Δ8 and Δ9 THC using SFC [24]. The structure of THC stereoisomers and their analytical separation is shown in Figure 12.4. The method was shown to be repeatable and could be used for qualitative and quantitative analysis of THC stereoisomers from natural and synthetic cannabis products.

SFC has also been evaluated for the separation of structural analogs and stereoisomers of synthetic cannabinoids. Runco et al. reported on the chiral analysis of synthetic cannabinoids including HU-210, HU-211, (±)-CP 47,497, (±)-epi CP 47,497, (±)-CP 55,940 and (±)5-epi CO 55,940 [25]. Breitenbach et al. evaluated chiral SFC for synthetic cannabinoid JWH018 and nine of its positional isomers [21]. The superiority of SFC over other separation techniques is demonstrated by the baseline separation of all ten positional isomers in the chromatogram shown in Figure 12.5.

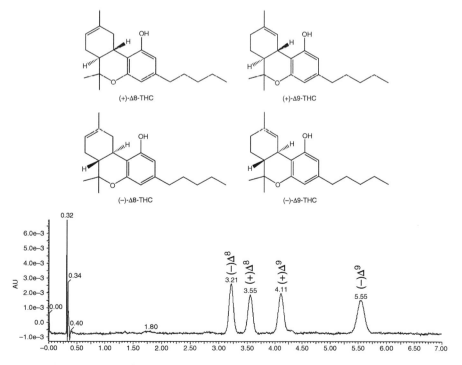

Figure 12.4 Structures of THC stereoisomers. Conditions: Trefoil AMY1, 2.5 μm, 3 × 150 mm, 10% ethanol/90% CO2, 2 mL/min. *Source:* Courtesy of Waters Corporation 2017.

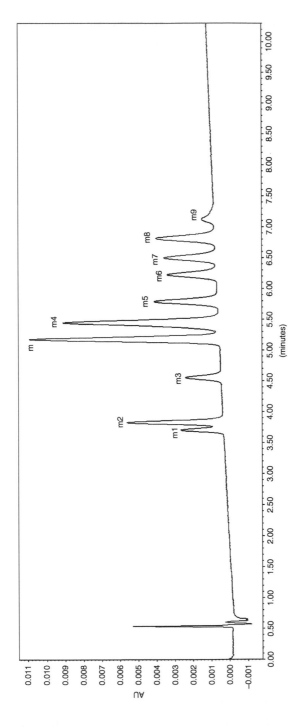

Figure 12.5 SFC separation of JWH 018 and nine of its positional isomers. Conditions: injection, 1 µL, column, Trefoil CEL1, 2.5 µm, 3 × 15 cm. Gradient conditions: 20% isopropanol/80% carbon dioxide to 31% isopropanol/69% carbon dioxide over 10.3 minutes linear gradient. Flowrate 1.25 mL/min, UV detection 273 nm. *Source:* Reprinted from Breitenbach et al. [21], with permission from Elsevier.

One problem faced in stereo-selective analysis of cannabis plant extracts is the lack of minor enantiomers or the racemates as reference standards. To overcome this limitation the "Inverted Chirality Columns Approach" (ICCA) can be utilized [26]. The method uses two chiral stationary phases having the same bound selector with an opposite configuration, to reverse the elution order of a given enantiomeric pair. Pirkle type chiral stationary phases (CSPs), being synthetic selectors available in both enantiomeric versions, are ideal for this application. Mazzoccanti et al. used the ICCA method in combination with UHPSFC to determine the enantiomeric excess (ee) of (-)delta-9-THC in medical marijuana to be high (99.73%), but the concentration of the (+)-enantiomer was not negligible (0.135%) [27].

12.2.5 Metabolite Analysis

To determine the impact of a drug on living systems it is important to perform metabolism studies that measure the breakdown of drugs upon administration. Only limited work has been published on metabolism of cannabis [28] and synthetic cannabinoids [29, 30], all of which used LC-MS/MS for analysis. Only recently has SFC been investigated for metabolism studies of cannabis and related products. Geryk et al. reported on SFC as a screening tool for natural and synthetic cannabinoids and their metabolites in biological samples [31]. The method has the advantages of short analysis time, high separation efficiency, and low solvent usage. It was demonstrated that the method was precise, selective, and robust with acceptable linearity within the calibration range. While the study used only UV detection, addition of a mass spectrometer may allow for systematic toxicology analysis.

Multiple studies have been performed to analyze urban wastewater as an indication of local drug use [32–34]. These studies use solid-phase extraction followed by separation and detection by LC-MS/MS. The low polarity of cannabinoids leads to problems with this approach, such as low recovery during SPE. Gonzalez-Marino et al. optimized and validated an SFC-MS/MS method to determine THC, three of its major human metabolites, and four JWH-type synthetic cannabinoids at low ng/L levels in wastewater [35]. The applicability of the method was demonstrated by the analysis of real wastewater where cannabis metabolites were positively quantified in all the samples analyze.

12.3 Preparative SFC

While Δ^9-THC is the main biologically active component of cannabis, more than 500 constituents have been isolated and/or identified from *Cannabis sativa* L. [36–38]. While a number of purification/isolation technologies are suitable for purification of cannabis related compounds, the majority of the work is performed using normal phase flash chromatography which uses larger particle packing materials (>40 µm) and large volumes of flammable organic solvents such as hexane, ethyl acetate, and/or toxic solvents such as dichloromethane [39, 40]. Preparative SFC, being ideally suited for purification of low polarity compounds, would appear to be a suitable alternative to existing purification technologies, as well as having advantages of higher productivities, higher separation efficiency, and reduced solvent usage.

The first step in cannabinoid isolation is processing of the plant material to enrich products of interest. This has been performed by crushing the plant material and extracting with organic solvents such as hexanes, ethyl acetate, and methanol, although there are safety concerns related to their toxicity and flammability [41]. Supercritical fluid extraction (SFE) is a technology used in large scale production of essential oils and a large number of bioactive compounds from plant materials [42–46]. The past five years have seen an increase in research of SFE for cannabis related materials including cannabinoids [47], aroma compounds [48], and production of hemp seed oil [49].

While SFE is useful for enriching materials, it is often not a highly specific purification process, often producing a mixture of compounds of approximately the same polarity and/or chemical class. To isolate individual components a high-performance separation process, such as preparative HPLC, is required. The disadvantages of preparative HPLC were discussed previously; it is not surprising that SFC is being explored for purification of cannabis components. The first SFC-based purification process for cannabis material was a 2008 US patent by Geiser et al. [50]. This patent discussed a process for purifying (-)-Δ^9-trans-tetrahydrocannabinol from a mixture of cannabinoids by preparative chromatography and carbon dioxide based mobile phases. The preferred embodiment of the invention utilized a two-step SFC purification process, the first with a 2-ethylpyridine achiral stationary phase followed by a second purification step using a polysaccharide based chiral stationary phase. This process is able to produce (-)-Δ^9-trans-THC with a purity of >99.5%.

While the first SFC purification of cannabinoids was reported in 2008, little additional work was performed due to limited legal uses for cannabis products in the United States. It was not until states began legalizing medical and recreational uses of cannabis that work in this field expanded. Enmark et al. presented a poster at SFC 2017 describing the isolation of CBD using preparative SFC [51]. The separation was scaled from a 4.6- to a 50-mm-i.d. column. No information on final purities or yields was discussed.

Denicola et al. presented a poster on cannabinoid isolation using immobilized chiral stationary phases [52]. The authors evaluated two columns (Chiralpak IB-N and DCpak P4VP) for the analytical separation of cannabinoid standards. The two phases were evaluated for preparative separation of real-world cannabinoid samples. The analytical and preparative chromatograms for these separations are shown in Figure 12.6. Of interest is the

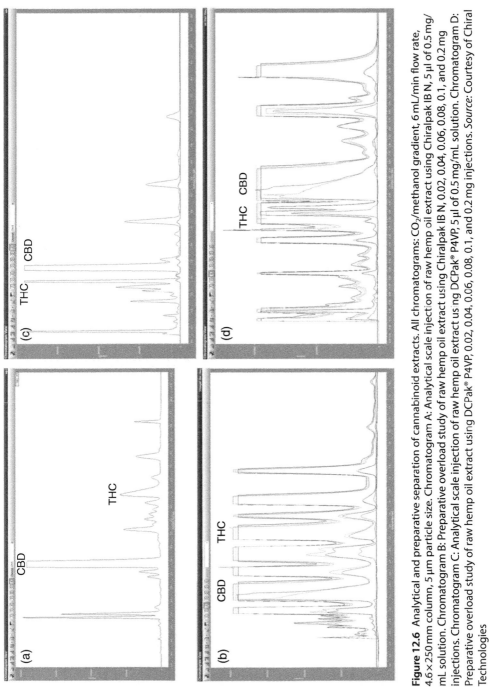

Figure 12.6 Analytical and preparative separation of cannabinoid extracts. All chromatograms: CO_2/methanol gradient, 6 mL/min flow rate, 4.6 × 250 mm column, 5 μm particle size. Chromatogram A: Analytical scale injection of raw hemp oil extract using Chiralpak IB N, 5 μl of 0.5 mg/mL solution. Chromatogram B: Preparative overload study of raw hemp oil extract using Chiralpak IB N, 0.02, 0.04, 0.06, 0.08, 0.1, and 0.2 mg injections. Chromatogram C: Analytical scale injection of raw hemp oil extract us ng DCPak® P4VP, 5 μl of 0.5 mg/mL solution. Chromatogram D: Preparative overload study of raw hemp oil extract using DCPak® P4VP, 0.02, 0.04, 0.06, 0.08, 0.1, and 0.2 mg injections. *Source:* Courtesy of Chiral Technologies

Table 12.2 Daily production parameters (normalized for 1 kg stationary phase over 24 hours run time).

	IB-N SFC	P4VP SFC	C18 Flash
Kilograms of CBD produced	1.13	1.41	0.77
Liters of solvent used	433	383	740
Solvent per kg CBD (L)	392	272	961

reversal of elution order between the two phases; depending on the purification need, either THC or CBD can be eluted first. The two methods were scaled to preparative loadings and compared to a C18 flash chromatography purification method often used for these types of purifications. A comparison of the purification methods can be found in Table 12.2. Purification using SFC had a productivity increase of approximately 50–90% over flash LC and required 60–70% less solvent per kg of CBD produced. While the poster did not discuss equipment cost differences between the two approaches (SFC and LC), it is obvious there would be significant costs savings in stationary phase and mobile phase transitioning from LC to SFC for cannabinoid isolation.

The use of preparative SFC for cannabinoid purification is a relatively new technique for cannabinoid processing but will only grow as additional research is performed. While few applications have been published, mostly from equipment vendors, there is a large amount of proprietary research occurring in cannabis companies. As increased research is performed on individual cannabinoids, preparative SFC will be an important technique for isolation of high purity compounds for biological testing. The expansion of this field is also evident by the number of preparative SFC vendors (both existing and new) that are producing equipment for the cannabis market (Thar Process, PIC Solutions, Waters, ExtraktLab). Preparative SFC equipment for cannabis processing is now available with column sizes up to 60 cm i.d. that, depending on cannabinoid levels in extracts, allow the production of greater than 50 kg of product per day.

12.4 Summary

SFC is a relatively new technology for the cannabis industry. The advantages of analytical and preparative SFC that have caused the technology to expand in pharmaceuticals and other areas will allow SFC to play a pivotal role in the burgeoning cannabis field for both analysis and purification. SFC has been identified as an ideal method for potency testing of various ingredients in cannabis such as THC, THCA, CBD, CBDA, and CBN, and for determining whether they are homogeneously distributed throughout the product. The high productivity, high resolving power and low solvent consumption of preparative SFC makes it an important technique for generation of high purity cannabinoids for research and manufacturing.

References

1 Mead, A. (2017). The legal status of cannabis (marijuana) and cannabidiol (CBD) under U.S. law. *Epilepsy & Behavior* 70: 288–291.
2 Lucas, P.G. (2008). Regulating compassion: an overview of Canada's federal medical cannabis policy and practice. *Harm Reduction Journal* 5 (1): 5.
3 Halford, B. (2013). Analyzing cannabis. *Chemical & Engineering News* 91: 32–33.
4 Izzo, A.A., Borrelli, F., Capasso, R. et al. (2009). Non-psychotropic plant cannabinoids: new therapeutic opportunities from an ancient herb. *Trends in Pharmacological Sciences* 30 (10): 515–527.
5 Halford, B. (2018). Medicine from Marijuana. *Chemical and Engineering News*: 28–33.
6 FDA Approves First Drug Comprised of Marijuana. American Pharmaceutical Review. 2018. https://www.americanpharmaceuticalreview.com/1315-News/351333-FDA-Approves-First-Drug-Comprised-of-Marijuana/
7 Hanus, L.O., Meyer, S.M., Munoz, E. et al. (2016). Phytocannabinoids: a unified critical inventory. *Natural Product Reports* 33 (12): 1357–1392.
8 Pertwee, R.G. (2014). *Handbook of Cannabis*. USA: Oxford University Press.
9 Lurie, I.S. (2016). The evolving role of SFC in forensic analysis. *The Column* 12 (14): 2–6.
10 Ciolino, L.A., Ranieri, T.L., and Taylor, A.M. (2018). Commercial cannabis consumer products part 1: GC–MS qualitative analysis of cannabis cannabinoids. *Forensic Science International* 289: 429–437.
11 McAvoy, Y., Bäckström, B., Janhunen, K. et al. (1999). Supercritical fluid chromatography in forensic science: a critical appraisal. *Forensic Science International* 99 (2): 107–122.
12 Richter, B.E., Knowles, D.E., Later, D.W., and Andersen, M.R. (1986). Analysis of various classes of drugs by capillary supercritical fluid chromatography. *Journal of Chromatographic Science* 24 (6): 249–253.
13 Bäckström, B., Cole, M.D., Carrott, M.J. et al. (1997). A preliminary study of the analysis of Cannabis by supercritical fluid chromatography with atmospheric pressure chemical ionisation mass spectroscopic detection. *Science & Justice* 37 (2): 91–97.
14 Raharjo, T.J. and Verpoorte, R. (2004). Methods for the analysis of cannabinoids in biological materials: a review. *Phytochemical Analysis* 15 (2): 79–94.

15 Wang, M., Wang, Y.H., Avula, B. et al. (2017). Quantitative determination of cannabinoids in cannabis and cannabis products using ultra-high-performance supercritical fluid chromatography and diode array/mass spectrometric detection. *Journal of Forensic Sciences* 62 (3): 602–611.

16 Wang, M., Wang, Y.H., Avula, B. et al. (2016). Decarboxylation study of acidic cannabinoids: a novel approach using ultra-high-performance supercritical fluid chromatography/photodiode array-mass spectrometry. *Cannabis and Cannabinoid Research* 1 (1): 262–271.

17 Castaneto, M.S., Gorelick, D.A., Desrosiers, N.A. et al. (2014). Synthetic cannabinoids: epidemiology, pharmacodynamics, and clinical implications. *Drug and Alcohol Dependence* 144: 12–41.

18 Brents, L.K. and Prather, P.L. (2014). The K2/Spice phenomenon: emergence, identification, legislation and metabolic characterization of synthetic cannabinoids in herbal incense products. *Drug Metabolism Reviews* 46 (1): 72–85.

19 Lewis, M.M., Yang, Y., Wasilewski, E. et al. (2017). Chemical profiling of medical cannabis extracts. *ACS Omega* 2 (9): 6091–6103.

20 Lurie, I.S., Marginean, I., and Rowe, W. (2016). Analysis of Synthetic Cannabinois in Siezed Drugs by High-Resolution UHPLC/MS and GC/MS. https://www.perkinelmer.com/lab-solutions/resources/docs/APP_Analysis-of-Synthetic-Cannabinoids-in-Seized-Drugs_012433A_01.pdf.

21 Breitenbach, S., Rowe, W.F., McCord, B., and Lurie, I.S. (2016). Assessment of ultra high performance supercritical fluid chromatography as a separation technique for the analysis of seized drugs: Applicability to synthetic cannabinoids. *Journal of Chromatography A* 1440: 201–211.

22 Jambo, H., Dispas, A., Avohou, H.T. et al. (2018). Implementation of a generic SFC-MS method for the quality control of potentially counterfeited medicinal cannabis with synthetic cannabinoids. *Journal of Chromatography B, Analytical Technologies in the Biomedical and Life Sciences* 1092: 332–342.

23 Gaoni, Y. and Mechoulam, R. (1966). Hashish – VII: the isomerization of cannabidiol to tetrahydrocannabinols. *Tetrahedron* 22 (4): 1481–1488.

24 Runco, J., Aubin, A., and Layton, C. (2016). The Separation of Delta 8-THC, Delta 9-THC, and Their Enantiomers by UPC2 Using Trefoil Chiral Columns. https://www.waters.com/webassets/cms/library/docs/720005812en.pdf.

25 Runco, J., Aubin, A., and Jablonski, J. (2016). Simple method development for the separation of chiral synthetic cannabinoids using ultra high performance supercritical fluid chromatography. *Chromatography Today*: 23–26.

26 Badaloni, E., Cabri, W., Ciogli, A. et al. (2007). Combination of HPLC "Inverted chirality columns approach" and MS/MS detection for extreme enantiomeric excess determination even in absence of reference samples. Application to camptothecin derivatives. *Analytical Chemistry* 79 (15): 6013–6019.

27 Mazzoccanti, G., Ismail, O.H., D'Acquarica, I. et al. (2017). Cannabis through the looking glass: chemo- and enantio-selective separation of phytocannabinoids by enantioselective ultra high performance supercritical fluid chromatography. *Chemical Communications (Cambridge)* 53 (91): 12262–12265.

28 Zhu, J. and Peltekian, K.M. (2019). Cannabis and the liver: Things you wanted to know but were afraid to ask. *Canadian Liver Journal*: 1–7.

29 Diao, X. and Huestis, M. (2017). Approaches, challenges, and advances in metabolism of new synthetic cannabinoids and identification of optimal urinary marker metabolites. *Clinical Pharmacology and Therapeutics* 101 (2): 239–253.

30 Presley, B., Gurney, S., Scott, K. et al. (2016). Metabolism and toxicological analysis of synthetic cannabinoids in biological fluids and tissues. *Forensic Science Review* 28 (2): 103.

31 Geryk, R., Svidrnoch, M., Pribylka, A. et al. (2015). A supercritical fluid chromatography method for the systematic toxicology analysis of cannabinoids and their metabolites. *Analytical Methods* 7 (15): 6056–6059.

32 van Nuijs, A.L., Castiglioni, S., Tarcomnicu, I. et al. (2011). Illicit drug consumption estimations derived from wastewater analysis: a critical review. *Science of the Total Environment* 409 (19): 3564–3577.

33 Zuccato, E., Chiabrando, C., Castiglioni, S. et al. (2008). Estimating community drug abuse by wastewater analysis. *Environmental Health Perspectives* 116 (8): 1027–1032.

34 Baker, D.R. and Kasprzyk-Hordern, B. (2011). Critical evaluation of methodology commonly used in sample collection, storage and preparation for the analysis of pharmaceuticals and illicit drugs in surface water and wastewater by solid phase extraction and liquid chromatography–mass spectrometry. *Journal of Chromatography A* 1218 (44): 8036–8059.

35 Gonzalez-Marino, I., Thomas, K.V., and Reid, M.J. (2018). Determination of cannabinoid and synthetic cannabinoid metabolites in wastewater by liquid-liquid extraction and ultra-high performance supercritical fluid chromatography-tandem mass spectrometry. *Drug Testing and Analysis* 10 (1): 222–228.

36 ElSohly, M.A. and Slade, D. (2005). Chemical constituents of marijuana: The complex mixture of natural cannabinoids. *Life Sciences* 78 (5): 539–548.

37 Radwan, M.M., ElSohly, M.A., Slade, D. et al. (2009). Biologically active cannabinoids from high-potency Cannabis sativa. *Journal of Natural Products* 72 (5): 906–911.

38 Pollastro, F., Taglialatela-Scafati, O., Allara, M. et al. (2011). Bioactive prenylogous cannabinoid from fiber hemp (Cannabis sativa). *Journal of Natural Products* 74 (9): 2019–2022.

39 Zulfiqar, F., Ross, S.A., Slade, D. et al. (2012). Cannabisol, a novel Δ9-THC dimer possessing a unique methylene bridge, isolated from Cannabis sativa. *Tetrahedron Letters* 53 (28): 3560–3562.

40 Ahmed, S.A., Ross, S.A., Slade, D. et al. (2015). Minor oxygenated cannabinoids from high potency Cannabis sativa L. *Phytochemistry* 117: 194–199.

41 Romano, L.L. and Hazekamp, A. (2013). Cannabis oil: chemical evaluation of an upcoming cannabis-based medicine. *Cannabinoids* 1 (1): 1–11.

42 Reverchon, E. and De Marco, I. (2006). Supercritical fluid extraction and fractionation of natural matter. *The Journal of Supercritical Fluids* 38 (2): 146–166.

43 Herrero, M., Mendiola, J.A., Cifuentes, A., and Ibáñez, E. (2010). Supercritical fluid extraction: recent advances and applications. *Journal of Chromatography A* 1217 (16): 2495–2511.

44 Rawson, A., Tiwari, B., Brunton, N. et al. (2012). Application of supercritical carbon dioxide to fruit and vegetables: extraction, processing, and preservation. *Food Reviews International* 28 (3): 253–276.

45 De Melo, M., Silvestre, A., and Silva, C. (2014). Supercritical fluid extraction of vegetable matrices: applications, trends and future perspectives of a convincing green technology. *The Journal of Supercritical Fluids* 92: 115–176.

46 King, J.W. (2014). Modern supercritical fluid technology for food applications. *Annual Review of Food Science and Technology* 5: 215–238.

47 Rovetto, L.J. and Aieta, N.V. (2017). Supercritical carbon dioxide extraction of cannabinoids from Cannabis sativa L. *The Journal of Supercritical Fluids* 129: 16–27.

48 Da Porto, C., Decorti, D., and Natolino, A. (2014). Separation of aroma compounds from industrial hemp inflorescences (Cannabis sativa L.) by supercritical CO2 extraction and on-line fractionation. *Industrial Crops and Products* 58: 99–103.

49 Aladić, K., Jarni, K., Barbir, T. et al. (2015). Supercritical CO2 extraction of hemp (Cannabis sativa L.) seed oil. *Industrial Crops and Products* 76: 472–478.

50 Geiser, F.O., Keenan, J.J., Rossi, R., Sanchez, A., and Whelan, J.M. (2008). Process for purifying (-)-Δ9-trans-tetrahydrocannabinol. US Patents 7449589B2. https://patentimages.storage.googleapis.com/07/e3/0e/ ca897914be8472/US7449589.pdf.

51 Enmark, M., Levy, K., Reid, B., and Seok Hur, J. (2017). Separation of natural cannabinoids by FC. *SFC* Rockville, MD (October 2017).

52 Denicola, C. and Barendt, J.M. (2018). Cannabinoid isolation models utilizing immobilized chiral stationary phases and SFC. *Emerald Conference* San Diego, CA (February 2018).

13

The Future of SFC

13.1   Introduction

The first 12 chapters of this book covered the history and current uses of SFC. The authors provided an overview of analytical and preparative SFC equipment, stationary phases, and mobile phases. The impact of SFC on several established industries including pharmaceutical, petrochemical, food, polymer, and personal care as well as newer industries such as cannabis was presented. This chapter focuses on the future of SFC.

What does the future hold for SFC? Is SFC a maturing technology, or a technology in its infancy? Is SFC being used in all industries where it is of use, or are there industries where SFC can have a major impact? Beyond the industrial setting, is SFC being used and researched in academia? Are there countries where SFC has yet to enter? What current and future technologies are disruptive to the increased use of SFC? Finally, what are the impediments to the expansion of analytical and preparative SFC? These questions will be addressed in the following pages.

Modern Supercritical Fluid Chromatography: Carbon Dioxide Containing Mobile Phases,
First Edition. Larry M. Miller, J. David Pinkston, and Larry T. Taylor.
© 2020 John Wiley & Sons, Inc. Published 2020 by John Wiley & Sons, Inc.

13.2 SFC Publication Record

The birth of modern SFC occurred in the mid- to late-1990s with the introduction of preparative SFC equipment targeting pharmaceutical purification. Preparative SFC is the predominant technique used for small scale chiral purifications in the United States and Europe, and more recently in China and India. The introduction of modern analytical SFC equipment in the 2010s facilitated the expansion of SFC from predominantly pharmaceutical discovery laboratories into areas requiring higher precision validated equipment and methods. The last 20 years have seen a resurgence of SFC after the unsuccessful incorporation in the 1980s and early 1990s of capillary SFC as an analytical technique. Publications by users and experts are necessary for a technology to prosper. They serve to introduce new users to the field, as well as demonstrate new areas of use for more experienced users. Figure 13.1 shows the number of publications (search performed with SciFinder) with "supercritical fluid chromatography" in the title since 1995. The late 1990s saw an average of 100 publications/year. Starting in 2000, the number of publications decreased until increasing around 2005. The introduction of modern analytical SFC equipment contributed to an increase in publications starting in 2012. Currently ~130 SFC articles are published per year. The number of publications does not accurately reflect the amount of SFC research being performed as the majority of the work performed in industry cannot be published due to patent and other confidentiality issues.

Examination of the last 10 years of publication shows a shift in the location of SFC research. This is illustrated in Figure 13.2, which lists publication by region

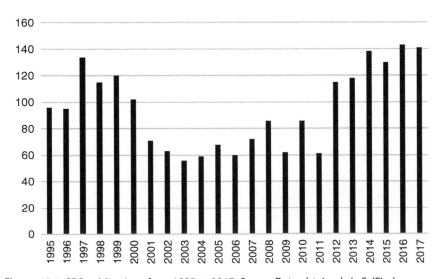

Figure 13.1 SFC publications from 1995 to 2017. *Source:* Data obtained via SciFinder.

for 2007, 2012, and 2017. In 2007, 75% of SFC publications originated from Europe and the Americas. By 2017, this dropped to 54%. Both Asia and Europe saw an increase in the percentage of SFC publications, at the expense of the Americas which saw its percentage decrease more than 50% from 2007 to 2017. This decrease is not caused as much by a decrease in publication numbers from the Americas as it is by a large increase in publications (>100%) from Asia and Europe. Figure 13.2 also demonstrates how SFC research has expanded; in

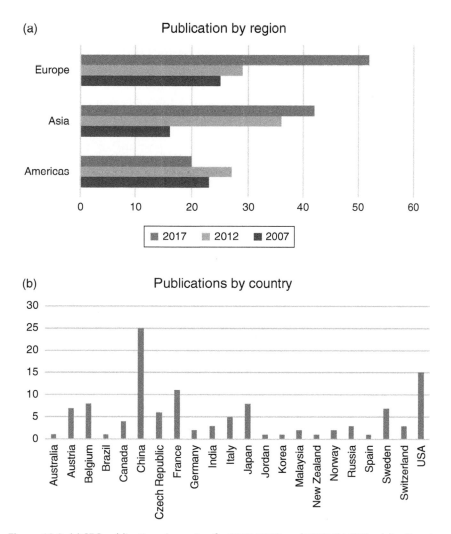

Figure 13.2 (a) SFC publications by region for 2007, 2012, and 2017. (b) SFC publications by country in 2017.

2007 and 2012 publications originated from 17 countries, in 2017 the number of countries publishing SFC research reached 22.

While the number of SFC publications are fewer than more established and utilized techniques, the steady increase shows an expanding technology base which is moving beyond pharmaceutical and into other industries. The expansion of SFC usage and research into additional countries is a positive sign for the future of the technology.

13.3 SFC Research in Academia

In 2008, Mukhopadhyay published a product review titled "SFC: Embraced by industry but spurned by academia" [1]. This article's main thesis was, while SFC was being used extensively in the pharmaceutical industry, it was faltering in academia. The author attributed the lack of excitement to a number of reasons, including overhype of SFC in the 1980s, academicians being accustomed to HPLC, and the high activation energy to using new methodologies. While not mentioned in by Mukhopadhyay, the increased cost of SFC vs. HPLC may be another factor preventing further use of SFC in academia. The impact of SFC not being widely used in academia has the consequence of students not being trained in the area, requiring training when they reach industrial positions.

While Mukhopadhyay's article suggest that academic is eschewing SFC, an evaluation of publications and presentations in the past 10 years offers some contradictory indications. Evaluation of published SFC literature shows an increase in university publications with 41 publications in 2007, 79 in 2012, and 93 in 2017, representing 59, 72, and 72% of SFC publications, respectively. The last 10 years have also seen a shift in oral and posters presenters on SFC at scientific meetings from mainly pharmaceutical scientists to a healthy mix of industry and academia. One area of concern is the reduction in SFC publications by US academicians. In 2017, only a handful of SFC publications originated from US universities. This corresponds with a decrease in all chromatography research at US universities. The impact on analytical chemistry in the United States is unknown at this time; will trained students from other countries be able to fill the hiring needs for skilled chromatographers in the United States?

13.4 SFC Conferences

Another avenue for sharing learnings and experiences with a technology is scientific conferences. Conferences serve to connect users with other interested scientists, bring users and vendors together, and allow presentation of research. The first series of conferences on SFC began in 1988 with 11 conferences held over the following 16 years. In 2004, this conference series was

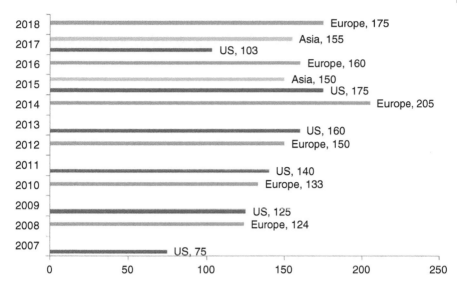

Figure 13.3 Attendance figures for Green Chemistry Group SFC conferences.

terminated due to lack of vendor and user interest. In 2007, a new series, sponsored by the Green Chemistry Group, a not-for-profit group consisting of industrial and academic scientists, began with a focus on packed column SFC. This conference series has shown better longevity, with a total of 14 conferences over the last 12 years. For the first 8 years, the conference alternated between the United States and Europe. Attendance figures for the SFC conference are shown in Figure 13.3. The figure shows a steady increase in attendance for the US and European meetings from 2007 to 2015. The 2016 and 2017 US/ European meetings saw a decline in attendance, but an increase was seen in 2018. Interest in SFC has exploded over the past 10 years in Asia. The increased interest and user base allowed biennial SFC meetings to begin in 2015, alternating between China and Japan. An additional benefit of the current SFC meeting series is a one-day SFC short course. The course introduces new users to the technology and serves as a refresher for more experienced chromatographers. In the past 10 years more than 400 students have attended courses at the SFC meetings and other courses taught by Green Chemistry Group scientists. These numbers indicate a steady influx of new scientists into the SFC arena.

13.5 Anticipated Technical Advances

There are still a number of "hoped-for" technical advances in SFC that these authors believe would help make SFC more widely applicable, more powerful and more user friendly. The first is a low-flow, low dead volume, reliable automated back-pressure regulator. This would allow easier coupling of SFC to detectors such as mass spectrometry, ELSD, and even flame-based detectors like the flame ionization detector (for pure CO_2 mobile phase). This would be especially useful when a fraction of the mobile phase flow is split and directed to the detector.

Some application areas would benefit from higher pressure limits and higher column temperature limits than are currently available from commercial instruments. These would be especially useful in the personal care products, polymer, and petrochemical fields. The higher temperatures and higher pressures would provide improved separations of complex, thermally stable mixtures.

Online SFE-SFC is commercially available and has provided far more flexibility in the direct application of SFC to solid samples. Yet the injection of samples dissolved in "strong" solvents (relative to the initial mobile phase) is still a challenge for SFC, as it is for all chromatographic methods. Technological advances in injection using strong solvents would provide more flexibility and a wider range of application for SFC. Online SFE-SFC is a powerful tool for analysis of lipophilic analytes such sa tocopherols, fat-soluble vitamins and fat-soluble naturan food colors.

Finally, these authors wish for a truly universal stationary phase for SFC. The advances in stationary phases described in Chapter 6 on achiral SFC method development have provided more widely applicable stationary phases, along with the ability to separate more polar and active analytes with little or no mobile-phase additive. But there is still no universal stationary phase such as the C18 column for reversed-phase HPLC. A truly universal stationary phase would greatly simplify method development, saving time, and money.

These technical needs attest to the need for further research in SFC by academic and industrial scientists and engineers.

13.6   Limits to SFC Expansion

Throughout this book the advantages of SFC over other chromatographic techniques in numerous industries has been discussed. Even with its advantages, the use of SFC pales in comparison to HPLC. This is evident by the number of publications/year (conservatively at least 10 times higher for HPLC) as well as the increased attendance at annual HPLC conference (~1000) compared to less than 200 for the annual SFC conference. What is keeping SFC from expanding in use?

Preparative SFC has become widely accepted over the past 20 years and is now the method of choice for sub-kg scale purifications. One purification area where SFC has not made inroads is as a replacement for flash purification performed in medicinal chemistry laboratories. Thousands of these purifications are performed per day, using a large volume of organic solvent. The main drawback to "flash SFC" has been commercially available equipment. Research in this area raises the hope this limitation will be eliminated in the near future [2–4].

Analytical SFC has not enjoyed the same success. What are some of the impediments to wider use of analytical SFC? The majority of analytical chromatographic analyses are performed in laboratories that require methods to be highly sensitive, validated, robust, and reproducible. These include, but are not limited to, clinical, environmental, food testing, and laboratories supporting pharmaceutical development and manufacturing. As discussed in Chapter 3, the first commercial SFC instruments were not sensitive and robust. Only the introduction of the latest SFC models in 2012 brought equipment with these attributes to the chromatography market. This compares with LC equipment that has been used in highly regulated, validated environments for more than three decades. Analytical SFC is working to attain a foothold against a well-established technology with thousands of users and tens of thousands of instruments across many industries throughout the world. In addition, in the 1980s capillary SFC was projected to be the next great chromatographic advance and would quickly supplant HPLC and GC. Unfortunately, the grandiose promises of capillary SFC were never achieved. Current state of the art packed-column SFC allows some, but not all, of the early promises of SFC to be realized. The adage of "once bitten, twice shy" applies and many scientists remember the hype and marketing of early SFC and are hesitant to wade back into SFC.

One advantage of SFC over HPLC is reduced solvent consumption. When packed column SFC was introduced as an analytical tool in the early 1990s the standard analytical method used 5 µm, 4.6-mm-i.d. columns, flow rates of 1 mL/min and on average consumed 20–30 mL of solvent. A comparable SFC method may require only 2–5 mL of solvent. This 50–80% solvent reduction added up to large savings in procurement and disposal costs and helped drive the incorporation of SFC in pharmaceutical discovery laboratories. In 2004, high performance UPLC systems using sub 2 µm, 2.1 mm i.d. columns, consuming at least 50% less solvent compared to 4.6-mm-i.d. columns were introduced. While high performance SFC instruments have now also been developed, as of 2018 the system dwell volumes were not as low as UPLC. The increased dwell volumes prohibit the use of 2.1-mm-i.d. columns; the smallest column to be used without system modification, is 3 mm i.d. The solvent reduction achievable with SFC vs. HPLC is significantly smaller than in the past. While solvent reduction can still be significant, especially in larger laboratories with multiple instruments for high throughput analysis, LC equipment advances of the past 20 years have reduced one of the main advantages of SFC.

HPLC is an analytical technique with a wide diversity of use, having the ability to analyze compounds with a wide range of polarities, from small hydrophobic molecules to highly polar, ionizable compounds and with a wide range of molecular weights, from small molecules to antibodies. SFC is used mainly as a technique for analysis of small molecular weight compounds of low to medium polarity. Recent research has expanded this range but the technique is still not suitable for larger molecular weight, higher polarity molecules such as larger peptides, proteins, and antibodies. This is a significant limitation to the expanded growth of SFC and could limit wider acceptance of the technology.

Another roadblock to increased SFC usage is the lack of SFC instruments in academia. While academic laboratories with SFC equipment has increased over past 10 years, especially in Europe and Asia, it is minimal relative to the number of HPLC instruments. For many scientists their first introduction to chromatography is through courses or independent research during their undergraduate or graduate education. Currently, this is nearly exclusively HPLC or GC equipment. SFC is not a routine component of instrumental analysis or chromatography courses. Equipment vendors should work to increase the number of SFC instruments in academia. This would lead to an increase in SFC use in industry down the road.

The final barrier to increased SFC use is also its main advantage, a carbon dioxide based mobile phase. Carbon dioxide is supplied through gas cylinders, whose use in a laboratory adds additional safety considerations such as storage, cylinder handling, leaks, and asphyxiation. While these concerns can be easily addressed, it is one more step to be undertaken before SFC can be introduced. Finally, when using CO_2 cylinders, it is difficult to determine the amount of gas remaining in the cylinder. The best approach to monitor cylinder levels is to place it on a scale. Many cylinders have a tare weight listed, allowing easy determination when it needs to be replaced. This compares to HPLC where it is easy to visually determine when additional mobile phase is needed.

13.7 Summary

Despite the challenges mentioned above, SFC has experienced tremendous growth in the past 10 years. It is a green analytical method with reduced solvent usage relative to HPLC [5, 6]. Advances in analytical equipment have allowed SFC to move into regulated and validated laboratories, greatly expanding the equipment footprint and user base [7, 8]. Research performed by both industrial and academic scientists has increased the theoretical understanding of SFC, allowing use by chromatographers without extensive additional training [9]. While SFC will never match the universality of HPLC, it is a valuable tool in the chromatography toolbox. SFC is orthogonal to reversed phase HPLC and is helpful to confirm peak coelution is not occurring [10]. Many experts agree that if you are performing analytical enantioseparations or normal phase separations, you should be using SFC. The high speed of SFC reduces method development time and has proven valuable in high throughput and 2-D analyses [11–16]. Analytical SFC has moved from just pharmaceutical analysis to the food, environmental, petrochemical, polymer, personal care, natural products, biomedical, and cannabis industries.

SFC is the technique of choice for sub kilogram scale preparative separations, especially for enantioseparations [17, 18]. The reduced solvent consumption makes preparative SFC "greener" and significantly reduces operating expenses [19]. Preparative achiral SFC have also begun to supplement most purifications currently performed using reversed phase chromatography [20, 21]. The increase in cannabis research has opened a new area for preparative SFC [22, 23].

While SFC has proven its value as an analytical and preparative tool, there are still advances that are required to increase the utilization rate. With the large number of talented academic and industrial scientists using and researching SFC, it is expected these challenges will be met and SFC will continue expanding into new application areas.

References

1 Mukhopadhyay, R. (2008). SFC: embraced by industry but spurned by academia. *Analytical Chemistry* 80 (9): 3091–3094.

2 Miller, L. and Mahoney, M. (2012). Evaluation of flash supercritical fluid chromatography and alternate sample loading techniques for pharmaceutical medicinal chemistry purifications. *Journal of Chromatography A* 1250: 264–273.

3 Ashraf-Khorassani, M., Yan, Q., Akin, A. et al. (2015). Feasibility of correlating separation of ternary mixtures of neutral analytes via thin layer chromatography with supercritical fluid chromatography in support of green flash separations. *Journal of Chromatography A* 1418: 210–217.

4 McClain, R., Rada, V., Nomland, A. et al. (2016). Greening flash chromatography. *ACS Sustainable Chemistry & Engineering* 4 (9): 4905–4912.

5 Schmidt, R.M., Ponder, C., Villeneuve, M.S., and Miller, L.A.D. (2012). Going greener in achiral & chiral separations: employing sustainable technologies to reduce our environmental impact. *10th International Symposium on Supercritical Fluids Proceedings*. San Francisco, CA, USA (13–16 May 2012).

6 Welch, C.J., Wu, N., Biba, M. et al. (2010). Greening analytical chromatography. *TrAC Trends in Analytical Chemistry* 29 (7): 667–680.

7 Dispas, A., Lebrun, P., and Hubert, P. (2017). Chapter 11 – validation of supercritical fluid chromatography methods. In: *Supercritical Fluid Chromatography* (ed. C.F. Poole), 317–344. Elsevier.

8 Hicks, M.B., Regalado, E.L., Tan, F. et al. (2016). Supercritical fluid chromatography for GMP analysis in support of pharmaceutical development and manufacturing activities. *Journal of Pharmaceutical and Biomedical Analysis* 117: 316–324.

9 Tarafder, A. (2016). Metamorphosis of supercritical fluid chromatography to SFC: an Overview. *TrAC Trends in Analytical Chemistry* 81: 3–10.

10 Wang, Z., Zhang, H., Liu, O., and Donovan, B. (2011). Development of an orthogonal method for mometasone furoate impurity analysis using supercritical fluid chromatography. *Journal of Chromatography A* 1218 (16): 2311–2319.

11 Zawatzky, K., Biba, M., Regalado, E.L., and Welch, C.J. (2016). MISER chiral supercritical fluid chromatography for high throughput analysis of enantiopurity. *Journal of Chromatography A* 1429: 374–379.

12 Pilarova, V., Gottvald, T., Svoboda, P. et al. (2016). Development and optimization of ultra-high performance supercritical fluid chromatography mass spectrometry method for high-throughput determination of tocopherols and tocotrienols in human serum. *Analytica Chimica Acta* 934: 252–265.

13 Ishibashi, M., Izumi, Y., Sakai, M. et al. (2015). High-throughput simultaneous analysis of pesticides by supercritical fluid chromatography coupled with high-resolution mass spectrometry. *Journal of Agricultural and Food Chemistry*.

14 Goel, M., Larson, E., Venkatramani, C.J., and Al-Sayah, M.A. (2018). Optimization of a two-dimensional liquid chromatography-supercritical fluid chromatography-mass spectrometry (2D-LC-SFS-MS) system to assess "in-vivo" inter-conversion of chiral drug molecules. *Journal of Chromatography B* 1084: 89–95.

15 Venkatramani, C.J., Al-Sayah, M., Li, G. et al. (2016). Simultaneous achiral-chiral analysis of pharmaceutical compounds using two-dimensional reversed phase liquid chromatography-supercritical fluid chromatography. *Talanta* 148: 548–555.

16 Stevenson, P.G., Tarafder, A., and Guiochon, G. (2012). Comprehensive two-dimensional chromatography with coupling of reversed phase high performance liquid chromatography and supercritical fluid chromatography. *Journal of Chromatography A* 1220: 175–178.

17 Miller, L. (2014). Pharmaceutical purifications using preparative supercritical fluid chromatography. *Chimica Oggi / Chemistry Today* 32 (2): 23–26.

18 Ventura, M.C. (2014). Chiral preparative supercritical fluid chromatography. In: *Supercritical Fluid Chromatography: Advances and Applications in Pharmaceutical Analysis* (ed. G.K. Webster), 171. Boca Raton, FL: CRC Press Taylor & Francis Group.

19 Welch, C.J., Leonard, W.R. Jr., Dasilva, J.O. et al. (2005). Preparative chiral SFC as a green technology for rapid access to enantiopurity in pharmaceutical process research. *LCGC North America* 23 (1).

20 Lazarescu, V., Mulvihill, M.J., and Ma, L. (2014). Achiral pPreparative supercritical fluid chromatography. In: *Supercritical Fluid Chromatography: Advances and Applications in Pharmaceutical Analysis* (ed. G.K. Webster), 97. Boca Raton, FL: CRC Press Taylor & Francis Group.

21 de la Puente, M.L., Soto-Yarritu, P.L., and Anta, C. (2012). Placing supercritical fluid chromatography one step ahead of reversed-phase high performance liquid chromatography in the achiral purification arena: a hydrophilic interaction chromatography cross-linked diol chemistry as a new generic stationary phase. *Journal of Chromatography A* 1250: 172–181.

22 Enmark, M., Levy, K., Reid, B., and Seok Hur, J. (2017). Separation of Natural Cannabinoids by FC. *SFC*, Rockville, MD (October 2017).

23 Denicola, C. and Barendt, J.M. (2018). Cannabinoid isolation models utilizing immobilized chiral stationary phases and SFC. *Emerald Conference*, San Diego, CA (February 2018).

Index

Modern Supercritical Fluid Chromatography: Carbon Dioxide Containing Mobile Phases,
First Edition. Larry M. Miller, J. David Pinkston, and Larry T. Taylor.
© 2020 John Wiley & Sons, Inc. Published 2020 by John Wiley & Sons, Inc.

Chemical Analysis

A SERIES OF MONOGRAPHS ON ANALYTICAL CHEMISTRY AND ITS APPLICATIONS

Series Editor

MARK F. VITHA

Editorial Board

Stephen C. Jacobson, Stephen G. Weber